MOLTEN SALTS

Characterization and Analysis

MOLTEN SALTS

Characterization and Analysis

Edited by GLEB MAMANTOV

DEPARTMENT OF CHEMISTRY
UNIVERSITY OF TENNESSEE
KNOXVILLE, TENNESSEE

1969

MARCEL DEKKER New York • London

MARCEL DEKKER, INC.
95 Madison Avenue, New York, New York 10016

United Kingdom Edition
published by

Marcel Dekker Ltd.
14 Craufurd Rise
Maidenhead
Berkshire, England

LIBRARY OF CONGRESS CATALOG CARD NUMBER 75-88481

PRINTED IN THE UNITED STATES OF AMERICA

PREFACE

Molten salts are of considerable importance in chemical technology. Applications range from the established ones, such as production of aluminum, sodium, magnesium and fluorine, to recent ones, such as plating of refractory metals, chemical synthesis and fuel cells. Of potentially great importance are the uses in nuclear technology. The experimental molten salt reactor at the Oak Ridge National Laboratory has operated successfully for more than 2 years; breeder reactors using molten salts appear promising.

From the fundamental point of view, molten salts represent a large and unique class of nonaqueous solvents, in which interesting redox and acid–base chemistry may be pursued.

This volume is a collection of papers, both of review and research nature, on most aspects of molten salt chemistry. These papers were presented at the Symposium on Characterization and Analysis in Molten Salts, American Chemical Society, Atlantic City, N.J., September, 1968. It is not surprising then that characterization of minor constituents by spectroscopic and electrochemical methods is discussed extensively in several of the chapters, and that solution chemistry in molten salt solvents is an important topic throughout the book. Other chapters stress the thermodynamic and transport properties of molten salts. The book is the first attempt to review advances in molten salt chemistry since the publication of two extensive monographs by Blander and Sundheim in 1964.

I would like to acknowledge gratefully the cooperation of the authors in submitting the manuscripts without undue delay, as well as to thank the Analytical Chemistry Division of the American Chemical Society for sponsoring the Symposium. The comments of several individuals, in particular G. P. Smith, J. Braunstein, D. L. Manning, and J. P. Young of the Oak Ridge National Laboratory, and M. Blander of the North American Science Center, were quite helpful to me in organizing the Symposium.

G. Mamantov

May, 1969
Knoxville, Tennessee

CONTRIBUTORS

C. A. ANGELL, Department of Chemistry, Purdue University, Lafayette, Indiana

MILTON BLANDER, Science Center, North American Rockwell Corporation, Thousand Oaks, California

CHARLES R. BOSTON, Metals and Ceramics Division, Oak Ridge National Laboratory, Oak Ridge, Tennessee

J. M. BRACKER, Nuclear Engineering Department, Brookhaven National Laboratory, Upton, New York

J. BRAUNSTEIN, Reactor Chemistry Division, Oak Ridge National Laboratory, Oak Ridge, Tennessee

M. A. BREDIG, Chemistry Division, Oak Ridge National Laboratory, Oak Ridge, Tennessee

JORULF BRYNESTAD, Metals and Ceramics Division, Oak Ridge National Laboratory, Oak Ridge, Tennessee

ELTON J. CAIRNS, Argonne National Laboratory, Argonne, Illinois

R. P. J. COONEY, Department of Chemistry, Oklahoma State University, Stillwater, Oklahoma

JOHN M. DALE, Oak Ridge National Laboratory, Oak Ridge, Tennessee

J. P. DEVLIN, Department of Chemistry, Oklahoma State University, Stillwater, Oklahoma

J. J. EGAN, Brookhaven National Laboratory, Upton, New York

L. F. GRANTHAM, Atomics International Division of North American Rockwell Corporation, Canoga Park, California

W. R. GRIMES, Reactor Chemistry Division, Oak Ridge National Laboratory, Oak Ridge, Tennessee

D. M. GRUEN, Argonne National Laboratory, Argonne, Illinois

R. J. HEUS, Brookhaven National Laboratory, Upton, New York

I. JOHNSON, Argonne National Laboratory, Argonne, Illinois

J. JORDAN, Department of Chemistry, The Pennsylvania State University, University Park, Pennsylvania

M. KRUMPELT, Argonne National Laboratory, Argonne, Illinois

P. C. LI, Department of Chemistry, Oklahoma State University, Stillwater, Oklahoma

W. B. MCCARTHY, Department of Chemistry, The Pennsylvania State University, University Park, Pennsylvania

P. B. MACEDO, Vitreous State Laboratory, The Catholic University of America, Washington, D.C.

G. MAMANTOV, Department of Chemistry, University of Tennessee, Knoxville, Tennessee

D. L. MANNING, Oak Ridge National Laboratory, Oak Ridge, Tennessee

VICTOR A. MARONI, Argonne National Laboratory, Argonne, Illinois

D. M. MOULTON, Reactor Chemistry Division, Oak Ridge National Laboratory, Oak Ridge, Tennessee

C. T. MOYNIHAN, Department of Chemistry, Purdue University, Lafayette, Indiana

NORMAN H. NACHTRIEB, Department of Chemistry, The University of Chicago, Chicago, Illinois

G. D. ROBBINS, Reactor Chemistry Division, Oak Ridge National Laboratory, Oak Ridge, Tennessee

J. H. SHAFFER, Reactor Chemistry Division, Oak Ridge National Laboratory, Oak Ridge, Tennessee

G. PEDRO SMITH, Metals and Ceramics Division, Oak Ridge National Laboratory, Oak Ridge, Tennessee

W. EWEN SMITH, Metals and Ceramics Division, Oak Ridge National Laboratory, Oak Ridge, Tennessee

R. E. THOMA, Reactor Chemistry Division, Oak Ridge National Laboratory, Oak Ridge, Tennessee

T. TIDWELL,* Brookhaven National Laboratory, Upton, New York

RICHARD J. TIVERS, Brookhaven National Laboratory, Upton, New York

JOHN D. VAN NORMAN, Brookhaven National Laboratory, Upton, New York

R. A. WEILER, Vitreous State Laboratory, The Catholic University of America, Washington, D.C.

J. K. WILMSHURST, Nuclear Engineering Department, Brookhaven National Laboratory, Upton, New York

S. J. YOSIM, Atomics International Division of North American Rockwell Corporation, Canoga Park, California

J. P. YOUNG, Oak Ridge National Laboratory, Oak Ridge, Tennessee

P. G. ZAMBONIN, Department of Chemistry, The Pennsylvania State University, University Park, Pennsylvania

* Present Address: North Texas State University, Denton, Texas.

CONTENTS

MOLTEN SALTS

Characterization and Analysis

SOME FUNDAMENTAL CONCEPTS IN THE CHEMISTRY OF MOLTEN SALTS

Milton Blander

Science Center
North American Rockwell Corporation
Thousand Oaks, California 91360

INTRODUCTION

In this paper I will review some of the concepts which have been important in molten salt chemistry. There will be special emphasis on those concepts which are an aid to physical intuition as well as those concepts which have a bearing on chemistry and chemical reactions. The evolution of a fairly sophisticated understanding of molten salts has taken place during the past fifteen years, and it is my purpose to outline some of these fundamental ideas without dwelling on details. In addition I will discuss the application of these ideas in making predictions on the physico-chemical properties of molten salts. Because of space limitations some of the significant ideas cannot be discussed and those which are discussed can only be covered superficially. More details are contained in several books on the subject.[1,2,3]

STRUCTURE AND ENERGETICS

From one point of view, molten salts, especially such salts as the alkali halides, have simpler structures than most liquids. They may often be considered as a collection of individual spherical ions and the fundamental interactions between the ions will govern the structure of any given system. The usual assumption made by theorists is that the total energy of a system is equal to the sum of the energies of the pair interactions. This is not strictly true and is a poor assumption when covalent bonds or ligand field interactions are significant. However, it is a reasonable first approximation to make for most cases. In a pure molten salt AX there are three kinds of pair interactions: cation-cation (A^+-A^+), anion-anion (X^--X^-), and cation-anion

(A^+-X^-). What characteristics of these three pair interactions have an influence on structure? The most important are the coulomb interactions ($\pm e^2/r$) which are of course repulsive for ions of like charge (cation-cation and anion-anion) and attractive for ions of opposite charge. A collection of ions will tend to sort themselves in such a way that cations will only "touch" anions and ions of like sign will not be very close. Thus, the repulsions of the ions due to their finite size (radii) which comes into play only when ions are close, are important only for the cation-anion pair. The other pairs are not close often enough for this short range core repulsion to be significant. Striking confirmation of this charge ordering is seen by comparison of the radial distribution functions (RDF) obtained from x-ray diffraction with those obtained from neutron diffraction by Levy and coworkers.[4,5] These RDF's for LiCl shown in Fig. 1 exhibit a positive value of the nearest neighbor peak in x-ray diffraction and a negative value in neutron diffraction of Li^7Cl due to the negative scattering amplitude of Li^7. These measurements are consistent with the predominance of nearest neighbors of opposite charge. Thus, of the three, only the ion size parameter for the cation-anion pair is

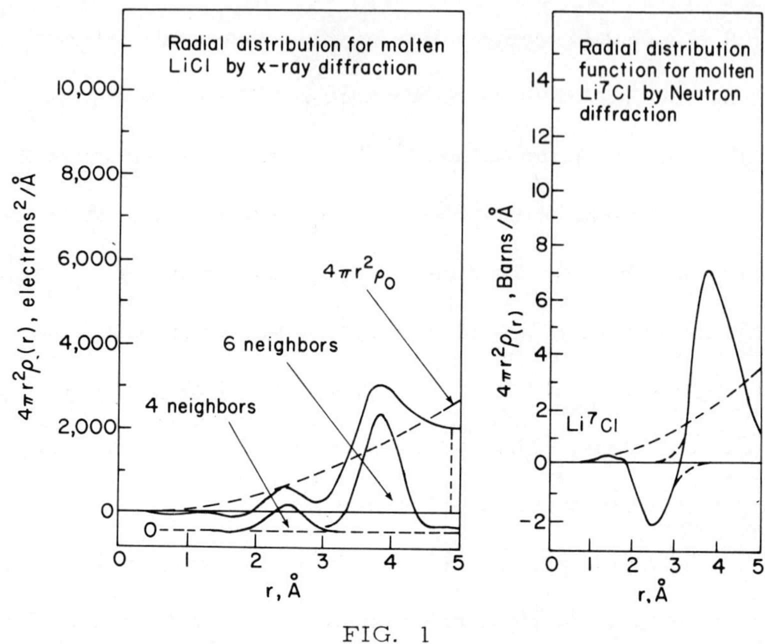

FIG. 1

Radial distribution functions for LiF calculated from x-ray and neutron diffraction data. The negative first peak in the RDF from neutron diffraction of Li^7F results from a combination of a negative scattering amplitude for Li^7 with a positive amplitude for Cl.

considered significant. This has been the single most important simplification in theories of molten salts.[6,7]

The second aspect of the pair potential which has an important bearing on structure is the softness of the core repulsions (i.e. those repulsions which define an ionic radius). To illustrate this for a hypothetical example: if, for the sake of illustration, the pair potential for a cation-anion pair is represented by the simple expression:

$$U = -\frac{e^2}{r} + \frac{k}{r^n}$$

where n is of the order of 8-12, then for a gaseous dipole

$$k = \frac{e^2 r_o^{n-1}}{n}$$

where r_o is the internuclear separation. For a solid with the NaCl structure

$$\frac{U}{n} = -\frac{1.7e^2}{r} + \frac{6k}{r^n}$$

From this one can show that the equilibrium value of r for the solid is about 19% larger than r_o for the vapor molecule if n = 8. Thus, lowering the coordination leads to significant foreshortening of cation-anion distances. This foreshortening is found in both vapor molecules and liquids. Of special note is the fact that despite the shorter cation-anion nearest neighbor separations in liquids their molar volume is greater than the solids at the same temperature. Of course, the importance of other interactions makes this simple calculation inexact for real salts. Table 1 gives some illustrations of this foreshortening.

TABLE 1

Nearest Neighbor Cation-Anion Separations in Alkali Halides (Taken from Levy[5] and Bauer and Porter[8])

LiF	2.10 (6)	1.95 (3.7)	1.545
LiCl	2.66 (6)	2.47 (4.0)	2.0207
LiBr	2.85 (6)	2.68 (5.2)	2.1704
LiI	3.12 (6)	2.85 (5.6)	2.3919
NaF	2.40 (6)	2.30 (4.1)	1.9259
NaCl	2.95 (6)	2.80 (4.7)	2.3606
NaI	3.35 (6)	3.15 (4.0)	2.7115
KF	2.80 (6)	2.7 (4.9)	2.171
KCl	3.26 (6)	3.10 (3.7)	2.6666
RbCl	3.41 (6)	3.30 (4.2)	2.7867
CsCl	3.57 (6)	3.53 (4.6)	2.9062
CsBr	3.86 (8)	3.55 (4.6)	3.0722
CsI	4.08 (8)	3.85 (4.5)	3.3152

( ) Coordination number is given in parentheses.

Another cause of foreshortening of a cation-anion distance for soft ions is an asymmetry of charge distribution about an ion. For example, in a one dimensional model it can be shown that the exchange of nearest neighbors in a binary system

$$\oplus\ominus\oplus + (+)(-)(+) \rightleftarrows 2\,\oplus\ominus(+) \qquad (1)$$

leads to a foreshortening of the shorter distance. In addition there is a lengthening of the long distance. Similar effects can be shown to exist if thermal motions introduced upon melting of a solid are taken into account as

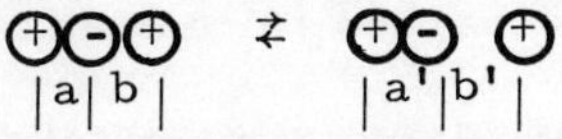

$$\underset{|a|\,b\,|}{\oplus\ominus\oplus} \rightleftarrows \underset{|a'|b'|}{\oplus\ominus\ \oplus} \qquad (2)$$

where an increase of b to b' by thermal motions of the cation leads, on the average to a decrease of a to a'. In addition, mixtures containing cations of different charge will have the average distances between the most highly charged cation and the anion foreshortened relative to that in the pure salts

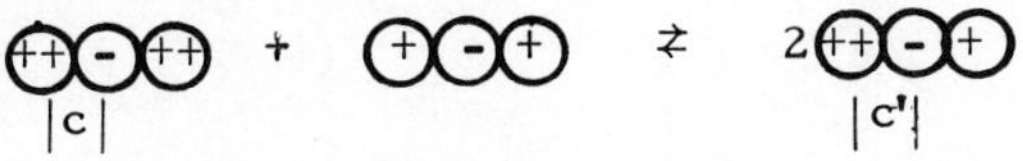

$$\underset{|c|}{(++)(-)(++)} + (+)(-)(+) \rightleftarrows 2\underset{|c'|}{(++)(-)}(+) \qquad (3)$$

so that the average value of c' is less than c. Such foreshortening is probably more significant in liquids than in solids.

In addition to these influences on structure other contributions to the pair interactions may have a significant effect. For example, if the anions are highly polarizable they are stabilized in regions of high field intensity. (The simplest classical interaction like this is the ion induced dipole interaction with $U = \frac{1}{2}\alpha \underline{\underline{E}}^2$ for weak fields where U is the energy, α the polarizability, and $\underline{\underline{E}}$ the field intensity.) This leads to a tendency for the charge environment of the anions to be asymmetric. The simplest illustration of this is the second configuration of (2). This effect will be greatest with small or highly charged cations and such stabilization by asymmetry of the cation environment of the anions should be easier to accomplish in liquids than in solids. Consequently, this tendency for the relative stabilization of liquids is greater for lithium salts than for sodium or potassium salts and may be one of the important reasons for the anomalously low melting points of lithium halides

TABLE 2

Melting Points of Alkali Halides (^{o}K)

	F	Cl	Br	I
Li	1121	883	823	742
Na	1268	1073	1020	933
K	1131	1043	1007	954
Rb	1068	995	965	920
Cs	976	918	909	899

Many other interactions have a profound influence on structure. Ligand field effects lead to strong tendencies for particular symmetries of anions about a transition metal cation which are strongly influenced by nearest neighbor anions as well as the next nearest neighbor cations. Steric effects are important, especially for small cations as Be^{+2} which cannot, for example, be closely surrounded by more than four F^- ions.

Another idea of interest is that of "significant structures" developed by Eyring and coworkers.[9] Although the basic concepts wilt under critical examination and cannot be justified from fundamentals, the theory has a direct intuitive appeal. Briefly, it is based on the fact that salts expand upon melting and the increased volume in the liquid is taken up by imperfections or passageways. Ions which border these openings are loosened and have gas-like degrees of freedom. The properties of the liquid are then calculated from the weighted product of partition functions for solid-like ions and gas-like ions. Thus, as a liquid is heated and expands, solid-like ions are converted to gas-like ions. Although it is easy to criticize such an artificial dichotomy which is inconsistent with structural data, this idea has produced a simple partition function which reasonably describes a large number of properties.

An analogue of this idea can be used to rationalize the significant measurements of Grantham and Yosim[10] who found that some salts exhibited maxima in a plot of specific or equivalent conductivities versus temperature (Fig. 2). They reasoned that at low temperatures the liquid was more dense and solid-like and the ions were relatively immobile.

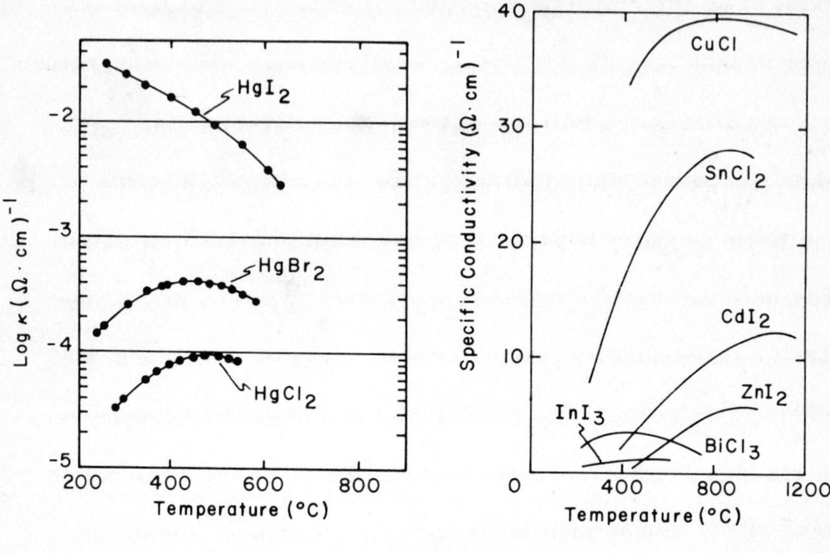

FIG 2

Specific conductivities of salts which exhibit maxima in κ versus T.

At increasing temperatures and decreasing densities, mobilities increase relative to the solid-like state and conductivity increases. At high temperatures and still lower densities, the ions become more gas-like and molecular-like groupings become more significant and although these molecular groupings may be mobile, the ions tend to move as uncharged or low charged units and contribute little to conductivity. The two states at opposite density extremes, the solid and gaseous salts are non-conducting and a maximum in conductivity must exist at an intermediate density. The influence of temperature is thus largely due to the density change with temperature.

That the mobilities of ions in a solid-like material should increase with melting can be understood in terms of the voids or holes which are presumably present in a liquid. The hole theory of Furth[11] is suggested by this and qualitatively, it can help to rationalize an increase of conductivity with temperature. Unfortunately, the attempts to apply the Furth hole theory to molten salts by Bockris and coworkers[12,13,14] have been incorrect.

Angell[15] has developed an important new idea which modifies this picture. He reasons by an extension of the work of Cohen and Turnbull[16] and Adam and Gibbs[17] that there exists a temperature related to the glass transition temperature, T_o, below which transport ceases and the ionic mobilities are zero. This temperature for simple salts is common for all transport properties. This idea justifies the representation of transport properties by simple equations of the form A exp $(B/(T-T_o))$ where A, B, and T_o are constants and T_o which is related to a Debye temperature should be the same for different transport properties.

If opening up the structure of a dense salt increases mobilities and conductivities what mechanism will rationalize the increase of conductivity of a vapor with an increase of density? My calculations (unpublished) can lend some insight into this. If we examine the dissociation of a salt dipole

$$AX \rightleftarrows A^+ + X^- \tag{4}$$

at reasonable temperatures (~ 1000-2000°K) the degree of dissociation and the number of free ions is very small. Since the mobility of ions in a gas generally decreases as the inverse power of the density, as the density of the gas increases the conductivity will decrease if dissociations as (4) are the only source of ions. However, my calculations indicate that at increased densities larger molecules and larger molecule ions become more important as

$$2AX \rightleftarrows A_2X^+ + X^- \tag{5}$$

$$3AX \rightleftarrows A_2X^+ + AX_2^- \tag{6}$$

$$4AX \rightleftarrows AX_2^- \tag{7}$$

This stems from the fact that the larger the molecule ion the easier it is (energetically) to form ions and ΔE for (5), (6), or (7) is much smaller than for (4). The decrease in energy for forming some of these molecule ions more than compensates for the estimated entropy effects which decrease with an increase of density. Once the average number of atoms in the molecule ions is greater than two, the conductivity will increase with density of the gas since mobilities decrease with $1/\rho$ and the number of ions will increase to a power greater than ρ. Superimposed upon this are the exchange mechanisms during collisions as

$$A'X + AXA^{+} \quad A'XA^{+} + XA \tag{8}$$

which will further increase the conductivity rise with density. This picture provides us with an intuitive feeling that conductivity in molten salts arises from the collective motions of clusters which have a net charge and which move through intervening voids. At very low densities there are too few ions to conduct. At somewhat higher densities the number of molecule ion clusters increases faster than mobility decreases and conductivity increases. This continues until the ions are fairly close packed and motion becomes increasingly hindered until ultimately the density increases to the point where mobility ceases.

EQUILIBRIUM PROPERTIES

Introduction

When kinetic factors are not important, then the extent of chemical reactions in solution is governed by the thermodynamic properties of the reactants and products

$$\Delta G^o = -RT\ln K$$

$$K = K(\mathbf{X}, \gamma)$$

and K is a function of the equilibrium concentrations X and the activity coefficients γ of the reactants and products. If ΔG^o and the γ's are known, then for known amounts of reactants we may calculate the extent of a chemical reaction by calculating the X's. (The dissolution of a slightly soluble reactant can be written as a chemical reaction which is here defined in its broadest sense.) The problem of calculating the extent of a reaction is then reduced to the problem of calculating ΔG^o or γ. The estimation or calculation of ΔG^o requires a knowledge of standard free energies of formation of the compounds involved in the reaction. Although there are extensive tabulations of these data there are still large gaps. The problems in obtaining these data are not primarily related to molten salt chemistry so I shall not discuss them here, but I will focus this discussion on the activity coefficients γ and other related quantities. These quantities, ΔH_m

the heat of mixing, μ^E the excess chemical potential, ΔG^E the excess free energy of mixing and their concentration dependence are interrelated and give us an insight into γ.

In this paper I will discuss some of the concepts which have been used to estimate γ and which have helped in making predictions concerning chemical equilibria in molten salts. Where possible, I will point out applications which are pertinent in analytical chemistry. In addition, I will point out those aspects which underpin the intuitive insights necessary to all chemists. Because of the limited length of this paper, no more than a superficial discussion can be presented.

Definitions

Perhaps the most important preliminary definition is that of an ideal solution. Although there is no unique definition, there is one that has been the most convenient in practice. For a mixture of monovalent ions the molar free energy of mixing per mole of an ideal mixture is given by

$$\Delta G^{id} = \Delta G^{o} + RT(\Sigma N_i \ln N_i + \Sigma N_j \ln N_j) \qquad (9)$$

where N_i is the ion fraction of cations and N_j of anions.

$$N_i = \frac{n_i}{\Sigma n_i} \qquad\qquad N_j = \frac{n_j}{\Sigma n_j}$$

where n are the number of moles of ions. This definition

of ideality was originally derived by Temkin[18] from a quasi-lattice model. The Temkin model consists of two interlocking sublattices, one of cations and the other of anions such that the nearest neighbors of the cations are anions and the nearest neighbors of the anions are cations. The cations mix on the cation sublattice and the anions mix on the anion sublattice. Equation (9) can also be derived from more general considerations without the use of a physical model.[19] The chemical potential of any component ij in an ideal solution is

$$\mu_{ij}^{id} = \mu_{ij}^{o} + RT\ln N_i N_j \tag{10}$$

The activity of a component is defined by

$$\mu_{ij} = \mu_{ij}^{o} + RT\ln a_{ij} \tag{11}$$

and consequently for an ideal solution

$$a_{ij} = N_i N_j \tag{12}$$

Real solutions deviate from ideal solution behavior, and we may define an activity coefficient, γ_{ij} which is a measure of this deviation from ideal behavior

$$a_{ij} = N_i N_j \gamma_{ij}$$

One can define excess thermodynamic functions as another expression of this deviation. The excess chemical potential of a component ij is

$$\mu_{ij}^{E} = \mu_{ij} - \mu_{ij}^{id} = RT\ln \gamma_{ij} \tag{13}$$

and the total excess free energy is

$$\Delta G^E = \Delta G - \Delta G^{id} = \sum_{\substack{\text{all} \\ \text{components}}} N \ln \gamma \tag{14}$$

Deviations from ideal behavior are positive when $\gamma > 1$ and negative when $\gamma < 1$. Another important thermodynamic function, the enthalpy (heat) of mixing, ΔH_m, generally parallels ΔG^E and does not differ greatly from it. Since:

$$\Delta G^E = \Delta H_m - T\Delta S^E \tag{15}$$

then another way to say this is that ΔS^E, although it may be significant, probably is relatively small. Consequently, measurements of ΔH_m can often give us considerable information on ΔG^E.

For solutions containing ions with valence higher than unity it has been convenient to use the same definitions of an ideal solution and of activity coefficients.

Binary Mixtures

The goal of much of the work on binary mixtures is to learn enough to make predictions of thermodynamic quantities. There has been a considerable amount of experimental data on binary systems which help in making correlations useful in such predictions. The measurements of Kleppa and coworkers[20-25] on the enthalpies of mixing have been among the most instructive measurements on binary mixtures. Kleppa found for one-one electrolytes with a common anion that the enthalpy of mixing could be expressed as a power series in the mole fractions X

$$\Delta H_m = X_1 X_2 (a + bX_1 + cX_1 X_2) \tag{16}$$

For alkali nitrates he found that the heats of mixing of a 50-50 mixture were negative and that they could be expressed as a simple function of the ion size parameters of the two components of a binary mixture

$$\Delta H_m (\text{50-50 mixture}) = -U\left(\frac{d_1 - d_2}{d_1 d_2}\right)^2 \tag{17}$$

where d is the sum of the radii of the cation and anion of the salt designated and U ~ 140 kcal/mole.

The experimental results could be rationalized in terms of the quasi-chemical theory[26] which leads to the equation for ΔH_m

$$\Delta H_m = X_1 X_2 \lambda \left(1 - X_1 X_2 \frac{\lambda}{ZRT} + \ldots\right) \tag{18}$$

where Z is a coordination number and λ is an energy parameter related to the exchange of next nearest neighbors in the mixing process (Fig. 3). This rationalization could only be made if it is assumed that λ is concentration dependent

$$\lambda = A + BX_1 \tag{19}$$

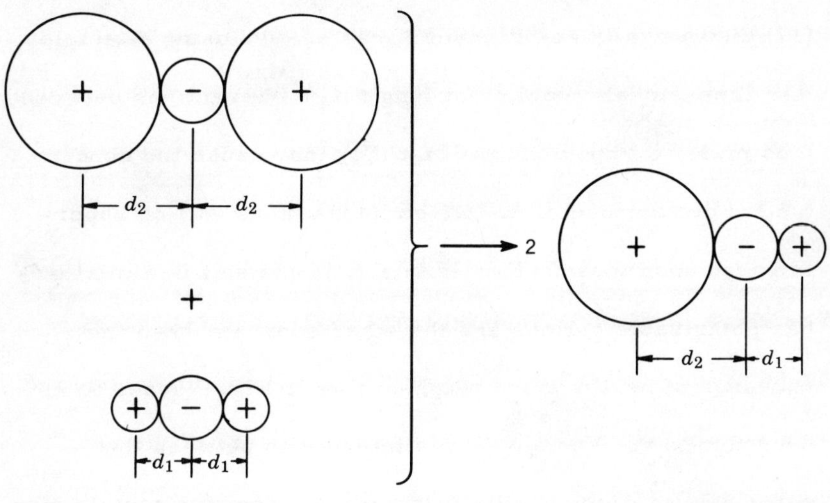

FIG. 3

Model for Førland's calculation of the energy change for the exchange of next nearest neighbor cations.

The sign and dependence of λ on ionic radii has been studied by Førland[27] who has shown that the decrease of the coulombic repulsion of next nearest neighbor cations (for hard sphere ions) led to an expression similar to (17). Lumsden[3,28] has shown an additional contribution of about the form of (17) (again for simple hard sphere ions) due to the change of the field intensity and the polarization of the anion in the asymmetric field of the cations. The real situation is much more complex. Although this picture is too simple for ionic systems, partial justification can be made on more rigorous grounds. Reiss, Katz, and Kleppa[7] have justified (16) and

(17) theoretically and Blander[29] has shown, using a simple one dimensional model, that long range interactions between ions must be important and that they may cause the concentration dependence of λ. Other interactions will be superimposed upon these. For example, if one salt in a mixture has polarizable cations and the other does not, then the exchange of next nearest neighbors as in Fig. 3 leads to a decrease in the Van der Waals interaction between the polarizable cations. This leads to a net positive contribution to λ which has been shown[28,30] to explain the differences in the values of ΔH_m of binary mixtures of the alkali salts and binary mixtures of an alkali salt and a silver or thallium salt.

Before going on to binary mixtures containing ions of higher valence, I shall digress to examine the structural implications for negative values of λ. A negative value of λ means that the like ions repel each other preferring the unlike ions as next nearest neighbors. Although an expression as (16) is generally adequate in concentrated solutions, it may be inadequate in dilute solutions for fairly negative values of λ. This comes about as a result of the strong repulsions between like ions which leads to a non-random mixing or ordering effect. The concentration dependence of ΔH_m

calculated from quasi lattice theory is given in Fig. 4 for different values of $(2\lambda/ZRT)$. For increasingly negative values of $(2\lambda/ZRT)$, the plot of $(\Delta H_m/\lambda)$ in Fig. 4 varies from a parabolic to a linear function of X_2. The linear dependence for $(-2\lambda/ZRT) = \infty$ means that in this extreme case $(\overline{H}_2 - H_2^o)$, the partial molar enthalpy of solution of component 2 is constant and is negative to a mole fraction of 0.5 and is zero for values of X_2 between 0.5 and 1. For finite large values of the parameter $(-2\lambda/ZRT)$, there will be rapid changes in $(H_2 - H_2^o)$ with composition in the vicinity of $N_2 =$ 0.5, with the steepness of the change being larger the more negative the value of $(2\lambda/ZRT)$. A physical insight into this concentration dependence of $(\Delta H_m/\lambda)$ can be obtained from an examination of the quasi-lattice model used for the calculations of the ordering or nonrandom mixing effect. The model is pictured two dimensionally in Fig. 5. The model consists of a sublattice of cations and a sublattice of anions with the cations having only anions and the anions having only cations as nearest neighbors. Although a lattice model is not an accurate representation for a liquid, most of the errors introduced are cancelled out in calculations since one calculates changes in thermodynamic properties between liquid mixtures

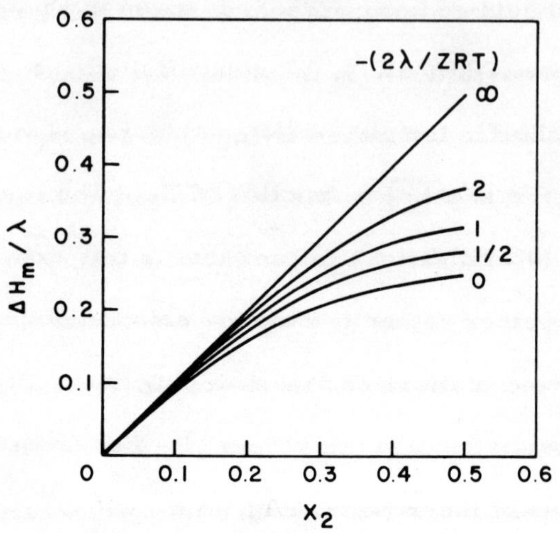

FIG. 4

Plot of $\frac{\Delta H_m}{\lambda}$ calculated from the quasi lattice theory for different values of $(2\lambda/ZRT)$.

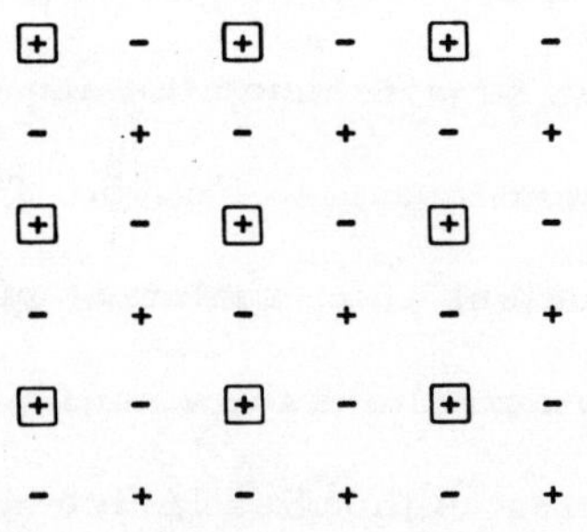

FIG. 5

Two dimensional representation of the quasi-lattice model for binary systems.

and the pure liquid components - each liquid being represented by a quasi-lattice. In the theory for binary mixtures the cation sublattice is further divided into two sublattices (designated as + and [+] in the two dimensional representation of Fig. 5) with cations in one set of sites having only cations on the other set as next-nearest-neighbors. In a binary mixture containing cations A^+ and B^+, these two cations will be distributed over these two sets of sites. Very negative values of the parameter λ lead to an ordering effect such that unlike cations tend to be next nearest neighbors and to be distributed on different sublattices. Thus for very negative values of λ, A^+ cations will populate one sublattice preferentially and B^+ cations the other. The maximum fraction of A^+-B^+ next nearest neighbors and the maximum ordering effect is at $X_2 = 0.5$. This tendency exists to a lesser degree for less negative values of λ. One aspect of the concept called complexing is this ordering effect which is not necessarily related to the interaction of the cations with anions but is related to the relative repulsion of like cations and attraction of unlike cations as next nearest neighbors. This is an important concept to which I will return later.

Kleppa and coworkers have also studied binary mixtures containing an alkali halide mixed with a salt having polyvalent cations.[24,25] One example, his measurements of ΔH_m for mixtures of $MgCl_2$ with monovalent halides are exhibited in Fig. 6 and $\Delta H_m/X_1X_2$ in Fig. 7. Obviously, equation (16) cannot be used to represent the data. In addition, Kleppa has shown that the interaction parameter, λ, at infinite dilution of the divalent salts varied linearly with change of the alkali cation according to the relation

$$\left(\frac{\Delta H_m}{X_1X_2}\right) = \lambda_\infty = A - B\left(\frac{d_1 - d_2}{d_1 d_2}\right) \tag{20}$$

a form justified theoretically by Davis.[31] The concentration dependence of $\Delta H_m/X_1X_2$ is most complex for the most negative values of λ_∞. This immediately suggests the discussion of ordering on the cation sublattice in the preceding paragraph. Thus, in terms of the simple model a very negative value of λ_∞ means that the divalent cations will tend to array themselves on one sublattice and the monovalent cations on the other with one divalent ion replacing two monovalent cations. This is supported by the linear dependence of ΔH_m on the mole fraction of $MgCl_2$ for mixtures with KCl, RbCl, and CsCl up to $X_2 \sim 0.25$. The model leads to the suggestion that the representation of excess free energies be made

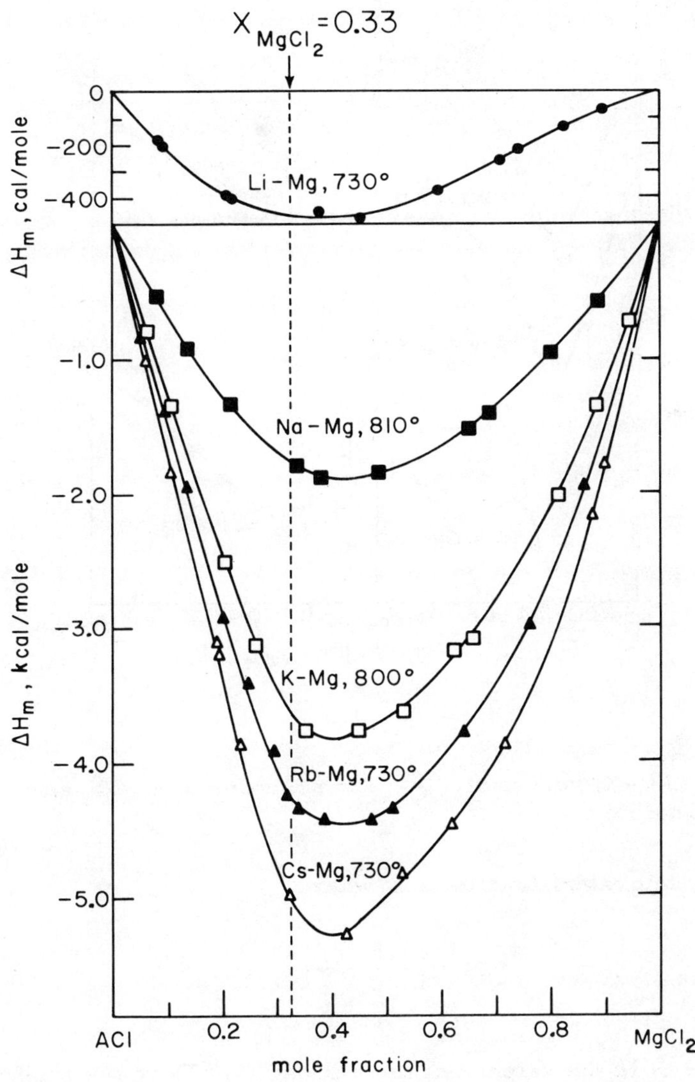

FIG. 6

Enthalpies of mixing of $MgCl_2$ with the alkali chlorides.

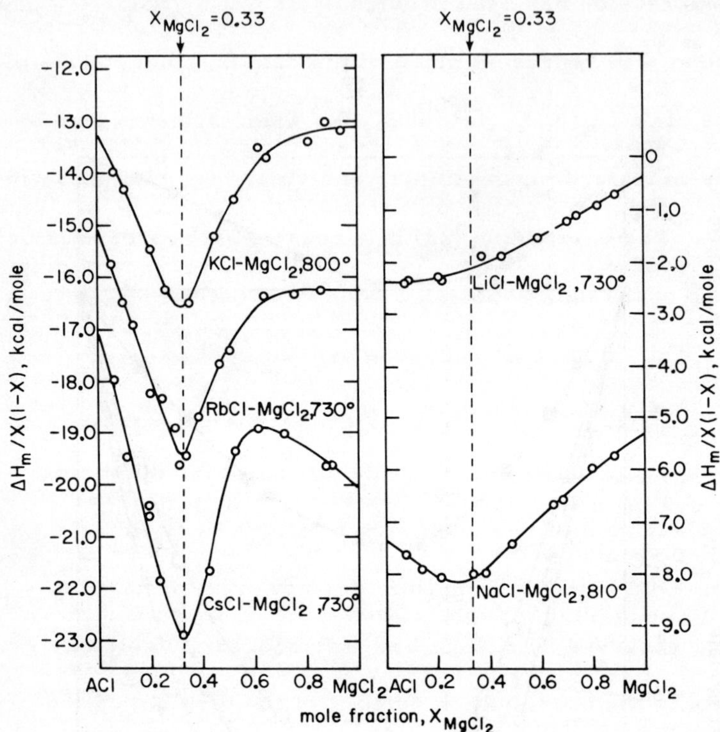

FIG. 7

$\Delta H_m/X(1-X)$ versus X_{MgCl_2} for solutions of $MgCl_2$ and alkali halides.

using equivalent fractions X_i' where

$$X_i' = \frac{z_i n_i}{\sum_i z_i n_i} \tag{21}$$

where z_i is the valence of the i th cation. Thus one might try the expression

$$\Delta H_m = X_1' X_2'(a + bX_1' + cX_1' X_2') \tag{22}$$

This expression has been previously shown by Forland[27] and Lumsden[32] to represent much of the data on binary systems. Values of $\Delta H_m/X_1'X_2'$ versus X_2' for these systems lead to curves which are much simpler and straighter than those in Fig. 7. The expression (22) is a reasonable representation of much of the data especially when λ is not very negative.

This picture is of course oversimplified as it does not take into account the major structural changes necessary and it glosses over the difficulty in visualizing the replacement of two monovalent cations by one divalent cation. One important implication, however, is that the minima observed by Kleppa are not necessarily related directly to the coordination number of the divalent cation but rather to the tendency to fill up one of the cation sublattices with polyvalent ions by the replacement of equivalent quantities of charge. I hope to investigate the applicability of this concept further.

This ordering phenomenon is only one of the many types of structural changes which are usually placed under the heading of complex ions. Other structural changes as the foreshortening of the average distance between the divalent cation and the anion, the close packed steric ordering of anions about a small highly charged cation which limits the coordination, and the tendency towards certain symmetries about transition metal cations all contribute to a structural complexity in these mixtures which defies an accurate description of mixtures of salts of different charge type without careful structural measurements. Some of these measurements will be discussed later in this volume.

Ternary Mixtures

Ternary molten salt mixtures contain four ions. The restriction of electroneutrality reduces the degrees of freedom to three. In this section two types of ternary systems will be discussed: reciprocal ternary mixtures containing two cations and two anions and additive ternary mixtures containing three cations and one anion. Much of the past theoretical work has been on reciprocal ternary systems and only recently has progress been made on additive ternary systems.

Perhaps the simplest significant predictions that have been made for reciprocal systems are obtained from a cycle first proposed by Flood, Førland, and Grjotheim[33] for the calculation of solubility products (standard free energies of solution at infinite dilution) and enthalpies of solution. For example, the dissolution of solid AgCl in $NaNO_3$ is broken up into the three steps

$$AgCl(solid) + NaNO_3(liquid) \rightleftarrows AgNO_3(liquid) + NaCl(liquid) \quad (A)$$

$$AgNO_3(liquid) \rightleftarrows AgNO_3(\infty dilution\ in\ NaNO_3) \quad (B)$$

$$NaCl(liquid) \rightleftarrows NaCl(\infty dilution\ in\ NaNO_3) \quad (C)$$

Thus

$$-RT\ln K_{SP} = \Delta G_A + \Delta G_B^{ST} + \Delta G_C^{ST} \quad (23)$$

and

$$\Delta H_{soln.} = \Delta H_A + \Delta H_B + \Delta H_C \quad (24)$$

where K_{SP} is a solubility product and the superscript ST signifies standard states. Thus, by using available thermodynamic data for (A) the calculation can be made with a knowledge of steps (B) and (C). The problem is thus reduced from understanding a very large number of ternary systems to understanding a much smaller number of binary systems. In general, the contribution from (A) is large and the contribution from (B) and (C) is relatively small. If data is not available for (B) and (C) one may approximate these from our general understanding of binary mixtures as discussed briefly earlier in this paper. Table 3 shows a comparison of data[34, 35] with predictions based upon this cycle. This cycle

TABLE 3

Measured Values of $-RT\ln K_{SP}$, ΔH_{soln}, and Their Comparison with Calculated Values

Solute	Solvent	$-RT\ln K_{SP}$, kcal./mole		ΔH, kcal./mole	
		Measured	Calcd.	Measured	Calcd.
AgI	$NaNO_3$	29.5	31.5	29.5	29.8
AgI	KNO_3	26.7	28.2	—	27.0
AgBr	$NaNO_3$-KNO_3	21.7	22.1	22.4	22.9
AgI	$NaNO_3$-KNO_3	28.4	29.9	27.9	28.4

allows one to make predictions of the solution properties and solubilities in dilute solutions.

Another prediction in dilute reciprocal salt systems concerns association constants for complex species. For example, association constants for the equilibria

$$Ag^+ + Cl^- \rightleftarrows AgCl \qquad K_1 \tag{25}$$

$$AgCl + Cl^- \rightleftarrows AgCl_2^- \qquad K_2 \tag{26}$$

can be derived from the quasi-lattice model.[36,37,38] The model is similar to Fig. 5 except that the anion sublattice now contains two different types of anions. In the simple model for very dilute solutions of Ag^+, only the influence of nearest neighbors is taken into account so that the association in (25) is characterized by an energy change ΔE_1 defined in terms of the lattice as shown in Fig. 8. The anion lattice is divided into two regions, one adjacent to Ag^+ and the other having no Ag^+ cations as a nearest neighbor. The interchange of the circled X^- and Y^- ions in Fig. 8 is accompanied by the energy change ΔE_1. Rather than discuss the detailed statistical mechanical calculations one can gain some insight into associations as (25) from an examination of the meaning of the association constant in terms of the model

$$K_1 = \frac{(AgCl)}{(Ag)(Cl)} = \frac{(AgCl)/(Ag)}{(Cl)} \tag{27}$$

If Henry's law holds for all species and in solutions so dilute

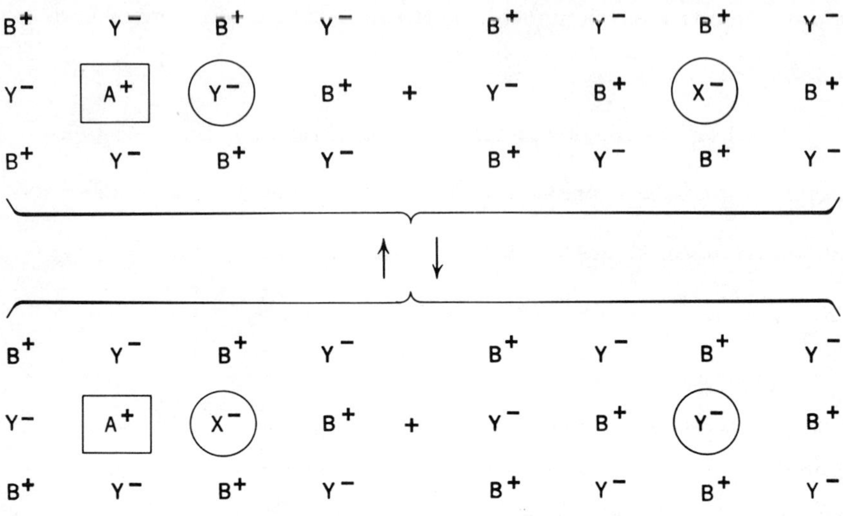

FIG. 8

Model for the definition of ΔE_1, the energy of AX ion pair formation for formation of the ion pair $A^+ + X^- \rightleftarrows AX$.

that AgCl pairs are the only significant species then K_1 is the ratio of the fraction of silver ions having one Cl^- as a nearest neighbor to the fraction of chloride anions in the anion sub-lattice not adjacent to Ag^+. If we define the fraction of Cl^- in the anion lattice adjacent to Ag^+ as $(Cl)'$ then in Zn_{Ag} positions about all the Ag^+ ions where Z is a coordination number, the number of Cl^- ions which are nearest neighbors to Ag^+ is $\sim Z(Cl)'n_{Ag}$ and dividing this by n_{Ag} yields $Z(Cl)'$ as the fraction of Ag^+ ions having a Cl^- as a nearest neighbor.

This fraction is also given by (AgCl)/(Ag), the numerator of the right hand side of equation (27). Consequently

$$K_1 = \frac{(AgCl)/(Ag)}{(Cl)} = \frac{Z(Cl)'}{(Cl)} = Z\beta_1 \qquad (28)$$

where the ratio (Cl)'/(Cl) is the ratio of concentrations of Cl^- in the two anion regions. For solutions dilute in chloride this ratio is a Boltzmann type factor $\beta_1 = \exp(-\Delta E_1/RT)$ which governs the distribution of Cl^- ions between these two regions (one adjacent to one Ag^+ cation and the other not adjacent to any Ag^+ cations). There is however a peculiarity in the definition of association constants such that K_1 goes to zero if there is no tendency for association. Thus (AgCl) in equation (27) is the number of pairs Ag-Cl in excess of those that would be present in a randomly mixed solution. Consequently

$$K_1 = \frac{Z\ (Cl)' - (Cl)}{(Cl)} = Z(\beta_1 - 1) \qquad (29)$$

where (Cl) is the concentration of Cl^- anions in the region adjacent to Ag^+ cations if mixing is random in this region (i.e., if $\Delta E_1 = 0$). Equation (29) is the result obtained from the statistical mechanical calculations.[36,37,38] The energy of association should be a <u>specific bond free energy.</u> For simple cases one might expect that ΔE_1 is a temperature independent quantity and equation (29) in this case constitutes a prediction of the temperature coefficient of K_1 and of the configurational contribution to the entropy of association. The constancy of ΔE_1 over wide ranges of temperature has been demonstrated for the association of many monatomic ions with halides in molten alkali nitrates.[19,37] Some of the data for the association of Ag^+ with Cl^- in alkali nitrates are given in Table 4. Values of ΔE_1 appear to be constant over large ranges of temperature for any reasonable choice of the coordination number (Z = 4, 5, or 6). The quasi-lattice model also leads to similar but more complex expressions for the higher association constants.[36,38] This example illustrates that molten salts are simpler than aqueous solutions where simple models and simple expressions such as (29) do not apply.

In concentrated reciprocal salt solutions there is no simple description of the structure of the large clusters and groupings of particular cation and anion pairs. From conformal solution theory[39] for a mixture of the ions $A^+, B^+/X^-, Y^-$ one obtains an expression for any component as BY of the form

$$RT\ln\gamma_{BY} = N_A N_X \Delta G^o + N_A N_X(N_X - N_Y)\lambda_{13} + N_X(N_A N_Y + N_B N_X)\lambda_{24} + N_A(N_A N_Y + N_B N_X)\lambda_{12} + N_A N_X(N_A - N_B)\lambda_{34} + N_A N_X(N_A N_Y + N_B N_X - N_B N_Y)\Lambda$$

= 1 metathetical term + 4 binary terms + 1 non-random mixing term (30)

TABLE 4

Values of ΔE_1 Obtained from the Comparison of Theory with Experimental Data

T, °K.	$-\Delta E_1$ (kcal.)			$K_1 = Z(\beta_1 - 1)$[a]
	Z = 4	Z = 5	Z = 6	
		$Ag^+ + Cl^-$ in KNO_3		
623	6.12	5.85	5.62	553
643	6.17	5.89	5.66	498
658	6.21	5.93	5.69	460
675	6.17	5.87	5.64	396
696	6.18	5.88	5.63	348
709	6.17	5.86	5.62	315
		$Ag^+ + Cl^-$ in $NaNO_3$		
604	5.10	4.83	4.62	277
637	5.12	4.84	4.62	226
658	5.17	4.88	4.65	205
675	5.10	4.81	4.57	176
696	5.13	4.83	4.59	160
711	5.12	4.81	4.56	146
773	5.14	4.82	4.55	110
		$Ag^+ + Cl^-$ in $(Na\text{-}K)NO_3$		
506	5.6	5.4	5.2	1050
551	5.57	5.33	5.13	644
658	5.67	5.38	5.15	302
752	5.72	5.40	5.13	180
801	5.62	5.28	5.00	133

[a] K_1 in mole fraction units.

ΔG^o is the free energy change for the metathetical reaction

$$AX \text{ (liquid)} + BY \text{ (liquid)} \rightleftarrows AY \text{ (liquid)} + BX \text{ (liquid)} \quad (31)$$

and the contribution of this term is positive when BY is a member of the stable pair and it contributes to positive deviations from ideal behavior. This term contributes negatively when BY is a member of the unstable pair. The terms λ_{ij} are binary interaction parameters for the indicated pairs of salts where AY is 1, BY is 2, AX is 3, and BX is 4 and are known from a knowledge of the four binary systems. The last term is a contribution due to non-random mixing or clustering of the pairs of ions of the stable pair of salts (AX + BY if ΔG^o is positive) and Λ can be approximated by $-(\Delta G^o)^2/2ZRT$ where a value of 6 for the coordination number Z is reasonable. Equation (30) leads to an insight into those interactions and parameters which are significant in understanding the chemical behavior of components. For a mixture of LiF with KCl the activity coefficients calculated from theory are in good agreement with values obtained from phase diagrams. Fig. 9 shows some of the calculations and measurements of the activity coefficients of LiF made for this system.[40] The "S" shaped character of the plot of RT log γ_{LiF} versus N^2_{KCl} is due to non-random mixing and clustering such that clusters of Li^+

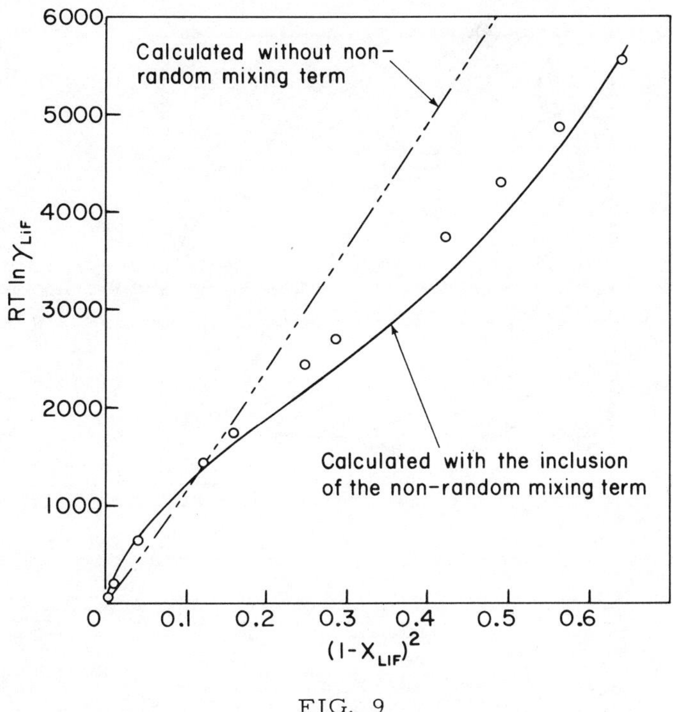

FIG. 9

Comparison of experimental values of $RT\ln\gamma_{LiF}$ vs. $(1-X_{LiF})^2$ in LiF-KCl with those calculated from theory.

with F^- and K^+ with Cl^- will tend to form. If there were no clustering, theory predicts a straight line for such a plot. Equation (30) enables one to understand the dependence of activities and activity coefficients upon concentration. This can be illustrated from the phase diagram for the Li^+, K^+/F^-, Cl^- system shown in Fig. 10.[41] The lines are isotherms for the components. The isotherms for LiF, which is a member

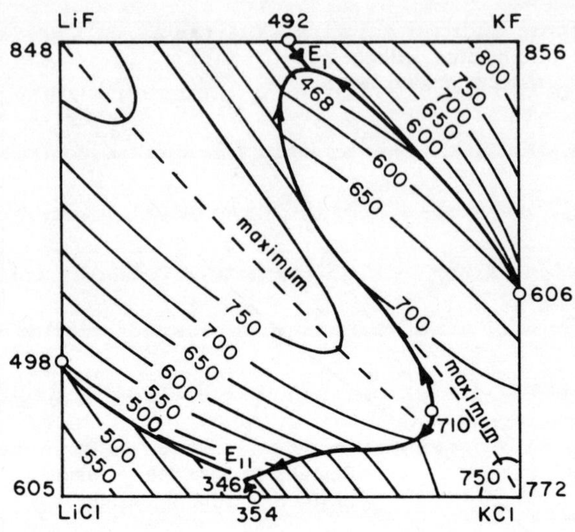

FIG. 10

Liquidus phase diagram for the Li^+, K^+/F^-, Cl^- system.

of the stable pair (ΔG^o = 16.3 kcal/mole), are bowed out away from the LiF corner. These isotherms are also isoactivity curves and change so that there is a minimum concentration and a maximum activity coefficient for LiF along the isotherm near the LiF-KCl diagonal. For temperatures above the liquidus temperature this means that for a fixed concentration of LiF the activity coefficient and activity of LiF goes through a sharp maximum as one varies the composition from the LiF-LiCl binary through the LiF-KCl diagonal to the LiF-KF binary. This phenomenon is also important in the chemical behavior of some additive ternary systems as will be discussed later in this paper. For KF, a member of the unstable pair, the isotherms are slightly bowed inwards towards the KF corner. From the theoretical equation (30) using reasonable values of the parameters it can be shown that one can predict all of the essential characteristics of the phase diagram of Fig. 10. Conversely, phase diagrams in conjunction with the theory may be utilized to deduce some of the values of the thermodynamic parameters in the theory.[41]

An important result of the theory is the calculation of an upper consolute temperature, T_c, below which a solution separates into two liquid phases. A simple theoretical approximation to T_c is

$$T_c = \frac{\Delta G^o}{5.5R} + \frac{\lambda_{12} + \lambda_{24} + \lambda_{13} + \lambda_{34}}{11R} \qquad (32)$$

which can be shown to correlate many of the large number of reciprocal molten salt systems which exhibit liquid-liquid immiscibility.[42,43] There has been very little experimental work in utilizing this phenomenon for liquid-liquid extractions although it shows promise as a practical tool. Kennedy has published two papers on this subject.[44,45] Utilizing the two phase molten systems, potassium nitrate-silver chloride or silver bromide, he measured the distribution of thallium between the two phases. These distribution coefficients depend on several factors such as the value of ΔG^o for the reaction

$$TlX + KNO_3 \rightleftarrows TlNO_3 + KX \qquad (33)$$

and the activity coefficients for the salts $TlNO_3$ in the nitrate phase and TlX and KX in the halide phases. Information on these factors is available or may be approximated. The activity coefficients of $TlNO_3$ change with the concentration of X in the KNO_3 phase largely due to the relatively weak association

$$Tl^+ + X^- \rightleftarrows TlX$$

and leads to a change of the distribution coefficients with the concentration of KX in the nitrate. Kennedy's results for the chloride system are plotted in Fig. 11 where the distribution coefficients are plotted as a function of the reciprocal of the chloride concentration in the nitrate.

Recently, Hagemark[46] has derived expressions for activities in additive ternary systems from the quasi-chemical approximation utilizing the quasi-lattice model. For dilute solutions of the salts BX and CX in the solvent AX he has shown[47] that some of the thermodynamic properties (e.g. the concentration dependence of the activity coefficients of

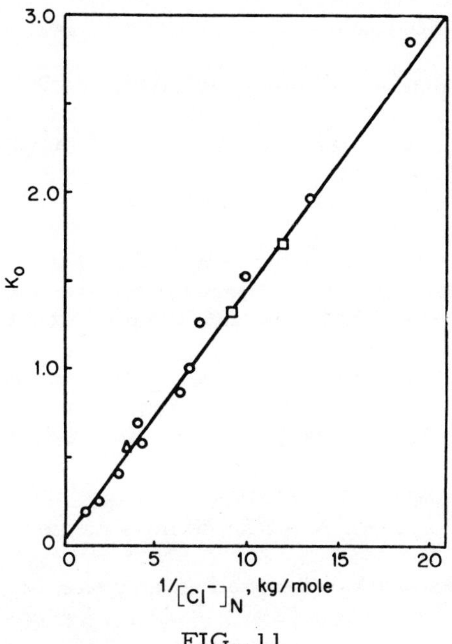

FIG. 11

Determination of true distribution constant and association constant in $AgCl$-KNO_3 system.

AX, BX, or CX) are related to associations of next-nearest-neighbor cations as shown in Fig. 12 for forming the simplest grouping, the B^+-C^+ pair.

The association constant has the same form as Equation (29) and the energy parameter is a combination of the values of the interaction parameters for the three binary systems AX-BX, AX-CX, and BX-CX. These associations can have a profound influence on activities in dilute solution.

$$\begin{array}{ccc} A^+ & X^- & A^+ \\ X^- & B^+ & X^- \\ A^+ & X^- & A^+ \end{array} \quad + \quad \begin{array}{ccc} A^+ & X^- & A^+ \\ X^- & C^+ & X^- \\ A^+ & X^- & A^+ \end{array}$$

$$\begin{array}{ccc} A^+ & X^- & C^+ \\ X^- & B^+ & X^- \\ A^+ & X^- & A^+ \end{array} \qquad \begin{array}{ccc} A^+ & X^- & A^+ \\ X^- & A^+ & X^- \\ A^+ & X^- & A^+ \end{array}$$

FIG. 12

In a binary system AX-BX which is dilute in the component CX, Hagemark has derived an equation for the activity coefficients of CX

$$\ln \gamma_{CX} = X_A \ln \gamma_{CX}(AX) + X_B \ln \gamma_{CX}(BX) - \Delta G^E(AX\text{-}BX) + f(N_A)$$

where $\ln \gamma_{CX}(AX)$ and $\ln \gamma_{CX}(BX)$ are the activity coefficients of CX in dilute solution in AX or BX. The last two terms on the right represent the deviations from linearity of a plot of $\ln \gamma_{CX}$ versus X_B. The third term on the right-hand side is the negative of the excess free energy of mixing of the AX-BX mixture. This function is zero at $X_A = 0$ or 1 and has a maximum or minimum at some intermediate composition. For stable solutions having negative excess free energies of mixing this leads to a maximum in the contribution of this term to the activity coefficients. This maximum in γ_{CX} can lead to minima in solubilities such as was observed for PuF_3 in the LiF-BeF_2 and NaF-BeF_2 binaries[48] or maxima in extraction coefficients such as was observed by Morrey and Moore[49] in the extraction coefficients of U by molten Al from 50-50 mixtures of KCl and $AlCl_3$ mixtures containing UCl_3. The last term $f(N_A)$ is related to the difference between the interactions of CX with AX and BX as measured by the difference $\ln\gamma_{CX}(AX)-\ln\gamma_{CX}(BX)$. When this is zero $f(N_A)$ is zero and f is negative for all other values of this difference.

Hagemark's calculations also justify the observation[50] that regions of ternary phase diagrams resemble reciprocal systems. Thus, the maxima in the activity coefficients of PuF_3 and UCl_3 in the systems mentioned in the last paragraph parallel similar maxima for a member of the stable pair in reciprocal systems as was discussed in an earlier section. Figs. 13 and 14 further illustrate this with the systems NaF-KF-BeF_2[51] and the important metallurgical system CaO-"FeO"-SiO_2.[52] The topology of the lower portions of these two ternary systems resemble reciprocal

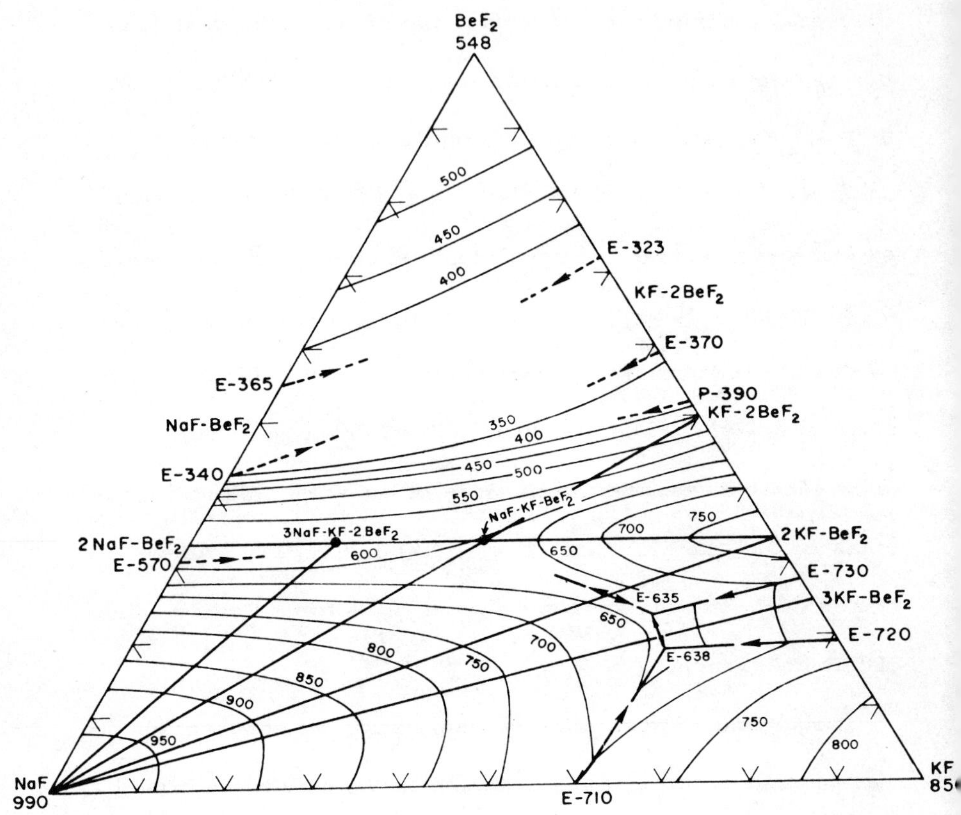

FIG. 13

Liquidus temperatures in the $NaF-KF-BeF_2$ system.

systems. (Compare Fig. 10 especially to Fig. 14.) The characteristics of the former system are related to the fact that molten $KF-BeF_2$ solutions have a somewhat more negative value of ΔG^E than $NaF-BeF_2$ and in the latter system it

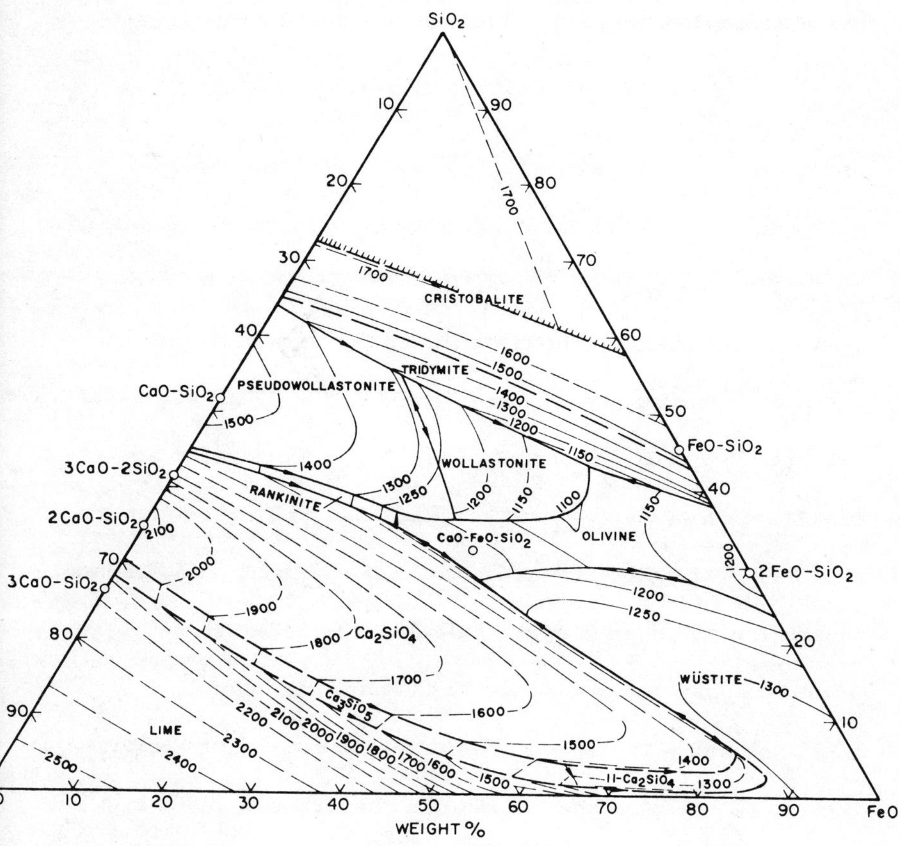

FIG. 14

Liquidus temperatures in the CaO-"FeO"-SiO_2 system.

is related to the fact that values of ΔG^E for CaO-SiO_2 binary solutions are much more negative than for "FeO-SiO_2" binary solutions. An essentially equivalent way of saying this is that ΔG^o for the reaction

$$\frac{1}{2}K_2BeF_4 + NaF \rightleftarrows KF + \frac{1}{2}Na_2BeF_4$$

has a somewhat positive value and ΔG^o for the reaction

$$\frac{1}{2}Ca_2SiO_4 + "FeO" \rightleftarrows \frac{1}{2}Fe_2SiO_4 + CaO$$

has a very positive value (~ 15 kcal.). The influence of ΔG^o parallels the effect in reciprocal systems as expressed by the first term in equation (30). The analogy with reciprocal systems can also be illustrated from the data of Taylor and Chipman[53] on the activities of "FeO" in the "FeO"-CaO-SiO_2 system. Figure 15 is a plot of log γ_{FeO} versus the equivalent fraction of Ca_2SiO_4 in the "FeO"-Ca_2SiO_4 quasi-binary system. The points are from the data of Taylor and Chipman and fit the "S" shaped curve calculated from equation (30) with the substitution of equivalent fractions for mole fractions and with the use of reasonable values for the parameters (Z = 4.6). The "S" shaped character is indicative of clustering of Fe^{++} with O^{2-} ions. This is consistent with expectations in simpler systems from calculations based on Hagemark's theory. A further expectation which can be understood for the simple systems treated by Hagemark is in ternary systems where one binary has a very negative value of ΔG^E. Analogous to reciprocal systems where ΔG^o is very large, immiscibility gaps which are completely enclosed within the ternary triangle should be found in these

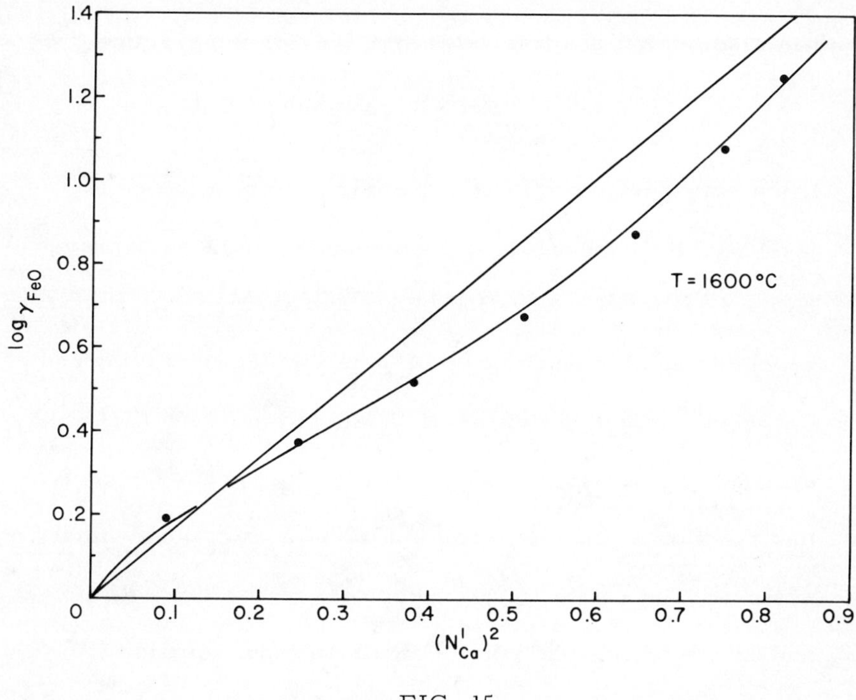

FIG. 15

Plot of log γ_{FeO} versus the square of the equivalent fraction of Ca_2SiO_4 in the quasi-binary system "FeO"-Ca_2SiO_4.

cases. The phase diagrams of two such systems, the LiCl-KCl-$AlCl_3$[54] and CaO-"FeO"-P_2O_5[52] ternary systems are illustrated in Figs. 16 and 17 where the KCl-$AlCl_3$ and CaO-P_2O_5 binary systems are characterized by very negative values of ΔG^E and the compounds $KAlCl_4$ and Ca_3PO_4 have very negative free energies of formation from the pure component end members (KCl + $AlCl_3$ and 3CaO + P_2O_5).

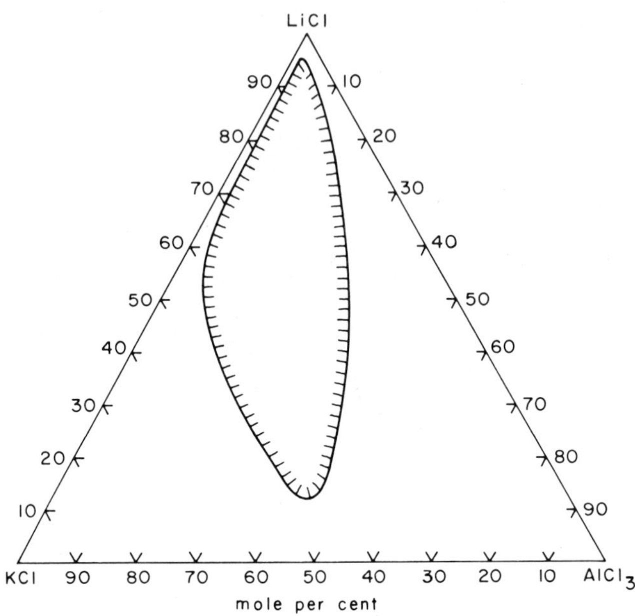

FIG. 16

Immiscibility gap in the LiCl-KCl-$AlCl_3$ system at 625°C.

Moore[55] has utilized the two phase region of the LiCl-KCl-UCl_3 system for extraction and has measured distribution coefficients for actinides between the LiCl and $KAlCl_4$ rich phases. His measurements indicate that separations of practical interest may be achieved.

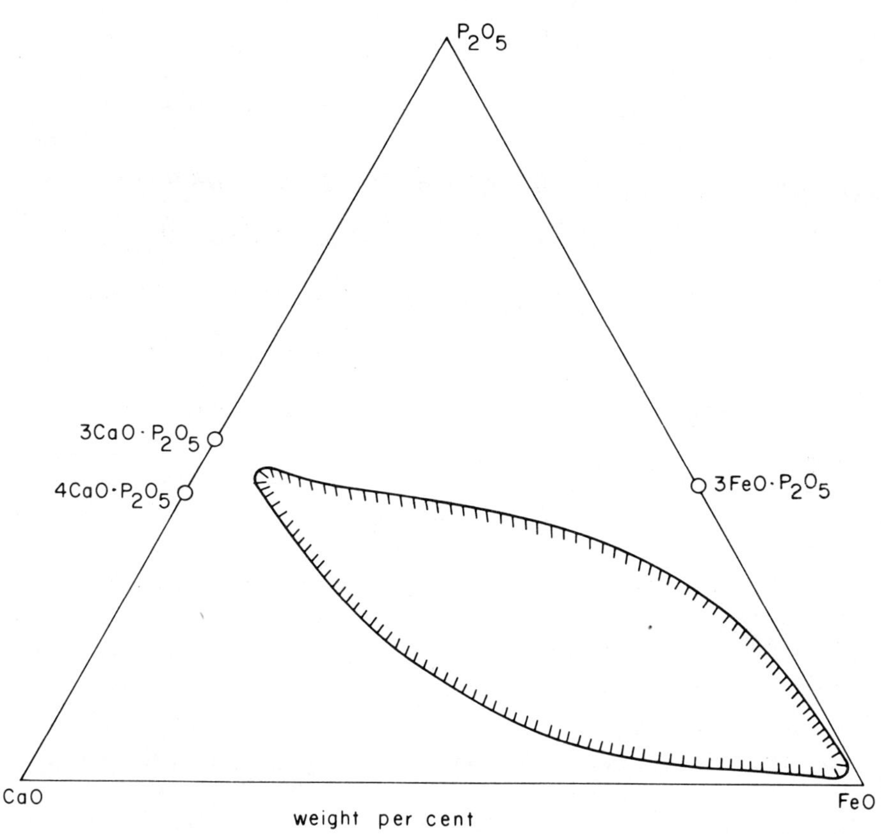

FIG. 17

Immiscibility gap in the CaO-"FeO"-P_2O_5 system.

CONCLUSIONS

Fundamental concepts of molten salts have given considerable insight into their chemistry. Theories based on simple structural and energetic concepts have helped in understanding chemical and phase equilibria, immiscibility gaps, extraction equilibria, complex ion formation in dilute solutions and many other phenomena relevant to analysis. It is hoped that this review, which in the interest of brevity has been superficial, will provide a useful reference for further study by practical chemists.

ACKNOWLEDGMENT

This work was partly supported by the Research Division of the U. S. Atomic Energy Commission.

REFERENCES

1. M. Blander, Ed., Molten Salt Chemistry, Interscience, New York, (1964).

2. B. R. Sundheim, Ed., Fused Salts, McGraw-Hill, New York, (1964).

3. J. Lumsden, Thermodynamics of Molten Salt Mixtures, Academic Press, New York, (1966).

4. H. A. Levy, P. A. Agron, M. A. Bredig, and M. D. Danford, Ann. N.Y. Acad. Sci., 79, 792 (1960).

5. H. A. Levy and M. D. Danford, "Diffraction Studies of the Structure of Molten Salts" in Molten Salt Chemistry, M. Blander, editor, Interscience, N. Y., London, (1964), p. 109.

6. H. Reiss, S. W. Mayer, and J. L. Katz, J. Chem. Phys., 35, 820 (1961).

7. H. Reiss, J. L. Katz, and O. J. Kleppa, J. Chem. Phys., 36, 144 (1962).

8. S. H. Bauer and R. F. Porter, "Metal Halide Vapors: Structures and Thermochemistry" in Molten Salt Chemistry, M. Blander, editor, Interscience, N. Y., London, (1964), p. 607.

9. H. Eyring, T. Ree, and N. Hirai, Proc. Natl. Acad. Sci., 44, 683 (1958).

10. L. F. Grantham and S. J. Yosim, J. Chem. Phys., 45, 1192 (1966).

11. R. Furth, Proc. Cambridge Phil. Soc., 37, 252 (1941).

12. J. O'M. Bockris and N. E. Richards, Proc. Roy. Soc. (London), A241, 44 (1957).

13. J. O'M. Bockris, N. E. Richards, and L. Nanis, J. Phys. Chem., 69, 1627 (1965).

14. J. O'M. Bockris, G. W. Hooper, Disc. Faraday Soc., 32, 218 (1961).

15. C. A. Angell, J. Chem. Phys., 46, 4673 (1967).

16. M. H. Cohen and D. Turnbull, J. Chem. Phys., 31, 1164 (1959).

17. G. Adam and J. H. Gibbs, J. Chem. Phys., 43, 139 (1965).

18. M. Temkin, Acta Physicochim. URSS, 20, 411 (1945).

19. M. Blander, "Thermodynamic Properties of Molten Salt Solutions" in Molten Salt Chemistry, M. Blander, editor, Interscience, N. Y., London (1964), p. 127.

20. O. J. Kleppa, J. Phys. Chem., 64, 1937 (1960).

21. O. J. Kleppa and L. S. Hersh, J. Chem. Phys., 34, 351 (1961).

22. O. J. Kleppa and L. S. Hersh, Disc. Faraday Soc., 32, 99 (1961); J. Chem. Phys., 36, 544 (1962).

23. O. J. Kleppa, R. B. Clarke, and L. S. Hersh, J. Chem. Phys., 35, 175 (1961).

24. O. J. Kleppa and F. G. McCarty, J. Phys. Chem., 70, 1249 (1966).

25. O. J. Kleppa and L. S. Hersh, Disc. Faraday Soc., 32, 99 (1961).

26. E. A. Guggenheim, Mixtures, Oxford Univ. Press, London, 1952.

27. (a) T. Førland, J. Phys. Chem., 59, 152 (1955).

(b) T. Førland, "Thermodynamic Properties of Fused-Salt Systems" in Fused Salts, B. R. Sundheim, ed., McGraw Hill, New York, 1961, p. 63.

28. J. Lumsden, Disc. Faraday Soc., 32, 138 (1961).

29. M. Blander, J. Chem. Phys., 34, 697 (1961).

30. M. Blander, J. Chem. Phys., 36, 1092 (1962).

31. H. T. Davis, J. Chem. Phys., 41, 2761 (1964).

32. J. Lumsden, Metallurgical J. (Univ. of Strathclyde) 18, 25 (1968).

33. H. Flood, T. Førland, and K. Grjotheim, Z. Anorg. Allgem. Chem., 276, 289 (1954).

34. M. Blander, J. Braunstein, and M. D. Silverman, J. Am. Chem. Soc., 85, 895 (1963).

35. M. Blander and E. B. Luchsinger, J. Am. Chem. Soc., 86, 319 (1964).

36. M. Blander, J. Chem. Phys., 34, 432 (1961).

37. (a) M. Blander, J. Phys. Chem., 63, 1262 (1959).

(b) J. Braunstein and R. M. Lindgren, J. Am. Chem. Soc., 84, 1534 (1962).

38. M. Blander and J. Braunstein, Ann. N.Y. Acad. Sci., 79, 838 (1960).

39. M. Blander and S. J. Yosim, J. Chem. Phys., 39, 2610 (1963).

40. M. Blander and L. E. Topol, Electrochim. Acta, 10, 1161 (1965).

41. M. Blander and L. E. Topol, Inorg. Chem., 5, 1641 (1966).

42. I. N. Belyaev, Usp. Khim., 29, 899 (1960); Russ. Chem. Rev., 29, 428 (1960).

43. J. E. Ricci, "Phase Diagrams of Fused Salts" in Molten Salt Chemistry, M. Blander, ed., Interscience, N.Y., London (1964), p. 239.

44. J. H. Kennedy, J. Phys. Chem., 65, 1030 (1961).

45. J. H. Kennedy, J. Chem. and Eng. Data, 9, 95 (1964).

46. K. Hagemark, J. Phys. Chem., 72, 2316 (1968).

47. K. Hagemark, private communication.

48. Reactor Chemistry Division Annual Progress Report ORNL 2931 (1960) Off. Tech. Serv., Dept. of Commerce, Washington, D.C.

49. J. R. Morrey and R. H. Moore, J. Phys. Chem., 67, 748 (1963).

50. M. Blander in Studies on Molten Salts, Euratom 2466 (1965), p. 39.

51. R. E. Thoma, Ed., *Phase Diagrams of Nuclear Reactor Materials,* ORNL 2548 (1959) Off. Tech. Serv., Dept. of Commerce, Washington, D.C.

52. A. Muan and E. F. Osborne, *Phase Equilibria Among Oxides in Steelmaking,* Addison-Wesley, Reading, Mass. (1965).

53. C. R. Taylor and J. Chipman, *Trans. AIME* *154*, 228 (1943).

54. R. H. Moore, *J. Chem. Eng. Data,* *8*, 164 (1963).

55. *Ibid.*, *9*, 502 (1964).

THE EXPERIMENTAL EVIDENCE FOR "COMPLEX IONS" IN SOME MOLTEN SALT MIXTURES[1]

M. A. Bredig

Chemistry Division, Oak Ridge National Laboratory
Oak Ridge, Tennessee 37830

INTRODUCTION

The interpretation of the thermodynamic, chemical, and physical properties of molten salt mixtures in terms of "complex ions" has been shown to present considerable conceptual difficulties when one deals with binary systems consisting of two simple salts with two kinds of cations and one kind of halide ion.[2] There are cases, to be sure, in which by various yardsticks and for good reasons complexing is definite enough for the sweeping objection to the concept to lose its force. Extremes of this variety include those systems in which one of the two components, e.g., $AlCl_3$, $BeCl_2$, or BeF_2, sometimes termed "Lewis acids," should in fact not be considered a "salt," certainly not in the sense of being an electrolyte. Rather, such substances actually occur in the form of volatile, molecular liquids, e.g., Al_2Cl_6, or as very viscous, polymeric network type of fluids, such as BeF_2 (a "weakened model" of SiO_2;[3] contrary to the belief of some[7] it differs greatly in its structure involving single fluoride bridges from $BeCl_2$ with its double chloride bridging). In molten mixtures of such substances with alkali halides, complex anions such as $AlCl_4^-$ or BeF_4^{2-} are hardly controversial and their stability can be readily rationalized in terms of the charges, sizes, and polarizabilities of the cations and halide ions involved. Systems of this type are comparable to oxide systems such as the silicate, phosphate, nitrate, or sulfate systems (M_xO_y - SiO_2, P_2O_5, N_2O_5, or SO_3), or to systems involving, with alkali fluorides or hydrides, a normally gaseous component such as BF_3 or B_2H_6. In the following we shall refer to such systems containing chloroaluminate or fluoroberyllate ions only briefly. Instead, this brief review will be confined to some critical comments on a number of examples of binary mixtures of truly salt-like components, e.g., KCl and $CdCl_2$. We shall sift some of the experimental evidence for a variety of proposed "complex anions" ranging from simple ones such as "$CdCl_3^-$" to larger and highly charged ones such as "$CdCl_6^{4-}$" or even "$MgCl_6^{4-}$."

The term "complex" or "complex anion" is used here with the implicit understanding that the type of interaction of the halide ion X^- with the cation B^{2+} within the complex "species" is usually not qualitatively ("covalent" vs. ionic) but just quantitatively distinguished from, that is, stronger than, the interaction outside with the cations of lesser charge. The emphasis is thus rather more on differences in "coordination number," or (average) number of nearest neighbors of negative charge, X^-, around the "central cation" B^{2+} than anything else.

THE INTERACTION PARAMETER λ AS EVIDENCE FOR A COMPLEX ANION

The question of complexing in molten binary mixtures of a monohalide such as one of the alkali metals with a dihalide of a variety of elements such as Zn, Cd, Hg, Pb, Ni, or Mg has attracted the attention of many. More recently, a special favorite among these, the system KCl-$CdCl_2$ has been studied calorimetrically by Metzger, Brenner, and Salmon[13] who presume to have succeeded for the first time in "an unambigious determination of the composition of the complex," namely, "$KCl.CdCl_2$." This claim seemed to refute the present writer's earlier conclusion[14] that the formation, not of $CdCl_3^-$, but of tetrahedral $CdCl_4^{2-}$ from 2 KCl and 1 $CdCl_2$ is the predominant or even the sole mode of complexing in this system. This conclusion was fundamentally based on structural considerations pertaining to the relatively large size of the central Cd^{2+} ion. Its field was thought not to be effectively shielded under the conditions of a condensed halide phase by a coordination of only three ions of the size of Cl^-. Moreover, certain experimental observations such as Raman spectra and electrical conductance measurements appeared to agree with this view. We wish to demonstrate here, that the more recent calorimetric data[13] also are preferably to be interpreted in terms of the 2:1 combination.

In discussing the experimental data[13] we follow the practice of O. J. Kleppa et al.[15] by considering the interaction parameter $\lambda \equiv \Delta H^M/N(1-N)$ as a function of the mole fraction, N. The recent study by Metzger et al.[13] gives the "partial molal heat effects" for the components KCl and $CdCl_2$, equal, but in sign opposite, to the more common "partial molal heats of mixing," $\Delta\bar{H}_1$ and $\Delta\bar{H}_2$, respectively. With $\Delta H^M \equiv N_1\Delta\bar{H}_1 + (1-N_1)\Delta\bar{H}_2$, we obtain, by substitution, $\lambda = \frac{\Delta\bar{H}_1}{(1-N_1)} + \frac{\Delta\bar{H}_2}{N_1}$. Fig. 1 shows λ thus computed from the data of Fig. 7, ref. 13 (KCl-$CdCl_2$, 780°C). For comparison, the data by Kleppa and McCarty[15] for RbCl-$MgCl_2$ are included.

It is very apparent that the λ data for KCl-$CdCl_2$ from partial molal heat effects (M., B. & S., full circles) are by several orders of magnitude less precise than those by Kleppa et al. from the total enthalpy of mixing ΔH^M (crosses connected by curve e). Of the two curves c and d, representing the KCl-$CdCl_2$ data of M., B. & S., the dotted one (d) is the best curve drawn through the mean values of λ for each composition. The smooth line (c), leading to no different conclusion, reflects an adjustment of the experimental data resulting from the observation that the curves for the experimental partial molal heat effects in Fig. 7, ref. 13, did not satisfy the required (Gibbs-Duhem) relationship between partial molal quantities. In other words, a curve for $\Delta\bar{H}_{KCl}$ obtained from the graphical evaluation of a plot of $\Delta\bar{H}_{CdCl_2}$ vs.

(N_{CdCl_2}/N_{KCl})(cf., e.g., Lewis-Randall, Thermodynamics, 1. ed. (1923) page 93) showed considerable disagreement with the direct experimental data for $\Delta\bar{H}$(KCl) (amounting to more than 100% for small $\Delta\bar{H}$ values). Violation of the Gibbs-Duhem relationship is also immediately apparent in the slopes for the experimental partial molal heat effect curves ($-\Delta\bar{H}_1$ and $-\Delta\bar{H}_2$) at the limits of $N_1 = 1$ and $N_2 = 1$ (Figs. 7 and 8, ref. 13), respectively. These slopes must be zero rather than finite. Furthermore, the values of the experimental points for $-\Delta\bar{H}$(KCl) at N(KCl) ≈ 0.73, 1.5 and 1.8 KCal/mole, and curve 4 drawn through them in Fig. 8, ref. 13 are by a factor 10 too high (error in decimal point?) when the mean of the corresponding values of 9.9 and 11.2 for $-\Delta\bar{H}(CdCl_2)$ on curve 4' at this concentration is taken to be nearly correct.

The adjustment made in our Fig. 1 for the smooth curve (c) consisted of using values of $\Delta\bar{H}_i$ intermediate between the experimental ones and those determined graphically from the experimental ones of the other component. While this simplified treatment is slightly less accurate than a more reiterative procedure, the inherent lack of higher accuracy of the data did not seem to warrant a greater effort.

EVIDENCE FOR "COMPLEX IONS"

Both the unadjusted and adjusted λ curves (c and d) go through their pronounced (negative) maxima not at the 1:1 composition but near 65-70 mole % KCl, clearly indicating the 2:1 complex, or $CdCl_4^{2-}$. This is in complete analogy to the $MgCl_4^{2-}$ species arrived at from the calorimetric measurements by Kleppa and McCarty.[15]

Interaction parameters derived from Fig. 9 of ref. 13 for 600°C after application of the adjustment above lead to the same conclusion as those for 780°. Thus, in spite of certain shortcomings, these calorimetric measurements must be taken strongly to support rather than refute the present writer's earlier proposal of $CdCl_4^{2-}$.

Also included in Fig. 1 are λ values (curves a and b) based on data for ΔH^M which Bloom and Tricklebank[16] obtained by a combination of drop calorimetry and solution calorimetry, for which the likelihood of a multiplication of experimental errors must be kept in mind. Except below 35 mole % KCl, the absolute values are somewhat smaller than those from ref. 13, yet they appear to be equally indicative of the absence of a 1:1, and of the presence of the 2:1 complex.

THE CONCENTRATION DEPENDENCE OF λ IN CHARGE-UNSYMMETRICAL SYSTEMS

In attempting to represent the interaction parameter or enthalpy of mixing by physically more or less meaningful expressions it must be realized as does not seem to have always been the case[16] that an equation of the form $\Delta H^M/N(1-N) = a + bN + cN(1-N) = \lambda$, which contains the term $cN(1-N)$, is capable of yielding only a symmetrical departure of λ from the linearity $(a + bN)$. In charge-symmetrical systems such symmetry is to be expected and has been observed and discussed by Hersh and Kleppa.[17] The expression is inherently incapable of reproducing the asymmetrical departure expected in charge-unsymmetrical systems $AX-BX_2$ and thus does not reflect the asymmetry in the deviation from linearity in $\lambda = \Delta H^M/N(1-N)$, that is indeed indicated in plots of λ vs. N for the experimental data of reference 16. In the $KCl-CdCl_2$ system (our Fig. 1, curves a and b, based on ref. 16), the more complicated shape of λ, with its strong (negative) crest near N(KCl) = 2/3 rather than 1/2, demonstrates the occurrence of the 2:1 complex beyond mere asymmetry in the deviation of λ from linearity.

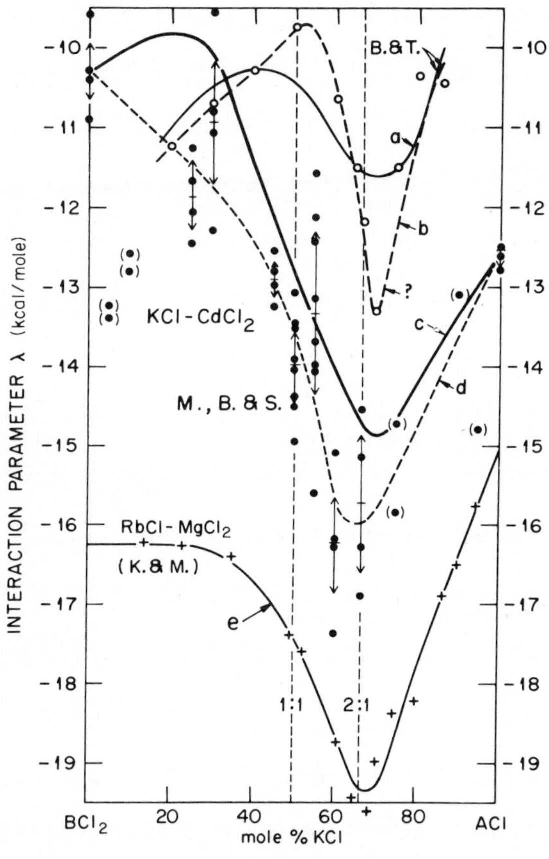

FIG. 1

Interaction parameter $\lambda \equiv \Delta H^M/N(1-N)$ for KCl-$CdCl_2$ (curves a, b, c, d) and RbCl-$MgCl_2$ (e) after data from ref. 13, 15, and 16. Curve a corresponds to ΔH^M curve adopted by Bloom and Tricklebank.[16] Full circles (arrows for mean deviation) and broken line d from data of ref. 13 not adjusted for Gibbs-Duhem relationship. Smooth curve c from adjusted data.

What the equation above, based on mole fractions, fails, in fact must fail, to do may be accomplished by an expression similar to it but based on equivalent fractions, f: $\lambda_f = \Delta H^M(\text{equiv})/f(1-f) = \alpha + \beta f + \gamma f(1-f)$; $f \equiv f(BX_2) \equiv 1 - f(AX)$. The use of equivalent fraction is not too different from the use of volume fraction, that important parameter in solution theory. The expression may hold in cases in which there is no pronounced complex formation. The actual occurrence of complexing is then indicated by the departure from, rather than agreement with, this equation.[18] Its application to the systems[15] $LiCl-MgCl_2$, $NaCl-MgCl_2$, and $AgCl-MgCl_2$ and comparison with some charge-symmetrical systems (LiCl-NaCl and LiCl-CsCl)[17] are shown in Fig. 2. The empirical parameters α, β, and γ were evaluated as follows: $\alpha = \gamma_f$ for $f = 0$; $\beta = -\alpha + \gamma_f$ for $f = 1$; $\gamma = [(d\lambda_f/df) - \beta]/(1 - 2f)$ for $f = 0$ or 1. For $LiCl-MgCl_2$, the agreement between the experimental and calculated curve is good. This is taken to signify very little complexing to $MgCl_4^{2-}$ in the system with the small lithium ion, as expected. For $NaCl-MgCl_2$ the deviations are considerably greater, with a maximum near $f = 1/2$. This is taken to indicate some formation of $MgCl_4^{2-}$, quite in agreement with Kleppa and McCarty.[15]

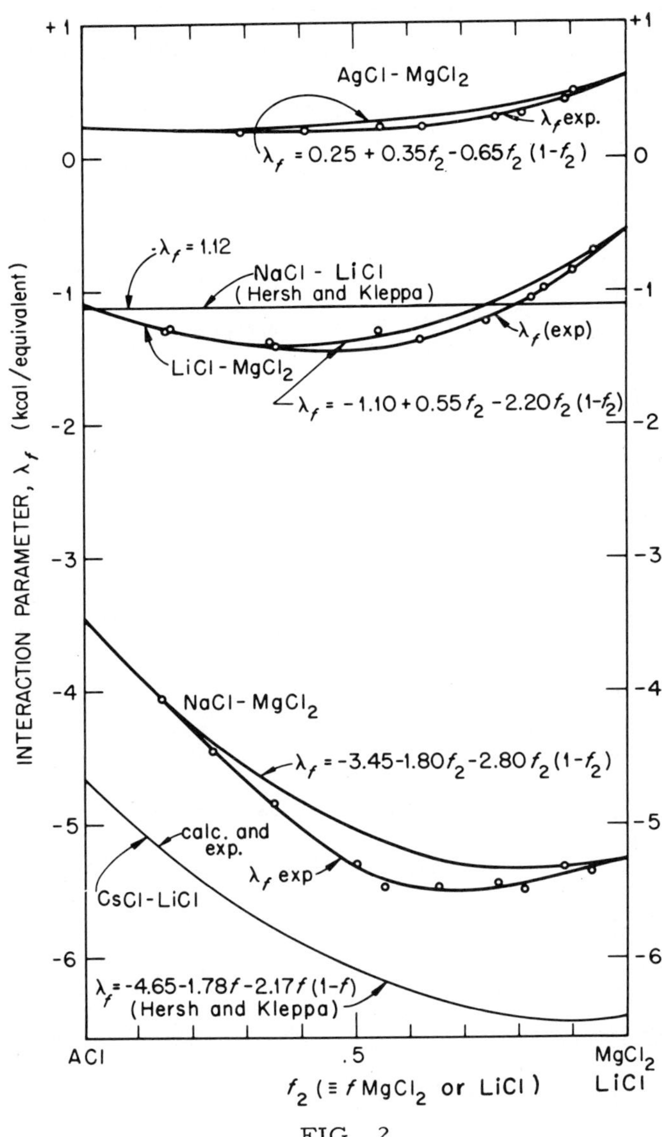

FIG. 2

Interaction parameter $\lambda_f \equiv \Delta H_f^M / f(1-f) = \alpha + \beta f + \gamma f(1-f)$ in kcal/equivalent compared with experimental data for charge-unsymmetrical and charge-symmetrical systems of Kleppa et al. ref. 15 and 17.

For other systems $AX - BX_2$ ($NaCl-CdCl_2$, $KI-CdI_2$, $KCl-PbCl_2$),[16] relatively low precision and lack of measurements above 86 mole % of alkali halide hamper the interpretation of the λ curves in terms of particular complexes. Very likely a decisive change to less negative λ would occur before approaching N(AX) = 1, but it is not apparent in the present data. In the case of $KCl-PbCl_2$ the maximum in $-\lambda$ might have appeared near N(KCl) = 0.75, in analogy to $CsCl-PbCl_2$ where instead of "$PbCl_3^-$," "$PbCl_4^{2-}$," or "$PbCl_6^{4-}$," a weak complex 3 $CsCl \cdot PbCl_2$, i.e., "$PbCl_5^{3-}$" is suggested by the data of McCarty and Kleppa.[19] This complex ion may conceivably have the symmetry C_4, i.e., its structure might be a tetragonally distorted octahedron constituted by the five Cl^- and the lone 6s electron pair in Pb^{2+}. It is further to be noted that for $KCl-PbCl_2$, at N(KCl) = 0.20, λ = -0.7 kcal/mole at 750°C (from ref. 16) is by more than a factor six smaller than λ = -4.3 at 655° (ref. 19), but it is doubtful that this is a valid indication of the temperature dependence of the interaction.

Recently, S. V. Meschel and O. J. Kleppa[20] have also represented their calorimetric data for two charge-unsymmetrical binary salt mixtures in terms of equivalent fractions. A simple linear equation was found to apply. However, it seems likely that for the full concentration range, i.e., above the present limits of 46 and 60 mole % cadmium salt in $TlNO_3$-$Cd(NO_3)_2$ and TlCl-$CdCl_2$, respectively, the inclusion of the additional term $\gamma f(1-f)$ will be required.

Mole fraction has thus far been used by Kleppa and McCarty[15] in discussing the energetic asymmetry in the systems ACl -$MgCl_2$ as represented by the constant b in the equation above, which is the difference in the limiting interaction parameters of the components ACl and $MgCl_2$. This may be amplified by a description in terms of equivalent fractions. It seems that this procedure may be advantageous if for no other than the following reasons:

1) In comparing a charge-unsymmetrical system such as CsCl-$MgCl_2$ with an analogous symmetrical one, CsCl-LiCl, one is inclined, because of the higher charge of the Mg^{2+} ion, in size similar to Li^+, to expect a larger <u>relative</u> asymmetry, $2b/(\lambda(CsCl) + \lambda(MgCl_2))$ than for $2b/(\lambda(CsCl) + \lambda(LiCl))$. Actually, the experimental values are in the reverse order, $(-3.0 \times 2)/(-20.0-17.0) = 16\%$ for the magnesium system[15] and $(-1.78 \times 2)/(-6.43-4.65) = 32\%$ for the lithium system.[17] In the scheme of equivalent fractions the relative asymmetry for the magnesium system becomes much

larger, $2\beta/(\lambda_f(CsCl) + \lambda_f(MgCl_2)) = (-11.5 \times 2)/(-8.5-20.0) = 81\%$.

2) The asymmetry parameter b for mole fractions changes sign in the middle of the range of the alkali metal ion sizes, even though all are either equal (Li^+) to, or increasingly larger than, Mg^{+2}(Fig. 4, ref. 15).

Our Fig. 3 shows β, the asymmetry parameter in the scheme of equivalent fractions, as a function of the ionic size parameter $\partial_{12} = (d_{ACl}-d_{MgCl_2})/(d_{ACl}\, d_{MgCl_2})$. Where b is a two-parameter linear function of ∂_{12}^3 (ref. 16), β is simply proportional to ∂_{12}^2, $\beta = C\,\partial_{12}^2$, $C = -780$ kcal/$\mathring{A}^2$ equivalent. It would appear that this simple relationship merits further attention.

MISCELLANEOUS EVIDENCE FOR COMPLEX IONS

For KCl - $MgCl_2$ melts, Raman effect measurements have been interpreted as demonstrating the "$MgCl_3^-$" ion, curiously shaped as a trigonal pyramid, to be prominently present.[21] The conclusion is mainly based on the alleged polarization of two out of an alleged total of four "Raman lines." However, the differences in the appearance of the two intensity curves in question (Fig. 3, ref. 21) are simply not large enough to seem significant. Moreover, especially

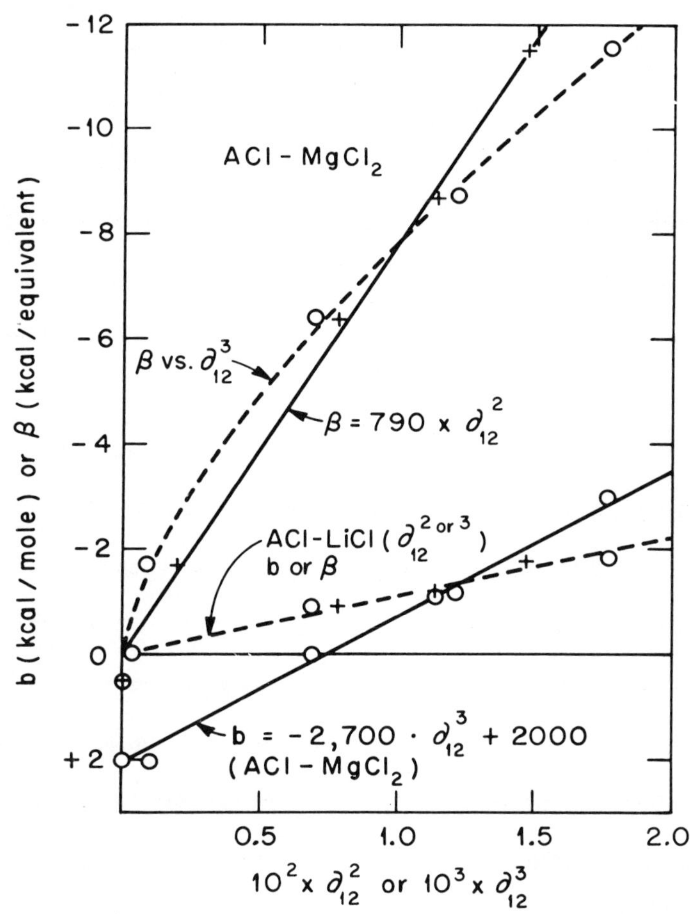

FIG. 3

Interaction asymmetry parameter $\beta = \lambda_f(MgCl_2 \text{ or } LiCl) - \lambda_f(ACl)$ for equivalent fractions and $b = \lambda(MgCl_2 \text{ or } LiCl) - \lambda(ACl)$ vs. cube or square of interionic distance parameter ∂_{12} from data by Kleppa et al.[15,17]

without a comparison of a blank run with pure molten KCl, it is extremely likely that many of the "Raman lines" reported in this and related studies are nothing but "noise" in the very high background (90% of the intensities of the peaks). The positions of the "lines" were not well reproducible, even though they were often far sharper (line half-width less than 10 cm^{-1}, cf., e.g. "line" near 150 cm^{-1}, Fig. 3, ref. 21) than compatible with a slit width of 15 cm^{-1} used or with the nature of a molten salt "complex" with poorly defined vibrational states. W. Bues[22] found a reasonable half-width as high as 40 cm^{-1} for the zinc chloride complex, $ZnCl_4^{2-}$, likely to be even more stable than $MgCl_4^{2-}$. It is further instructive to compare line widths shown for the far better defined nitrate ion, in molten $AgNO_3$ ranging from 40 or 60 cm^{-1} on up.[23] (In this connection, it would also be far more reasonable to interpret the spectrum presented for pure molten $MgCl_2$ (Fig. 1, ref. 21) in terms of one Raman line only, with a peak intensity of 25% above background, at 220 cm^{-1} and a half-width of 45 cm^{-1}.) The proposed pyramidal shape of a "$MgCl_3^-$" (and of a "$CdCl_3^-$," ref. 24) has as little foundation in chemical theory or intuition, for the elements involved, as has the very large difference of 20%

between the two pairs of Al - Cl distances in an allegedly "sphenoidal" $AlCl_4^-$ ion of KCl - $AlCl_3$ melts, deduced similarly from Raman effect measurements.[25] Again it is the abundance of what simply cannot be true Raman lines but must be artifacts such as electronic noise, namely, the far too narrow and poorly reproducible "lines" at 136, 155, 168, 182, 196, 562, and 592 cm^{-1}, which makes the interpretation given unacceptable. The reported data are much more realistically consolidated into 3 lines of reasonable width ($>30\ cm^{-1}$) as follows: 180(vw)depol. (in the foot of the exciting line), 353(I)pol., and 510(m)depol. A fourth and very weak one, to complete the number of fundamental modes expected for a regular tetrahedron $AlCl_4^-$ in $K(AlCl_4)$, just as in $Na(AlCl_4)$ with lines 145(m), 183(m), 349(I) and 580 (?, vw),[26] may be hidden near 150 cm^{-1} in the large foot of the exciting line (Figs. 1 and 3, ref. 25; cf. also Figs. 2 and 4, ibidem). This analysis also avoids the extremely tenuous explanation of alleged different structure in terms of "compressed bonds" as a result of the difference in size between Na^+ and K^+.

For mixtures of $HgCl_2$ with KCl or NH_4Cl,[27] the interpretation given to minute details in the Raman spectra observed, which included even estimates of amounts of "$HgCl_3^-$" presumed to be present seems open to question for reasons similar to those just discussed. One of various objections is that the demonstration of observed polarization effects documented by Fig. 1 and listed in Table II of ref. 27 is not at all convincing. Although the width of the slit used was not reported, seven out of a total of nine alleged Raman bands, two of them, 180 and 192 cm^{-1} outside the range depicted in Fig. 1,[27] must very likely again be considered spurious and be consolidated into just two bands of reasonable width (40 cm^{-1}) with peaks at 314 and 270 cm^{-1}, for pure $HgCl_2$ and, probably, tetrahedral $HgCl_4^{2-}$, in the mixtures, respectively. The appearance or disappearance of shoulders on either band indicate the growing-in of the other with change in melt composition.

In a report on electrical conductance of KCl - $MgCl_2$ melts[28] it was recently stated that "the presence of $MgCl_3^-$ complex ions can be taken as established." The evidence from electrical measurements is far from satisfactory. The activation energy for electrical conductance, said to increase sharply above 50 mole % $MgCl_2$ actually is seen to be constant up to 70 mole % (cf. points in their Fig. 5). Yet a rather sudden change in $d\kappa/dN(KCl)$ from 0 to 2.5 ohm^{-1} cm^{-1} in the specific conductivity (κ) isotherm occurs below 35 mole % $MgCl_2$ (points in their Fig. 3), or in other words, maximum deviation from additive conductivity occurs at this concentration, suggesting K_2MgCl_4 in agreement with the calorimetric data.[15] The definite evidence for the occurrence of $KMgCl_3$ in the vapor which is cited has no bearing at all upon the molecular structure of the melt.

For the corresponding cadmium system KCl-$CdCl_2$ we were unable in careful measurements[29] to confirm the sharp dip in the specific conductance of KCl-$CdCl_2$ melts at the composition 1:1 that was reported by N. M. Tarasova[30] as evidence for the existence of "$CdCl_3^-$" ions.

Electrical conductivity was cited[31] to support also the existence of the complex ion "$CdCl_6^{4-}$" in KCl-$CdCl_2$ at 20.4 mole % $CdCl_2$. Sharp drops in conductivity either with decreasing temperature or increasing KCl concentration, much sharper than would seem to be reasonable for a dissociation equilibrium of a complex, were reported to occur 30-40° above the liquidus (Fig. 13, ref. 31 and Fig. 6, ref. 32). It was implied that it was impossible to assume as had been suggested[14] that the close vicinity of the liquidus temperature per se was responsible for this observation through solidification of salt on at least one of the electrodes. The upper one especially (cf. Fig. 2, ref. 32) partly as the result of high metallic thermal conductivity, might have been at a somewhat lower temperature than the melt. No drop in conductivity had occurred, presumably as close as 10° above the alleged liquidus temperature of "584°" in $NaCl-PbCl_2$ (1:1) where complex formation is not indicated otherwise. This argument, however, is fully invalidated by the fact that the $NaCl-PbCl_2$ liquidus temperature assumed in Fig. 13 (ref. 31) is in error and actually lies near 548° (not 584) according to two different sources (Treis, 1914, quoted in ref. 31 as "Landolt-Bornstein, 1958"; and Ill'yasov and Bergman[33]). This places the lowest reported temperature of the conductivity measurements in $NaCl-PbCl_2$ (594°) at least 45° (not 10°) above the liquidus. It seems highly probable that at a temperature, as measured in the bulk liquid, just slightly closer to, but perhaps even as far as $30-40^{o}$ above the true liquidus temperature in $NaCl-PbCl_2$ a sharp apparent drop in "conductivity" similar to that in $KCl-CdCl_2$ would have been observed, merely as a result of the particular experimental conditions.

EVIDENCE FOR "COMPLEX IONS"

A presumed maximum in the partial molal volume of $CdCl_2$ (ref. 34, figure 4) was also taken to indicate the complex 4:1, with $(CdCl_6)^{4-}$ ions. Actually the significance of a maximum in the partial quantity corresponding merely to an inflexion in the total one (experimentally not even too well supported in these systems) is very minor compared with that of an inflexion and a rapid change in the partial quantity, which correspond to the maximum deviation in the total quantity from linearity. (Cf. e.g. Kleppa and McCarty[15] fig. 6). A large rise and inflexion in $\overline{V}(CdCl_2)$ occurs in KCl-$CdCl_2$ between 50 and 80 mole % KCl, i.e. centered around 65% KCl,[32] suggesting again the predominance of the complex 2 KCl·$CdCl_2$.

The assumption has been made that a "building up of octahedral complex ions began as KCl, RbCl or CsCl were added to $CdCl_2$," and that "these would reach a maximum concentration at 80 mole per cent alkali metal chloride..."[31] Contrary to this suggestion the following model would seem to be in better agreement with general notions about molten salt mixtures. In most AX-BX_2 systems one already starts with an environment ("coordination") of about six halide anions X^- around B^{2+} cations in the largely ionized pure dihalide melt (reasonably high conductivity, with a few exceptions such as in the highly polymeric BeF_2, $BeCl_2$, or $ZnCl_2$ where the coordination number is more likely four). Then, the addition of AX with large A cations of low charge, such as K^+ and especially Cs^+, initiates a structure change towards lower, but much more intense (tetrahedral) coordination of X^- around B^{2+}, also involving asymmetric polarization of the X^-. This structural change proceeds further and further with increasing alkali halide concentration. No reliable evidence exists that even with a large excess of X^- ions over and above the composition A_2BX_4 this coordination number of 4 is again exceeded to form "BX_6^{4-}." Possibly an exception occurs when a highly polarizable, relatively large B^{2+} cation such as Pb^{2+} is involved, but even then the resulting complex ($PbCl_5^{3-}$? cf. page 10) is a rather weak one.[19]

It is also of great interest to note that spectroscopic observations[35] on molten mixtures of $NiCl_2$ with CsCl which are reasonably analogous to the $CdCl_2$-KCl mixtures show the tetrahedral complex $NiCl_4^{2-}$ to be stable up to the highest CsCl concentrations, N(CsCl) ≈ 1.

CONCLUSIONS

The preceding discussion has attempted to bring out the fact that the experimental evidence for the existence of BX_3^- or BX_6^{4-} complex ions in binary molten halide systems, AX-BX_2, based on a variety of techniques as applied by various investigators, is not satisfactory. Rather, measurements that at present appear to be by far the most trustworthy ones, such as calorimetric determinations of the integral heats of mixing, indicate a predominance of the BX_4^{2-} species in such systems possibly even to the exclusion of all other configurations.

REFERENCES

1. Research sponsored by the U. S. Atomic Energy Commission under contract with the Union Carbide Corporation.

2. Cf., e.g., M. Blander, "Molten Salt Chemistry," Interscience Publ., New York 1964, p. 186 ff.

3. V. M. Goldschmidt, Geochem. Verteilungsgesetze VIII, p. 131, Skr. Norske Vidensk.-Akad. Oslo, I. Matem. Naturvid. Kl. 1926, No. 8. With the polymeric nature of highly viscous molten BeF_2, analogous to that of SiO_2, the adoption of an entropy of fusion, $\Delta S_m = 12$ e.u. as derived from phase diagrams[4] was in sharp disagreement. A more recent and reasonable value of 1.4 e.u. determined calorimetrically[5] closely resembles that for SiO_2.[6] Large excess free energies of mixing polymeric SiO_2 or BeF_2 with their silicates or fluoroberyllates explains the discrepancy, above.

Another case of an unrealistic entropy of fusion from incomplete or erroneous phase diagram data is that of lithium hydride, "9.3" e.u.,[8] as against 5.75 measured calorimetrically well in line with all alkali metal halides and with other, reliable phase equilibria.[9] The interpretation of the LiH-LiCl phase diagram leading to "9.3" e.u. was no more satisfactory than an earlier one[10,11] employing the improper notion of strong ion association to a $LiCl_2^-$ species even in dilute solution. A much more rational interpretation was attained by considering solid solution formation.[12] This applies also to the LiH-Li system (ref. 8, p. 298).

4. T. Førland in "Fused Salts," B. R. Sundheim, Ed., McGraw Hill, 1964, p. 157.

5. A. R. Taylor and T. E. Gardner, U. S. Bur. Mines Rept. No. RI-6644 (1965).

6. Natl. Bur. of Stand., Bulletin 500.

7. F. A. Cotton and G. Wilkinson, "Advanced Inorgan. Chem.," 2nd Ed., Interscience Publ., 1966, p. 249.

8. J. Lumsden, "Thermodynamics of Molten Salt Mixtures," Academic Press, 1966, pp. 133, 224, and 298.

9. C. E. Messer, J. Mellor, J. A. Kroll, and I. S. Levy, J. Chem. Eng. Data, 6, 328 (1961).

10. C. E. Johnson, S. E. Wood, and C. E. Crouthamel, Inorg. Chem., 3, 1487 (1964).

11. C. E. Johnson, S. E. Wood, and E. J. Cairns, J. Chem. Phys., 46, 4168 (1967).

12. M. A. Bredig, J. Chem. Phys., 46, 4167 (1967).

13. William H. Metzger, Abner Brenner, and Harry I. Salmon, J. Electrochem. Soc., 114, 132 (1967).

14. M. A. Bredig, J. Chem. Phys., 37, 451 (1962), and earlier references there.

15. O. J. Kleppa and F. G. McCarty, J. Phys. Chem., 70, 1249 (1966), and earlier publications by Kleppa et al.

16. H. Bloom and S. B. Tricklebank, Australian J. Chem. 19, 187 (1966).

17. L. S. Hersh and O. J. Kleppa, J. Chem. Phys., 42, 1309 (1965).

18. This is somewhat different from the procedure of Kleppa and McCarty (15) who used mole fraction in referring to the asymmetric deviation from linearity.

19. F. G. McCarty and O. J. Kleppa, J. Phys. Chem., 68, 3846 (1964).

20. S. V. Meschel and O. J. Kleppa, J. Chem., Phys., 46, 1853 (1967).

21. K. Balasubrahmanyam, J. Chem. Phys., 44, 3270 (1966).

22. W. Bues, Z. anorg. allgem. Chem., 279 104 (1955).

23. David W. James, Fig. 5, p. 520, "Molten Salt Chemistry," M. Blander, Ed., Intersc. Publ., New York, 1964.

24. K. Balasubrahmanyam, J. O'M. Bockris, and M. Tanaka, Electrochim. Acta, 8, 621 (1963).

25. K. Balasubrahmanyam and L. Nanis, J. Chem. Phys., 42, 676 (1965).

26. Cf. also D. E. H. Jones and J. L. Wood, Spectrochim. Acta, 23A, 2695 (1967), who propose 485 cm^{-1} for this vibration.

27. G. J. Janz and D. W. James, J. Chem. Phys., 38, 902, 905, (1963).

28. E. A. Ukshe and E. B. Kachina-Pullo, Russ. J. Inorg. Chem., 11, 638 (1966).

29. H. R. Bronstein and M. A. Bredig, ORNL-3320, p. 101 (1962).

30. N. M. Tarasova, *Zh. Fiz. Khim.*, *21*, 825 (1947), discussed in "Electrochemistry of Fused Salts," IU. K. Delimarskii and B. F. Markov, Engl. Transl., Sigma Press, Washington, D. C., 1961.

31. H. Bloom, *Pure and Appl. Chem.*, *7*, 398 (1963).

32. H. Bloom and E. Heymann, Proc. Roy. Soc. (London), *188*, 392 (1946).

33. I. I. Ill'yasov and A. G. Berman, Russ. *J. Inorg. Chem.*, *7*, 181 (1962).

34. H. Bloom, P. W. D. Boyd, J. L. Laver, and J. Wong, *Australian J. Chem.*, *19*, 1591 (1966).

35. G. P. Smith, C. R. Boston, and J. Brynestad, *J. Chem. Phys.*, 45, 829 (1966).

THE ROLE OF PHASE EQUILIBRIA IN MOLTEN SALT RESEARCH*

R. E. Thoma

Reactor Chemistry Division
Oak Ridge National Laboratory
Oak Ridge, Tennessee

INTRODUCTION

Much of the interest which has appeared in molten salt chemistry over the last few years has arisen from the needs of high-temperature technologies, primarily those related to extractive metallurgy, molten salt breeder reactor development, and reactor fuel reprocessing. As a special class of liquids, one which is comprised entirely of positively and negatively

*Research sponsored by the U.S. Atomic Energy Commis- under contract with the Union Carbide Corporation.

charged ions undiluted by a weak-electrolyte supporting medium, molten salts have found increasing use in many different types of research. In much of the chemical research which is carried out with molten salt systems, the conditions of temperature and composition are varied far more extensively than is routine with aqueous or organic systems. Drastic changes in equilibrium phase relationships are thus commonplace. In many aspects of molten salt research and development, the most economic and time saving approach therefore begins with a general understanding of the heterogeneous equilibria of the systems involved. Pragmatically, this approach need not entail a complete nor even extensive study of the systems involved, but may be quite useful as a framework which prescribes the conditions which can be employed in related investigations of molten salts. In this connection, a frequently useful result of the characterization of phase equilibrium relationships in a system is the recognition, through experiment, of the metastable equilibria which are likely to be encountered as well as the relative tendency for their occurrence.

The investigation of phase equilibrium relationships in mixtures of salts is a very old-fashioned area of chemistry, and one which seems to have lost some of its appeal to the research scientist because of the limited extent to which accurate generalizations and predictions can be made concerning systems other than those investigated. I wish to show today that the pursuit of such investigations is intrinsically worthwhile and, in addition, provides an aproach to the fruitful evolution of many kinds of research with molten salt mixtures.

For many years scientists in this country and elsewhere expended a great deal of effort in attempts to predict the phase relationships of uninvestigated systems, but the necessary thermochemical data which were required, such as heat capacities of the constituents, the variation of this property with temperature, the heats of fusion, and the free energies of mixing, were not available. It soon became evident that their acquisition was at least as arduous a task as was the determination of the actual equilibrium behavior, and that quantitatively, the phase relationships predicted from the thermochemical data were too imprecise to make the method attractive. As a result, it has not yet become routine to estimate with practical accuracy the behavior of untested systems. Accordingly, the determination of equilibrium phase diagrams has remained principally in the region of experimental science. To be sure, numerous correlations are available which can assist the experimentalist in minimizing his tasks, such as inter-relationships of crystal structure, thermodynamic quantities, and the predictable degree of ideality of some ionic systems. Such information makes it possible for phase studies to be conducted with greater efficiency and rapidity, and therefore tends to maintain the field as experimental. The largely experimental character of molten salt phase equilibria today arises from the development over the last few decades of experimental techniques which have greatly increased the capability of the experimental scientist to delineate the phase relationships in molten salt systems far more accurately and rapidly than was previously possible.

The role of phase equilibria in molten salt research is multifaceted, ranging from that work that is done because of its attraction as intrinsically interesting phenomena to highly pragmatic investigations which seek to establish process controls or to devise a new or improved method of analysis. It is the purpose of this paper therefore to indicate several aspects of the role of phase equilibria in molten salt research and development as well as to describe some of the techniques which form a basis for this role.

TYPICAL EQUILIBRIA IN MOLTEN SALT SYSTEMS

In order to discuss the role of phase equilibria in molten salt research we must first survey the principal kinds of behavior which are most frequently encountered in heterogeneous phase equilibria. I will confine description of the varieties of behavior to a minimum in order to indicate the role of such studies to molten salt research and development. A number of excellent treatises[1-4] are available for more specific description.

Molten salt systems which find most frequent use as reaction media and for technological application are those which have sufficiently low liquidus temperatures that containment is not a problem, which are stable in the molten state, and which have low vapor pressures. To some extent, these requirements are conflicting, because salts with an appreciable covalent character which are useful in obtaining low freezing points also tend to give increased vapor pressure. The phase diagrams presently available for the binary and ternary systems of molten salts show that these systems are distinguished from those of other classes of substances by significant absences of retrograde solubility, appreciable solubility in the solid state among intermediate compounds, of immiscibility in the liquid state, and of solid solutions containing maximum temperature singular points. These are the qualities which, together with chemical stability, are principally responsible for the increasing use of molten salt media in industrial and research applications.

Characterization of salt phase equilibria is simple and straightforward with binary and ternary systems, but becomes increasingly intricate and of decreasing utility as the effect of additional components must be considered. For purposes of the present discussion, considerations are therefore limited to condensed equilibria in binary and ternary systems.

Let us now consider the binary system A-B. As drawn (Fig. 1) the phase diagram indicates that the components A and B react to form the intermediate compounds A_2B and AB. Limited miscibility in the solid state occurs when mixtures of A and A_2B are present at equilibrium at temperatures between T_1-T_2, and T_3 and T_4. The compound AB melts incongruently to A_2B and liquid. Crystals of AB are dimorphic; at temperatures above T_5 they exist in a crystalline

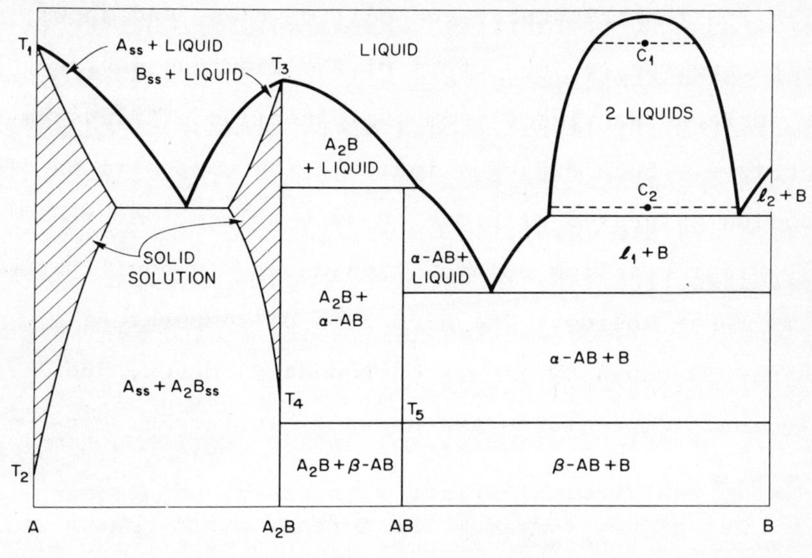

FIG. 1
Typical phase equilibrium relationships in condensed binary systems of molten salts.

state which transforms on cooling below temperature T_5 to a less symmetric (lower energy form) crystalline state. Liquids formed from mixtures of AB and B tend to be immiscible at compositions such as C_1 and C_2, which separate into immiscible liquids of compositions described by horizontal tie lines in the two-liquid region shown in Figure 1.

The temperature-composition relations in condensed ternary systems are generally represented by a polythermal projection of the liquidus surfaces. This is a projection parallel to the temperature axis, on the triangular composition plane and shows therefore the various parts of the liquidus surface or surfaces as viewed from the direction of high temperature. Such diagrams indicate the compositions of liquids saturated with one or more solids. At the invariant reaction points, liquids are in equilibrium with three solids. The direction of temperature change is shown by arrows on boundary curves, and liquidus temperatures are shown by isothermal contours.

Let us now consider the ternary system shown in Figure 2 in order to appreciate the kind of information which is described by such diagrams. We shall assume that invariant equilibrium points in the phase diagram A-B-C will ultimately be located as shown in Figure 2. No intermediate compounds are formed from components B and C; a single intermediate compound AC_2, melting congruently, occurs in the system A-C; the congruently melting compound AB_2 and the incongruently melting compound A_2B are formed from components A and B. A careful examination of several crystalline mixtures cooled from the liquid state can be expected to furnish much information about equilibria in the system A-B-C.

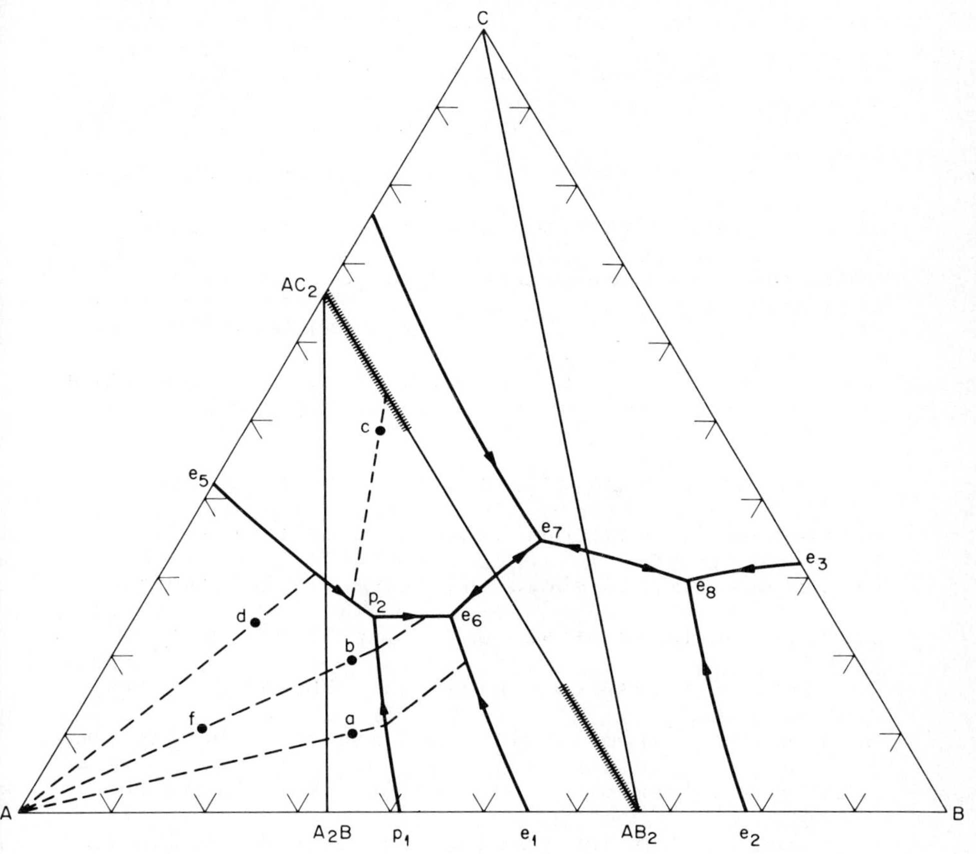

FIG. 2
Typical phase equilibrium relationships in condensed ternary systems of molten salts.

Examples

1. All crystallized specimens having compositions within the triangle AB_2-B-C contain only the pure solid phases AB_2, B, and C. These results indicate that the compounds AB_2, B, and C form a subsystem, that AB_2 probably melts congruently, and that

the section AB_2-C is a quasi-binary system. Under these circumstances, a single invariant point, the eutectic e_8 can occur in AB_2-B-C. It may be possible to estimate its approximate composition by examination of the crystallized mixtures. Morphological data, obtained from petrographic or metallographic methods, may reveal the domains of each of the primary phases AB_2, B, and C, as well as the composition of the eutectic.

2. Assume that AC_2 and AB_2 form extensive but limited solid solutions exhibiting maximum solubility at the solidus, as shown in Figure 2. In general, lattice constants of AB_2 and AC_2 will change as a function of solute concentration. Therefore, x-ray data obtained from crystallized specimens having compositions intermediate between AB_2 and AC_2 can be examined for lattice-spacing changes and correlated employing the parametric method. Some exsolution is to be expected as temperatures are lowered. The exsolution curve often exhibits considerable slope and precludes the possibility of obtaining data which indicate concentrations of maximum solubility of AB_2 and AC_2 except by static methods.

3. Occurrence of solid solutions of AB_2 and AC_2 indicates that inferences drawn regarding the subsystem AB_2-AC_2-C can be made from phase analyses of

mixtures having compositions in triangle AB_2-B-C in addition to those in AB_2-AC_2-C. Some mixtures in triangle AB_2-AC_2-C which crystallize primary-phase AC_2 solid solutions will solidify before the liquid fraction attains the composition e_7 and may not contain detectable concentrations of AB_2 or C. Mixtures which crystallize primary-phase AB_2 solid solutions will behave similarly, producing possibly undetectable amounts of AC_2 or C. The identity of the subsystem AB_2-AC_2-C may thus be equivocal without the conclusion based on other data that AB_2-C is quasi-binary.

4. Much information about phase behavior with the triangles A_2B-AC_2-AB_2 and A-A_2B-AC_2 can be obtained from x-ray analyses of specimens crystallizing within triangle A_2B-AC_2-AB_2. The solid-solution formation along AB_2-AC_2 and the melting relations of A_2B will cause mixtures within A_2B-AB_2-AC_2 to undergo a variety of crystallization reactions. Depending on the character and extent of the solid solution and the melting-freezing behavior of A_2B, x-ray analysis may furnish very informative phase data. Crystallization in polycomponent systems often results in the formation of final products which are in disequilibrium. Identification of the crystalline phases

in the final crystallized mixture provides useful inferences as to the nature of the equilibrium-phase diagram. Assume that in the system A-B-C the following crystallization mechanisms occur:

a. Liquids with original composition a reach curve p_1p_2, then e_1e_6, and will solidify before reaching e_6. On examination of the crystallized mixture, one will find present the phases A, A_2B, AB_2 solid solution, and possibly some AC_2 solid solution.

b. Liquids with original compositions of b or f will reach curve p_1p_2, then p_2e_6, and solidify at or near e_6. The final crystalline solid will contain the solid phases A, A_2B, AC_2 solid solution, and AB_2 solid solution.

c. Liquids with original compositions of c may or may not reach curve e_5p_2, depending on the behavior and composition range of the AC_2 solid solution. If the liquid reaches the transition curve e_5p_2, it may or may not reach p_2e_6, again depending on the solid solution. If curve p_2e_6 is reached, solidification will occur before the liquids reach e_6. Solid phases AC_2 solid solution, A, A_2B, and possibly small amounts of AB_2 solid solution will be present in the crystallized mixture.

d. Liquids with original compositions of d will reach curve e_5p_2. Crystallization from that point on will be essentially the same as that from composition c.

On identification of the solid phases produced during the crystallization of liquids of compositions a, b, c, d, and f, along with rough estimates of the relative quantities of each phase, one could deduce several characteristics of the system:

(1) The presence of compound A in crystallized mixtures within A_2B-AB_2-AC_2 indicates that the compound A_2B melts incongruently to A + liquid and therefore that the section A_2B-AC_2 is not quasi-binary. It also indicates the corollary that no invariant equilibrium points are to be found in the composition triangle A-A_2B-AC_2.

(2) Two invariant points p_2 and e_6 occur in the composition triangle A_2B-AB_2-AC_2.

(3) If it is established that AB_2 and AC_2 exhibit limited mutual solubility and therefore precipitate from liquids in two separate primary-phase domains, the identity of the solid phases present at the invariant points p_2 and e_6 is established. For p_2 they are the solid phase A, A_2B, and AC_2 solid

solution, while for e_6 they are A_2B, AB_2 solid solution, and AC_2 solid solution.

Much of the early work on solution theory in molten salts was carried out with systems in which anionic as well as cationic interactions could be examined Such systems, with two or more anions and two or more cations, are classed as reciprocal systems, and, for this simplest class, ternary reciprocal systems. The characteristics of such systems, as well as methods of representation were discussed recently by Blander[5] and need not be discussed here. In contrast to the representation of additive ternary systems by points in equilateral triangle as described above, ternary reciprocal systems may be represented by points in a square.

Reciprocal systems are characterized by the tendency of the two salts which form the stable pair to precipitate in preference to the two salts of the unstable pair. The stable pair is related to the ionic radii and theoretical lattice energies of the ions in such a way that there is a strong tendency to separate the salt consisting of the small cation and small anion, and the salt containing the large cation and large anion (the two salts which constitute the stable pair). In the phase diagram, this tendency is manifested by liquidus temperatures which are high relative to those of binary systems in which there is a common anion. If the stability of the stable pair relative to that of the unstable pair is high enough, the positive deviations from ideality are manifested not only by relatively high liquidus temperatures but also by the formation of a region of two liquid layers. It is with such systems that the most notable advances in solution theory with molten salts have been made. Current aspects of these advances are reviewed by Flood[6] and Blander[7-10].

CHARACTERIZATION AND ANALYSIS BY PHASE STUDIES

A comprehensive study of the heterogenous phase equilibria in a system of salts consists essentially of characterization and analysis. The capabilities for performing the relevant determinations have development rapidly through several advances in the experimental methods which have been introduced in the last several years. As a consequence, numerous systems of salts, particularly the fluorides, are so well established that the phase diagrams and properties of the crystalline phases may frequently find use as the basis for analytical methods. Classically, phase equilibrium diagrams are derived from two general types of experiments, those in which deductions are made from measurements of thermal effects occurring while heating and cooling mixtures, and those which permit a direct or indirect identification of the numbers and compositions of phases occurring at all temperature-composition points. Commonly, fused-salt diagrams are based on information from cooling curves. Changes in the slope of the temperature of the sample, when plotted as a function of time, reflect phase changes which occur on cooling. This technique is generally adequate for determining all except the steepest liquidus curves; steep curves represent small changes in saturation concentrations with temperature and hence small heat effects to influence the cooling rate. Cooling curves also provide information on the solidus and subsolidus phase changes, but are prone to give misleading indications because of the impossibility of maintaining equilibrium during cooling.

The determination of tie lines in ternary fused-salt systems using Schreinemaker's wet-residue method, commonly employed in studies involving equilibria in aqueous systems, consists of separating and identifying the composition of the liquid accompanying the precipitation of a solid phase. It is not used frequently in high-temperature salt systems because of the experimental difficulties imposed by the high-temperature filtrations and the requirements of chemically analyzing a large number of samples.

A much more effective method than either of these two is the quenching of equilibrium samples and identification of the phases by crystallographic examination with microscopic and x-ray diffraction techniques. Attainment of equilibrium can generally be assured by the use of long equilibration periods, and effective quenching to freeze the high-temperature equilibrium is accomplished by using small samples that can be cooled rapidly. Samples of 10-20 mg are ample for analysis with petrographic microscope and x-ray diffraction examinations.

Recent applications of the thermal-gradient quenching method in phase studies of ternary salt systems has brought the petrographic microscope into very effective use for locating fractionation or crystallization paths when used with this method. This is accomplished by determining the identity of a precipitating primary phase by measuring its optical properties. A single thermal-gradient quenching experiment can provide 25-30 samples of a fused-salt mixture quenched from as wide a range of temperatures as 150°C. The measurements from one such group of samples can provide a large amount of definitive data for the determination of tie lines.

Innovations in the original means of study have brought about the application of a wide variety of experimental techniques to studies of heterogeneous phase equilibria. These include differential thermal analysis, methods for observing melting and freezing materials visually, filtration of partially molten mixtures, high-temperature centrifugation, hot stage microscopy and high-temperature x-ray diffraction studies. Since 1956, electromotive-force measurements and vapor-pressure determinations have been used with success in the study of heterogeneous equilibria in salt systems. A review of the application of a number of these techniques has been made by Porter.[11] These advanced techniques are frequently more applicable to the studies of molten salt phase equilibria than to other types because of the generally low vapor pressures of salts, and the fact that salt mixtures are generally lower-melting than oxides, refractories, and metals.

Until recently, the definition of composition - temperature relationships in liquid-liquid immiscibility regions in salt systems was considered to be either intractable or too tedious to warrant extensive exploration. Friedman's[12] high-temperature centrifuge has simplified this problem greatly and affords a method for establishing the tie-lines in regions of liquid-liquid immiscibility. In operation, the apparatus rotates four sealed metal capsules containing 5-10 g specimens at a maximum speed of 2.4 x 10^4 rev min^{-1} and at a maximum temperature of approximately 1000°C. After equilibration, the rotation of the centrifuge is stopped slowly; the capsules are then quenched by rapidly removing the furnace and immersing the capsules in water. The ends of the capsules are cut off and the salts contained therein are subjected to composition analysis.

Perhaps the most important recent advance among the experimental methods which contributes to solid state phase studies is the development of automatic x-ray diffractometers for single crystal structure studies. With automatic diffraction equipment, the average period for determination of new structures is about one tenth that required previously. With such equipment, it is now commonplace to measure 1200 to 2000 reflections and to deduce the structures with significantly greater accuracy than was obtained previously. This development has a special function in phase studies, for by concurrently determining the equilibrium relations in a system and the structure of the crystalline solids which are found in the system, the stoichiometry, compositional variability, coordination chemistry, etc., are established unequivocally. An interesting example of the significance of concurrent determinations of phase equilibria and structure is shown by the CrF_2-CrF_3 system, which was reported by Sturm[13] and Steinfink and Burns.[14] Their inter-related investigations showed that the stoichiometry of the intermediate phase formed from CrF_2 and CrF_3 is Cr_2F_5, but that the crystalline phase of this stoichiometry is not produced on crystallization of CrF_2-CrF_3 melts. Instead, the intermediate phase crystallizes with a deficiency of Cr^{3+} ions and with a stoichiometry which is limited to $CrF_{2.4}$ to $CrF_{2.45}$.

APPLICATION OF PHASE EQUILIBRIA IN MOLTEN SALT RESEARCH AND DEVELOPMENT

The principal efforts in molten salt research and development today seem to be shared about equally between the United States and Russia. Two general categories encompass the principal activities in phase equilibrium investigations, those which center about technological applications, and those which are essentially fundamental science. In the United States, the quality of research has proved to be consistently excellent, so good, in fact, that the mission-oriented goals which give source to practically all of the work, are rarely discerned as impeding the accomplishment of comprehensive and penetrating investigations.

The principal applications in which molten salts are currently employed in technology and research are listed below. As is implied in the foregoing discussion, the division between technology and research for a number of the listed categories is artificial.

<u>Technology</u>

1. Reductive extraction processes for metals production.

   a. electrochemical processes

      (1) Sacrificial anode methods: Ta, Nb, W, and Mo

      (2) Electrowinning methods: lanthanides, Al, transition metals

   b. pyrochemical processes: lanthanides and actinides

2. Working fluids for Molten Salt Breeder Reactor Development.

   a. fuels, based on UF_4, ThF_4, PuF_3

b. blankets for breeder reactors

c. coolant salts

3. Solvents for use in volatility processes for recovery of uranium from spent fuel elements.

4. Non-nuclear heat exchange media - nitrates, fluoroborates, carbonates.

5. Application to regenerative fuel cells.

6. Metal alloys - salt compatibility.

7. Development of analytical chemical procedures.

Research

1. Coordination chemistry

   a. lanthanide-actinide model relationships

   b. complex ions in solutions

   c. electrochemical research

   d. transition metal chemistry

   e. structure studies

2. Thermodynamics of solution behavior

   a. freezing point depression

   b. liquid-liquid immiscibility

   c. reciprocal systems

3. Optically active crystals

   a. lasers

   b. electronic theory

4. Crystal growth
   a. single crystals of oxides and fluorides for structure studies
   b. zone refining
   c. phonon scattering
5. Metal-metal halide systems
6. Source of new types of molten salt research

Perhaps the principal factor which distinguishes technological and research applications of molten salt phase equilibrium studies is that the materials problems in technology often necessarily involve multicomponent systems, whereas the number of components in fundamental research studies is limited by choice. In technology the processes frequently involve highly complex reactions involving metal, gas, salt, and oxide or other phase reactions as well as oxidation-reduction equilibria at the same time. It will be advantageous to limit the discussion of the applications of phase equilibria in technology to a few samples, and for this purpose we will choose those related to the selection of a fertile-fuel mixture for application in a large molten salt breeder reactor.

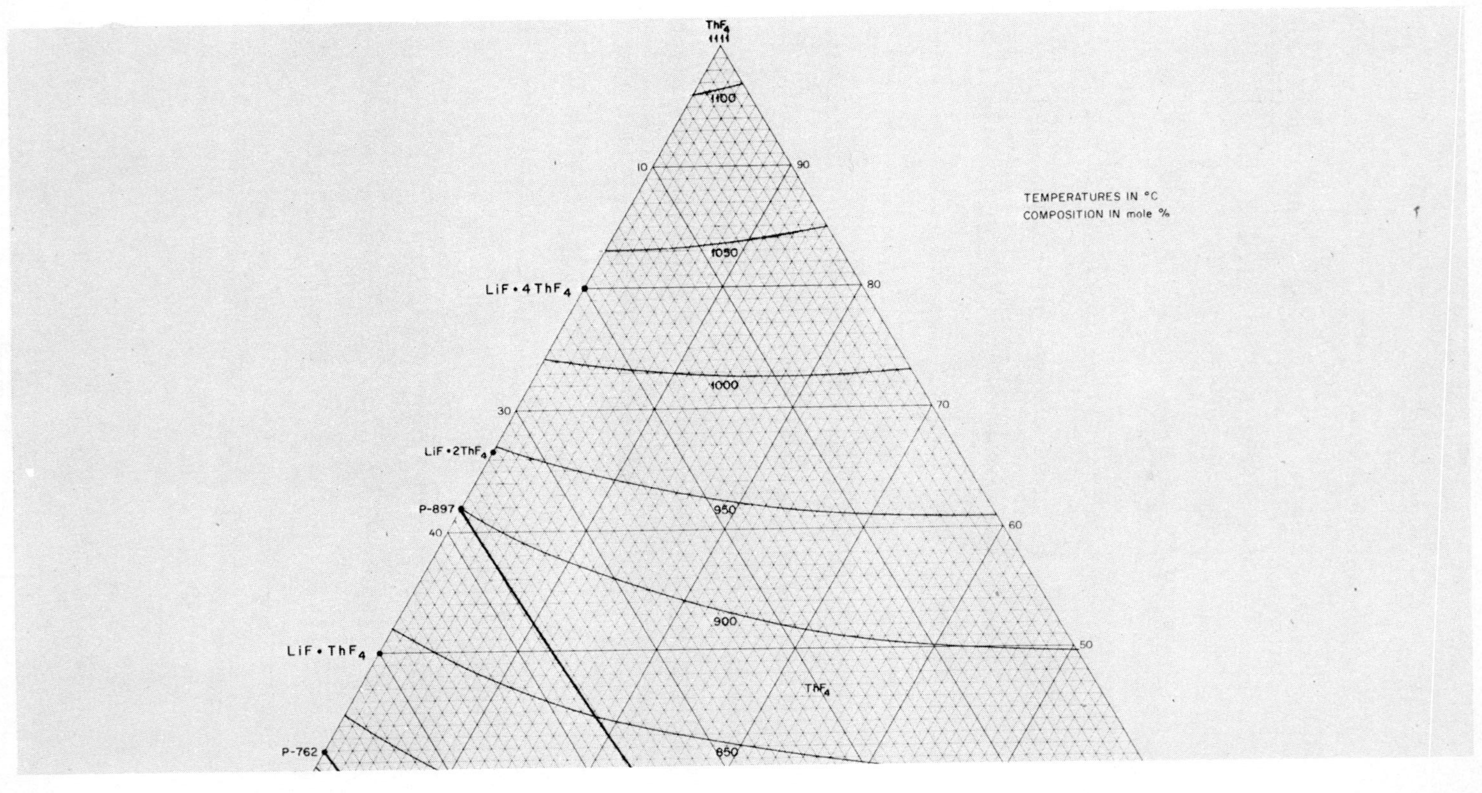
ThF_4
1100
10
90
TEMPERATURES IN °C
COMPOSITION IN mole %
1050
$LiF \cdot 4ThF_4$
80
1000
30
70
$LiF \cdot 2ThF_4$
P-897
950
40
60
900
$LiF \cdot ThF_4$
50
ThF_4
P-762
850

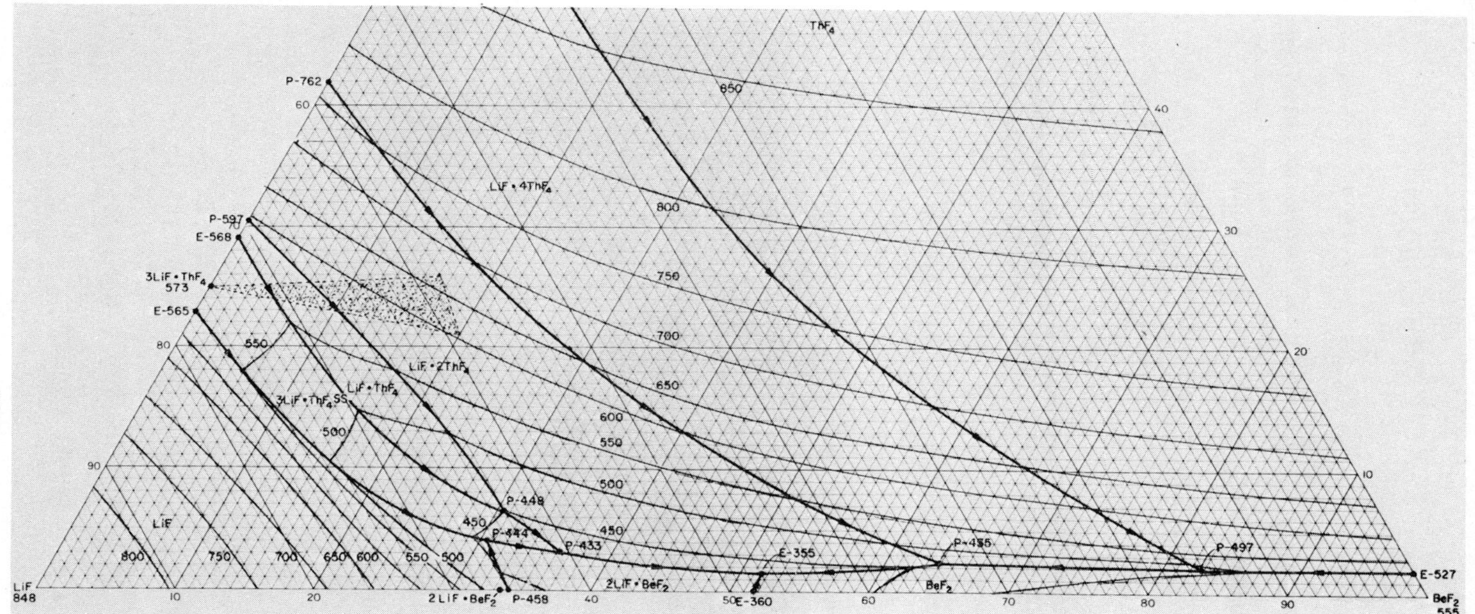

FIG 3

Equilibrium phase diagram for the system LiF-BeF_2-ThF_4. (Peppered area designates solid solution of Li_3 (Th, Be)F_7.)

Within the last year, an advance in the chemistry of processing molten salts to remove protactinium and the lanthanides makes the development of a large, i.e., 1000 MW(e), single-fluid molten salt breeder reactor feasible. Such a reactor would circulate its fertile-fissile salt through a graphite moderator in the core. The energy generated in the reactor fluid would then be transferred to a secondary coolant-salt circuit, which couples the reactor to a supercritical steam cycle. The results of design studies show that the salt system should be comprised of ^{7}LiF, BeF_2, ThF_4 and $^{233}UF_4$, and should contain about 12 mole % of ThF_4 and about 0.2 mole % $^{233}UF_4$. The phase diagram of the ^{7}LiF-BeF_2-ThF_4 system[15] (Fig. 3) then dictates the choices which are available in selecting the optimal composition of the salt mixture. For minimum liquidus temperatures, minimum viscosity of the molten salt, and optimal nuclear performance, the composition ^{7}LiF-BeF_2-ThF_4-$^{233}UF_4$ (68-20-11.8-0.2 mole %) was selected provisionally. In the future development of reprocessing technology, it is possible that further optimization of the ^{7}LiF-BeF_2 composition ratio will take place. If so, primary consideration will necessarily be made on the basis of the phase behavior in the four component system.

ROLE OF PHASE EQUILIBRIA

It has become quite clear that the technology related to molten salt reactors which rely on thermal or epithermal neutron energies will be based on the solvent system $^7LiF-BeF_2$ (Fig. 4). The physicochemical behavior of this system has therefore been the subject of rigorous investigations for some time. As a result, the heterogeneous phase equilibria,[16] viscosity,[17] crystal structures,[18] and activity coefficients,[19] have been determined with an uncommon degree of accuracy. The results of such detailed characterization now constitute a body of independent analytical methods for analysis of $LiF-BeF_2$ materials. Equilibrium phase behavior forms the basis for a very large number of analytical procedures, since solubility phenomena at or near equilibrium conditions depend on the principles of the Gibbs phase rule. All of classical gravimetric analysis is based on such behavior. The special relations to phase solubility techniques were described in detail by Higuchi and Conners,[20] who pointed out that within the last decade a phase titration technique called heterometry has been developed largely by Bobtelsky and coworkers.[21] In practice, a stirred titration system is monitored photoelectrically. At critical points

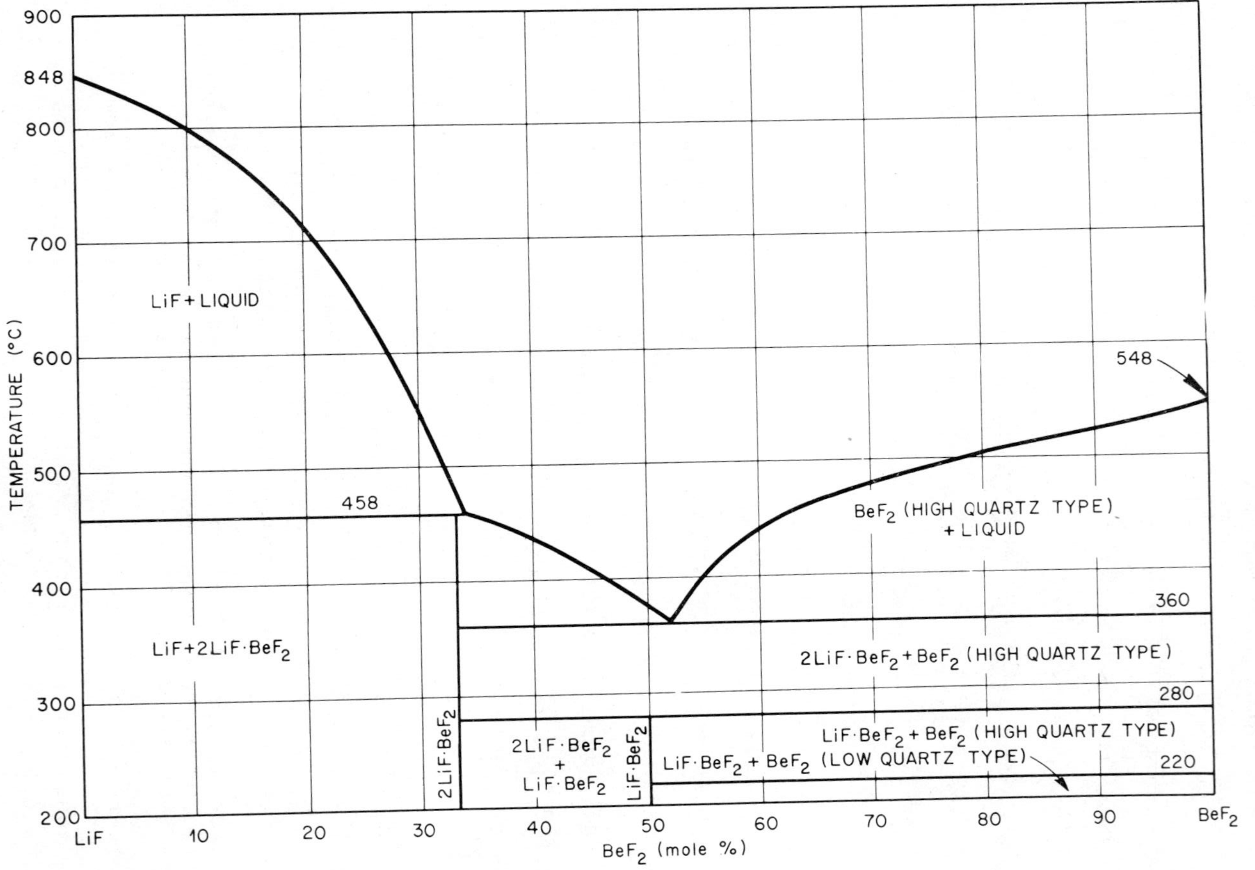

FIG. 4
Equilibrium phase diagram for the system $LiF-BeF_2$.

which correspond to end points of various titrant-titrand reactions, the apparent absorbance of the system changes sharply and is evidenced by more or less sharp breaks. Such analyses have been applied to the determination of inorganic substances by titration with organic reagents. One application is the titration of calcium with dipotassium phthalate in the presence of magnesium.[22] A general review of the application of such phase relationships to chemical analysis was given by Case.[23]

An earlier significant application of complex phase behavior in fluoride systems involved determination and use of the LiF-NaF-ZrF_4 phase diagram[24] (Fig. 5). In reprocessing Zircaloy-base fuel elements by the Volatility Process,[25] it was desirable to dissolve the spent fuels in a low melting fluoride solvent by reaction with HF, followed by extraction of the uranium, as UF_6, by purging the melt with fluorine. The LiF-NaF-ZrF_4 phase diagram proved to be extremely useful in choosing the operating variables for this process, for application of the salts as solvents was limited only by the liquidus temperature and dissolution kinetics. In practice, a solvent of the approximate composition, LiF-NaF-ZrF_4 (35-35-30 mole %) with a liquidus temperature of 550°C was used. As is evident from Figure 5, this solvent could dissolve up to a total of 55 mole % ZrF_4 before the liquidus temperature of the solution exceeded that of the original solvent.

Among the research efforts in salt chemistry which entail characterization and analysis of salt systems by phase studies, few are more demanding of a variety of techniques than those which are devoted to studies of coordination chemistry and to solution theory.

To cite one example of coordination chemistry, the application of the results of phase and structure studies of the alkali fluoride-lanthanide fluoride systems as models for actinide analogs is a particularly rewarding activity. Except for the heaviest of the lanthanides trifluorides of the lanthanides and actinides crystallize from the melt as hexagonal crystals of the space group $P6_3mcm$. Similarly, the tetrafluoride of the two groups, as well as zirconium and hafnium crystallize as monoclinic crystals of the space group I2c. The similar ionic size of the cations of the two series causes their intermediate compounds with the alkali fluorides to exhibit crystalline properties which, although similar, differ slightly with increasing atomic number of the heavy

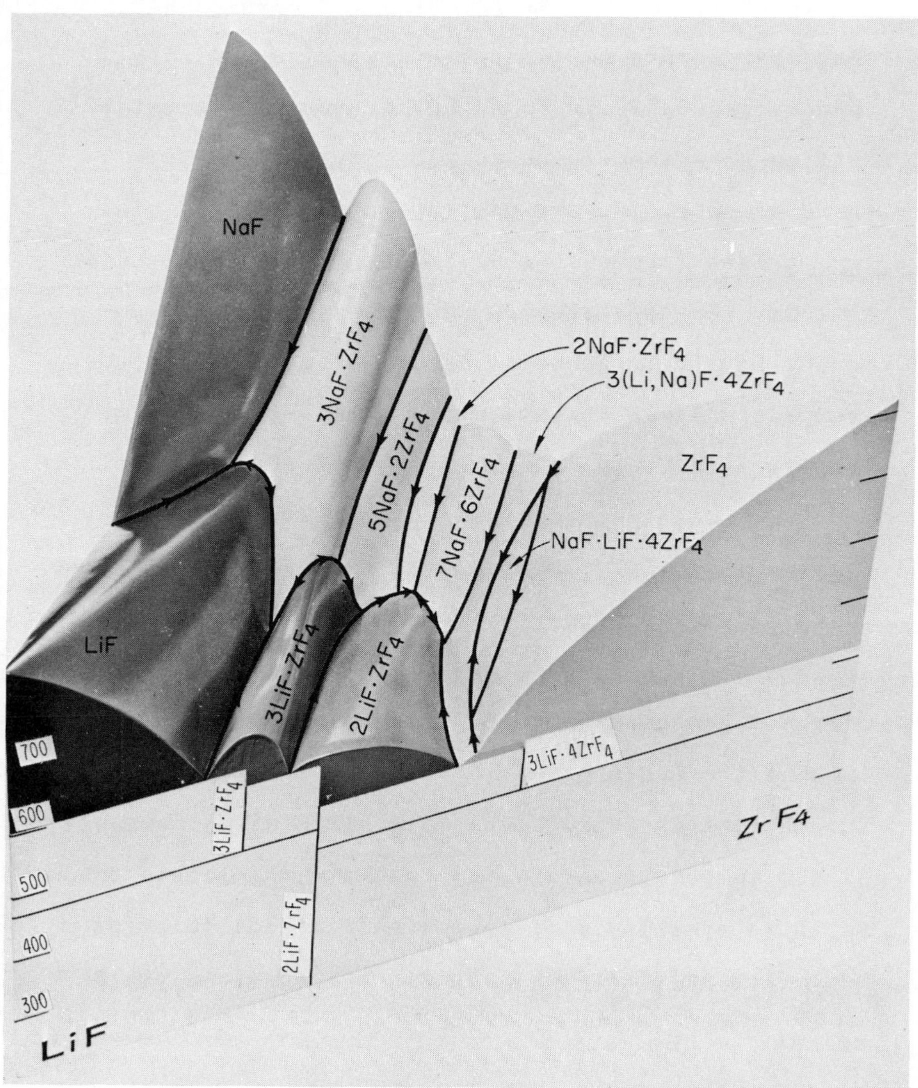

FIG. 5
Equilibrium phase diagram for the system $LiF-NaF-ZrF_4$.

metal. With increasing disparity in size of the alkali and heavy metal ions coordinations of the ions change, gradually increasing the order and crystal lattice energies of the complex compounds. This trend is seen, for example, in the complex crystalline phases formed in the various alkali fluoride-LnF_3 systems. In the LiF-LnF_3 systems, the single compound, $LiLnF_4$ is formed. It is tetragonal,[26] space group $I4_1/a$. In this structure the lanthanide ions are highly ordered and have ninefold coordination. Two intermediate phases characterize the NaF-LnF_3 systems (Fig. 6)[27]; a highly disordered fluorite-like phase of variable composition which prevails in these systems at high temperatures (Fig. 7) and a low-temperature hexagonal phase of equimolar composition. In that the cubic phase is isostructural with fluorite, the cations experience complete mixed site occupancy. At low temperatures the intermediate equilibrium phase in these systems is of essentially fixed stoichiometry and crystallizes as hexagonal crystals of $NaLnF_4$.[28] The cation sites in these crystals and also in $KLaF_4$ (Fig. 8)[29] are partially ordered and also have ninefold coordination. The mixed site occupancy is a result of the nearly identical ion sizes of the alkali and lanthanide ions in these

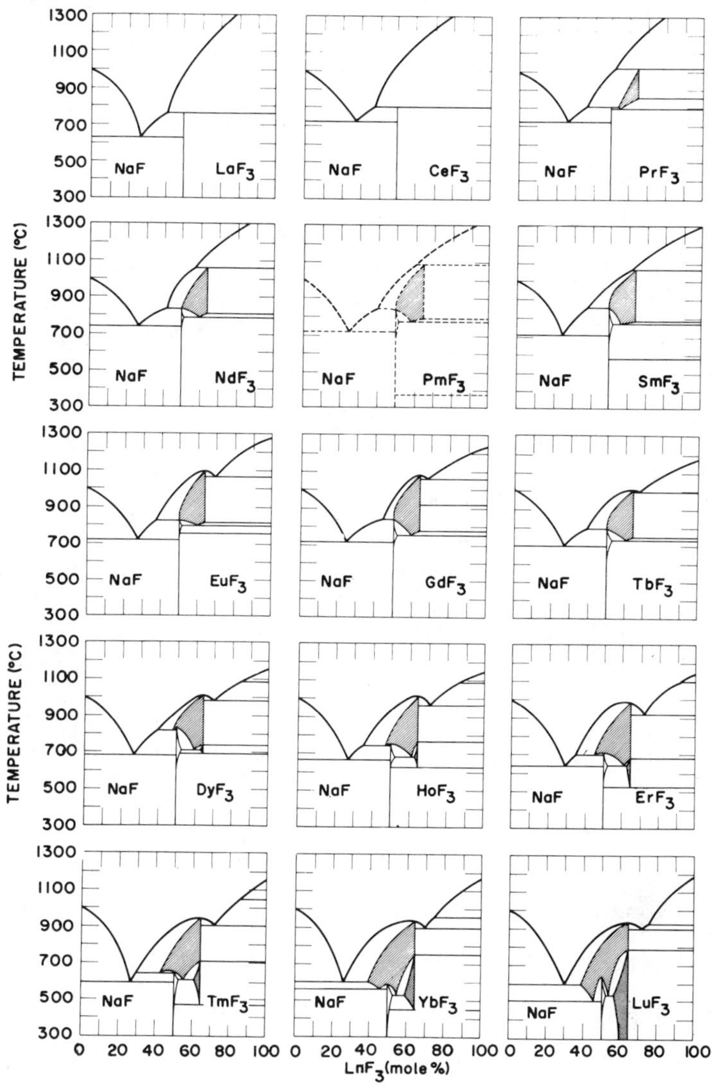

FIG. 6
Phase equilibria in the binary systems of sodium fluoride and the lanthanide trifluorides. Diagonally hatched area designates the region of single phase cubic solid solution. Peppered area designates the region of the ordered $5NaF \cdot 9LnF_3$ solid solution phase.

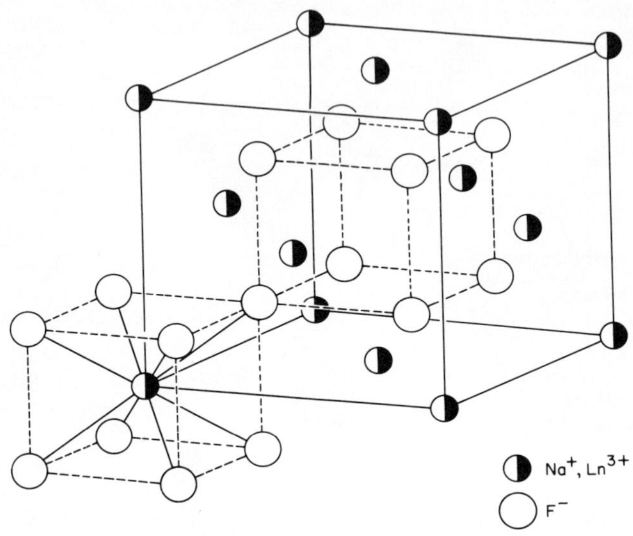

FIG. 7
Crystal structure of the sodium fluoride - lanthanide trifluoride cubic solid solutions.

cases, and ceases to occur with slightly increasing disparity in the sizes of the alkali and lanthanide ions. Current studies[30] have shown that in crystals of the compound $KCeF_4$, the cations, unlike those in $KLaF_4$, occupy unique positions, again with ninefold coordination. The systems RbF and CsF-LnF_3 have not been investigated nearly so thoroughly as have those of the lighter alkali fluorides. It is anticipated that with increasing ratios of the alkali ion/lanthanide ion radii, congruently melting phase, $3MF \cdot LnF_3$ will appear to increase in stability and impose a pronounced effect on the liquidus in these systems.

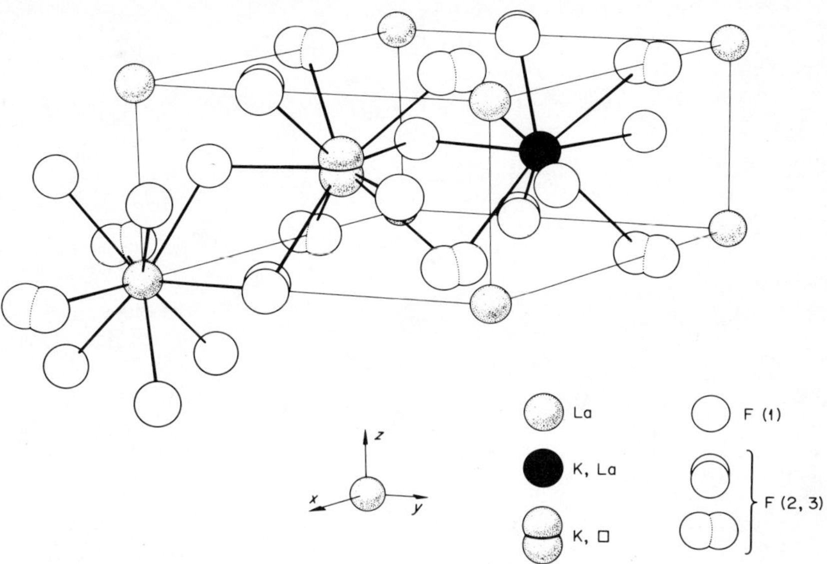

FIG. 8
Structure of crystalline potassium tetrafluorolanthanate, β-$KLaF_4$.

Some preliminary evidence of this trend has been given by Dergunov[31] which shows increasing lattice energies of a 3:1 compound as suggested by the relative differences in the melting temperatures of those compound as compared with those of the eutectics formed from the alkali fluorides and the 3:1 compounds (Fig. 9).

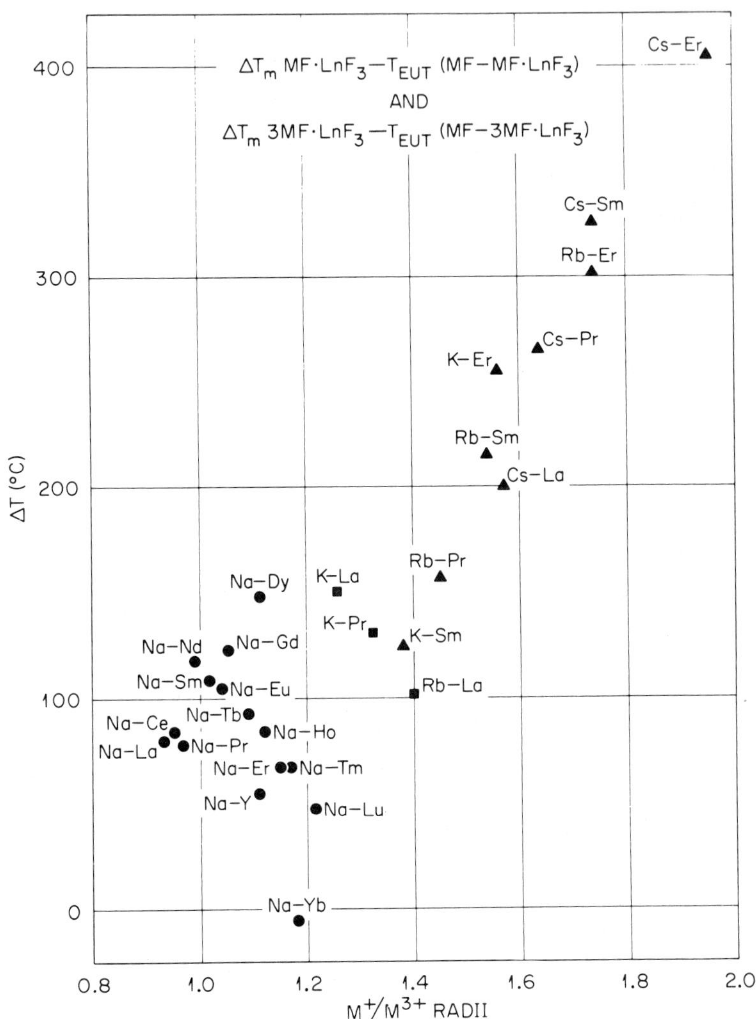

FIG. 9
Melting temperatures of the alkali fluoride - lanthanide trifluoride compounds, $3MF \cdot LnF_3$ and $MF \cdot LnF_3$, relative to their respective $MF\text{-}3MF \cdot LnF_3$ and $MF\text{-}MF \cdot LnF_3$ eutectic temperatures.

CONCLUSION

Advances in the methods for determination of phase equilibrium relationships in molten salt systems have greatly enhanced characterization and analysis in such systems. Application of the information available from the precise characterization which is a part of such determinations is both straightforward and highly useful in molten salt technology and research.

REFERENCES

1. J. E. Ricci, "The Phase Rule and Heterogeneous Equilibrium," Van Nostrand, New York, 1951.

2. J. Zernike, "Chemical Phase Theory," A. E. Kluwer, Deventer, 1956.

3. G. Masing, "Ternary Systems: Introduction to Three Component Systems," Translated by B. A. Rogers, Dover Publications, New York, 1960.

4. E. M. Levin and H. F. McMurdie, "Phase Diagrams for Ceramists," American Ceramic Society, Columbus, Ohio, 1960.

5. M. Blander, Chem. Geo., **3**, 33 (1968).

6. H. Flood, T. Førland, and K. Grjotheim, Z. Anorg. Allgem. Chem., **276**, 289 (1954).

7. M. Blander, J. Phys. Chem., **63**, 1262 (1959).

8. M. Blander and J. Braunstein, Ann. N. Y. Acad. Sci., **79**, 838 (1960).

9. M. Blander and L. E. Topol, Electrochim. Acta., **10**, 1161 (1966).

10. M. Blander and L. E. Topol, Inorg. Chem., 5, 1641 (1966).

11. R. F. Porter in Annual Review of Physical Chemistry, Eyring, H., Ed., Vol. 10, George Banta Co., Samford, California, 1960, p. 219.

12. H. A. Friedman, J. Sci. Instr., 44, 454 (1967).

13. B. J. Sturm, Inorg. Chem., 1, 665 (1962).

14. H. Steinfink and J. H. Burns, Acta Cryst., 17, 823 (1964).

15. R. E. Thoma, H. Insley, H. A. Friedman, and C. F. Weaver, J. Phys. Chem., 64, 865 (1960).

16. R. E. Thoma, H. Insley, H. A. Friedman, and G. M. Hebert, J. Nucl. Mat., 27, (1968).

17. C. T. Moynihan and S. Cantor, AEC Report ORNL-4076, Oak Ridge, Tennessee, December 1966, p. 25.

18. J. H. Burns and E. K. Gordon, Acta Cryst., 20, 135 (1966).

19. A. L. Mathews and C. F. Baes, Inorg. Chem., 7, 373 (1968).

20. T. Higuchi and K. Conners in Advances in Analytical Chemistry and Instrumentation, Reilley, C. N., Ed., Vol. 4, Interscience Publishers, New York, 1965.

21. M. Bobtelsky, "Heterometry," Elsevier, New York, 1960.

22. M. Bobtelsky and I. Bar-Gadda, Anal. Chim. Acta, 9, 168 (1953).

23. L. O. Case in Treatise on Analytical Chemistry, Part 1, Theory and Instrumentation, Kolthoff, I. M., Ed., Interscience Publishers, New York, 1961.

24. R. E. Thoma, H. Insley, H. A. Friedman and G. M. Hebert, J. Chem. and Eng. Data, 10, 219 (1965).

25. G. I. Cathers, R. L. Jolley, and E. C. Moncrief, Nucl. Sci. and Eng., **13**, 391 (1962).

26. G. D. Brunton, 1968, unpublished work.

27. R. E. Thoma and R. H. Karraker, Inorg. Chem., **5**, 1222 (1966).

28. J. H. Burns, Inorg. Chem., **4**, 881 (1965).

29. D. R. Sears and G. D. Brunton, Acta Cryst., in press.

30. G. D. Brunton and D. R. Sears, AEC Report ORNL-4229, Oak Ridge, Tennessee, December 1967, p. 64.

31. E. P. Dergunov, Dokl. Akad. Nauk. SSSR. **60**, 1185 (1948), and op. cit., **85**, 1025 (1952).

IONIC INTERACTIONS IN MOLTEN SALTS

Norman H. Nachtrieb

Department of Chemistry
The University of Chicago
Chicago, Illinois

INTRODUCTION

One of the central problems in molten salt chemistry is the characterization of the interactions that exist among the constituent ions. It is an important problem because the structural, thermodynamic, and transport properties of pure molten salts and their mixtures depend upon the nature and range of such interactions. These vary widely, from the idealized extreme of point-centered coulomb forces between positive and negative charges on the one hand, to those of the directed but delocalized electronic charge distributions of covalent bonds on the other. Such extremes are rarely, if ever, realized in actual condensed systems of ions above $0^{o}K$.

Near to the ionic extreme, typified by alkali halide-like salts in which the free ions are in a ^1S_o ground electronic state, it is feasible to measure small deviations from ideal ionic behavior. One of the techniques that lends itself to this purpose is nuclear magnetic resonance. Nuclei having non-zero spins may serve as probes of their local magnetic environment, sensing distortions of their surrounding electron distributions and manifesting them through a shift in the resonance condition. Such chemical shifts are designated by

$$\delta = -\left[\frac{(\nu_S - \nu_R)}{\nu_R}\right]_H = \left[\frac{(H_S - H_R)}{H_R}\right]_\nu , \qquad (1)$$

where the subscripts S and R refer to the nucleus in the system of interest, relative to a system taken as a suitable reference state. A dimensionless scalar for isotropic systems such as molten salts, δ takes on negative values (downfield chemical shift) for nuclei that experience a larger local magnetic field in the system of interest than in the reference system. Such local field enhancements may arise from magnetic moments of electronic states having non-zero angular momentum which are mixed into the diamagnetic (^1S_o) ground state by a variety of mechanisms: polarization, charge transfer, and the overlap of wave functions of neighboring ions. These are aspects of the particle interactions that are included in the concept of the covalent bond.

The ideal reference system to which chemical shifts should be related is one of free ions in a pure ^1S_o state, unperturbed by the influence of neighboring ions. Since this is, in general, not experimentally feasible it is necessary to select as the reference a salt in which the nucleus of interest exhibits its smallest magnetogyric ratio. Such a state, in which the nuclei are surrounded by the most nearly spherically symmetric electron distributions, is most "ionic." Relative to it, the downfield chemical shifts of the same nuclei in other systems provide a measure of the extent of covalent interaction of their ions with other ions.

According to Ramsey's theory[1] the chemical shift is given by

$$\delta = -\frac{8}{3}\mu_B^2 \left\langle \frac{1}{r^3} \right\rangle \frac{Z\Lambda}{\Delta E}, \tag{2}$$

where μ_B is the Bohr magneton, $\langle \frac{1}{r^3} \rangle$ is an average over the radial part of the wave function of the excited state, Z is the coordination number of the ion, ΔE is an average over the energies of low-lying paramagnetic states of the ion above its ground state, and Λ is the fraction of such paramagnetic states that are admixed with the ground state of the ion. Because p states are energetically the lowest-lying above the ^1S_o ground state, $Z\Lambda$ may be looked upon as the degree of p character in incipient σ-bonds between the ion and its Z nearest neighbors of opposite sign[2].

Some difficulties arise in the use of Eq. 2 that center about the correct values that should be selected for the quantities $<\frac{1}{r^3}>$ and ΔE. These difficulties have their origin in the approximations made for the electronic configuration of the mixed ground state. It may be regarded as arising from the transfer of charge from a neighboring anion to a cation, or from the overlap of the wave functions of proximal cations and anions. For the former model, proposed by Yosida and Moriya[3], the first order approximation to $<\frac{1}{r^3}>$ is the value for the neutral atom and ΔE is a charge transfer energy which may be approximated by the Von Hippel formula:

$$\Delta E = \frac{e^2(2\alpha - 1)}{R} + E - I \text{, where } \alpha \text{ is the} \qquad (3)$$

Madelung constant, R is the internuclear separation, E is the anion electron affinity, and I is the ionization potential of the metal atom. For the second model $<\frac{1}{r^3}>$ is an average over the radial part of the outermost p orbital of the ion and ΔE is the excitation energy of that state for the free ion. Both are crude approximations to the actual electronic state of the system, and higher order approximations for the two approaches should converge. As will be seen, however, the agreement is surprisingly good even at this level of approximation.

EXPERIMENTAL

Measurements of the chemical shifts of molten salts require some modification of commercially available NMR instrumentation for use at high temperatures. The principal changes involve the probe, which is shown schematically in Fig. 1. It employs a single coil with an RF bridge, rather than the more conventional crossed 2-coil design. The NMR signal is recovered through the bridge, which is adjusted to a slightly unbalanced condition, and displayed on either an oscilloscope or as the derivative of an absorption line on an X-Y recorder after amplification and demodulation. The essential features of the probe are the RF coil consisting of about 13 turns of silver wire wound internally in grooves of a cylindrical Lava shell, a surrounding furnace of non-inductively wound platinum wire, and a water-cooled outer jacket. The salts were contained in Pyrex or Vycor ampoules of about

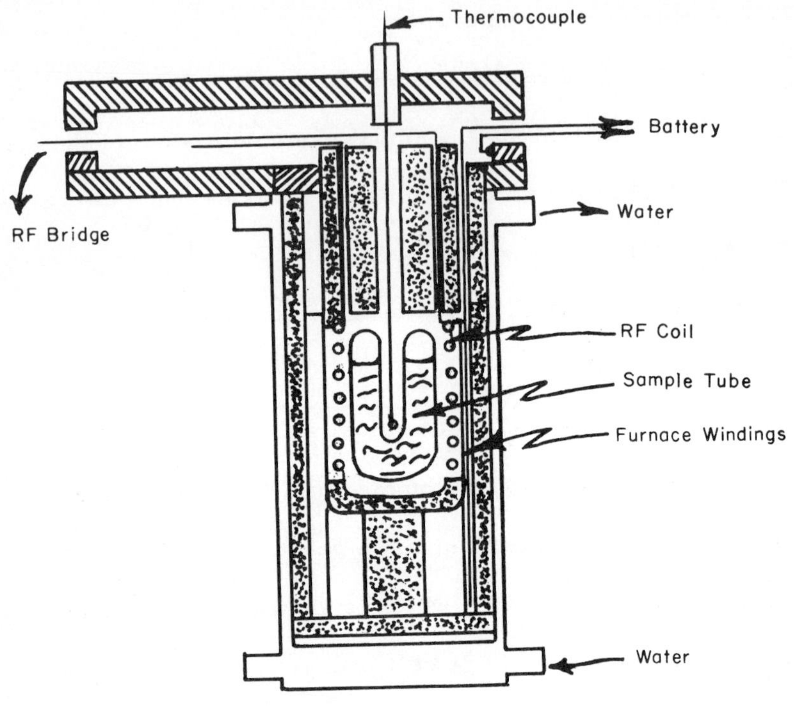

FIG. 1

Probe and furnace design.

1 cm^3 volume, sealed in vacuo, and equipped with an axial reentrant well to accept a thermocouple. A second probe was constructed to contain the working reference, that could be interchanged with the sample probe in the magnetic field in a matter of moments by means of a lateral sliding motion. It was separately fed by a conventional marginal oscillator, and the resonance frequency was measured with a second counter. Further details are given in earlier publications[4, 5, 6].

MEASUREMENTS ON PURE THALLIUM SALTS

Fig. 2 shows the results of chemical shift measurements for a series of pure thallium salts over a range of temperature from 20° to 700°C. Thallium salts were chosen for study because of the favorable nuclear properties of Tl^{205} (large magnetic moment and spin, $I = \frac{1}{2}$), cubic structures in the

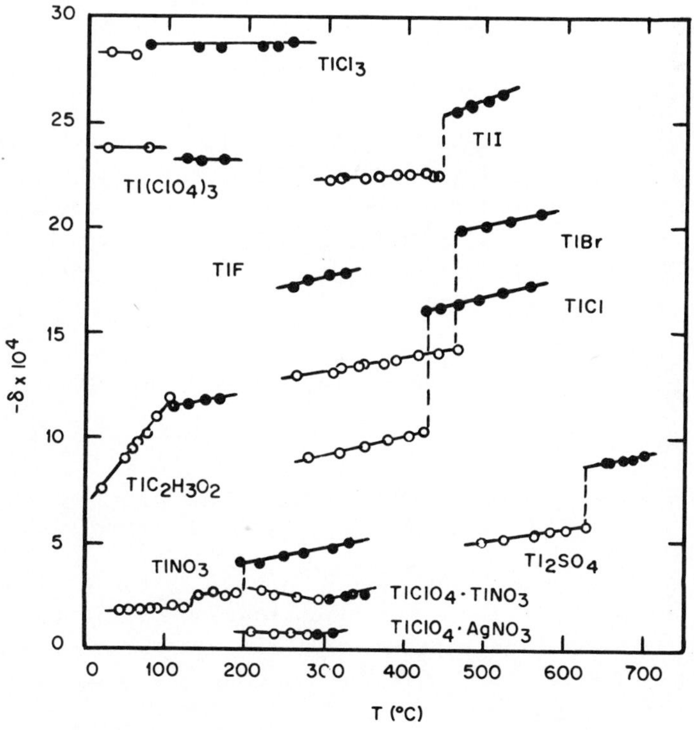

FIG. 2

Chemical shifts for crystalline and molten salts versus temperature.

solid state, and chemical similarity to alkali metal salts. Further studies have been carried out on salts of other nuclei (Br^{81} and I^{127}) that show the feasibility of extending the work to other systems.

Four interesting qualitative observations are immediately evident in Fig. 2:

1. The chemical shifts of pure thallium salts are well separated from one another and span a range of almost 3000 ppm.

2. The order of the downfield chemical shifts is qualitatively in accord with intuitive predictions of the degree of covalent interaction between thallium and various anions; i.e., increasing from highly ionic salts such as $TlClO_4$ and $TlNO_3$ to poorly conducting covalent salts such as $TlCl_3$.

3. With increasing temperature in both the crystalline and molten state there is in general a linear downfield chemical shift.

4. As a rule there is a discontinuous downfield shift at the melting point (solid $\longrightarrow$ liquid).

These observations may largely be understood in terms of Ramsey's theory of the chemical shift, and some further inferences may be drawn from Fig. 2 about the interactions of the thallium ion with its environment. Table 1 gives the values of the covalency parameter, Λ, calculated from both the charge transfer model with Eq. 3, and the overlap model of excited ionic states.

As noted above, both approximations give remarkably concordant results for the degree of covalent interaction. The absolute values are subject to the uncertainty in the definition of the zero chemical shift reference state, which would introduce a constant, but presumably small, additive correction. The values calculated for the molten state are undoubtedly an underestimate, because the coordination number is probably closer to 4.5 than to $Z = 8$, the number of nearest neighbors for the crystalline solids [7]. The chemical shift is proportional to $Z\Lambda$, and the covalency parameter would be even larger if the decrease in average coordination number were taken into account.

An important conclusion to be drawn from Fig. 2 and Table 1 is that the chemical shift of Tl^{205} is not due to the polarization of Tl^{+} by the halide ion. Although the Tl^{+} ion is large and polarizable, its deformation should increase with decreasing anion radius, and the observed order of the chemical shift is just the opposite. Nor is the effect to be attributed to "ion-

TABLE 1

Covalency of Thallous Halides Based on Chemical Shifts

	Solid						Melt[c]			Vapor[d]
	20° C.			350° C.			500° C.			
	$-\delta \times 10^4$	Λ^a	Λ^b	$-\delta \times 10^4$	Λ^a	Λ^b	$-\delta \times 10^4$	Λ^a	Λ^b	
TlCl	6.3	0.032	0.037	9.4	0.048	0.055	16.4	0.084	0.095	0.17
TlBr	11.0	0.056	0.059	13.1	0.067	0.071	19.8	0.101	0.107	0.20
TlI	21.7	0.103	0.104	22.0	0.113	0.106	26.0	0.133	0.124	0.28

Λ^a calculated from Eq. 2, with ΔE from Eq. 3, and $\langle \frac{1}{r^3} \rangle = 11.8$ a.u.$^{-3}$

Λ^b calculated from Eq. 2, with $\Delta E = 7.42$ eV, and $\langle \frac{1}{r^3} \rangle = 18.8$ a.u.$^{-3}$

[c] Coordination No. taken to be Z = 8

[d] From microwave spectra (A. H. Barrett and M. Mandel, Phys. Rev. 109, 1572 (1958)).

pairing" in the usual sense of that term, since the coulomb interaction of point-centered charges does not induce a magnetic moment.

Temperature Dependence of the Chemical Shift

It is the temperature dependence of the effect that provides a clue to the nature of the ionic interactions responsible for the chemical shift. It is seen to be negative (downfield) and linear. The interaction therefore increases with temperature, and is associative rather than dissociative in nature. Moreover, it behaves unlike a system of magnetic spins that follow a $1/T$ law. Since the chemical shift is linear in T, it is clearly affected by the kinetic energy of the system. For the interionic distances that prevail in melts this is largely vibrational. To a good approximation, the ions oscillate with linear harmonic motion about their mean positions with a characteristic period of the order of 10^{-12} sec. Superimposed upon this vibrational motion, the ions undergo cooperative anharmonic displacements that result in macroscopic diffusion. It is the former, "solid-like" harmonic motions that are relevant in the chemical shift. They provide the thermally-driven overlap of the electron distributions of neighboring ions that mix excited states into the ground states of both cations and anions.

Equation 2 may be differentiated logarithmically, giving:

$$1/\delta_o(d\delta/dT) = 1/\Lambda_o(d\Lambda/dT) \quad (4)$$

where δ_o is the chemical shift and Λ_o is the corresponding covalency parameter at some arbitrary reference temperature. It is convenient to use the overlap model at this point, and to assume that Λ is proportional to the square of the overlap integral for the normalized wave functions of a cation with a neighboring anion: $\Lambda = \Sigma(\int \psi_c \psi_a d\tau)^2$. The covalency parameter may then be expanded in a power series with the vibrational amplitude giving, to first order:

$$\Lambda = \Lambda_o + \frac{\partial^2 \Lambda}{\partial x^2} \cdot \langle x^2 \rangle \quad , \quad (5)$$

where the higher order terms are neglected.

Equation 4 may now be rewritten as:

$$1/\delta_o(d\delta/dT) = 1/\Lambda_o\left(\frac{\partial^2 \Lambda}{\partial x^2}\right)\cdot\left(\frac{\partial\langle x^2\rangle}{\partial T}\right) \quad . \quad (6)$$

The mean square amplitude along a particular direction (e. g. the line of centers of a cation-anion pair) is one-third the mean square amplitude, $\langle x^2\rangle = 1/3 \langle u^2 \rangle$, for an isotropic system, and above the Debye temperature may be approximated as

$$\langle x^2 \rangle = 6kT/M\omega_D^2 \quad . \quad (7)$$

Substitution of its temperature derivative into Eq. 6 gives:

$$1/\delta_o(d\delta/dT) = (6k/M\omega_D^2)(1/\Lambda_o)\left(\frac{\partial^2 \Lambda}{\partial x^2}\right) = \frac{6\hbar^2}{Mk\theta_D^2}\cdot\frac{1}{\Lambda_o}\frac{\partial^2 \Lambda}{\partial x^2}. \quad (8)$$

The factor $\frac{1}{\Lambda_o}\frac{\partial^2\Lambda}{\partial x^2}$ could be evaluated from the overlap integrals at the equilibrium interionic separation, given accurate wave functions for the anion and cation in their ground and excited states. However, Löwdin[8] has tabulated values of the overlap integrals for the s and p orbitals of the alkali halides which yield the average value, $\frac{1}{\Lambda_o}\frac{\partial^2\Lambda}{\partial x^2} = 1.3 \times 10^{17}\ cm^{-2}$. This quantity is largely independent of the particular ion or choice of orbital because of the exponential decrease in wave functions at large distances. Table 2 gives the observed and calculated values of $1/\delta_o(d\delta/dT)$, based upon Eq. 8 and the assumption that the thallous halides have the same fractional overlap variation with separation as the alkali halides.

TABLE 2

Fractional Change in Chemical Shift with Temperature

	$\theta_D(^oK)$	$\frac{1}{\Lambda_o}\frac{\partial^2\Lambda}{\partial x^2}\ (cm^{-2})$	$\frac{6\hbar^2}{Mk\theta_D^2}\ (cm^2deg^{-1})$	$\frac{1}{\delta_o}\frac{d\delta}{dT}\ (deg^{-1})$ (calc.)	(obs.)
TlCl	126	1.3×10^{17}	2.7×10^{-20}	1.2×10^{-3}	1.4×10^{-3}
TlBr	113	1.3×10^{17}	3.4×10^{-20}	1.5×10^{-3}	0.55×10^{-3}

The agreement between the observed and calculated temperature coefficient is satisfactory, and supports the conclusion that the covalent contribution to the interaction of ions increases with temperature by a coupling of vibrational modes with electronic states. The coupling has presumably both a static component, associated with the loss of cubic site symmetry of the ions when the salt melts, and a dynamic component which is associated with non-centric vibrational modes. A transfer of electron density from anion to cation presumably accompanies both the static and dynamic spects of the chemical shift, and it is conjectured that the transfer of charge may be proportional to the square of the overlap integral.

Composition Dependence of the Chemical Shift

The presence of foreign cations or anions might be expected to affect the chemical shift of a given nucleus, and this proves to be true. The simplest case is that of a solution of two salts that have a common ion. If the NMR nucleus belongs to a cation (e.g. Tl^{205}), its chemical shift is a function of the mixed anion fraction. Figure 3 illustrates the linear dependence of $\delta_{Tl^{205}}$ upon the mole fraction of TlBr in a binary solution of thallous chloride and thallous bromide. This is what would be expected if strong local electroneutrality obtains and the probability of finding a given anion species in the first coordination sphere of a cation is proportional to the mole fraction of that anion.

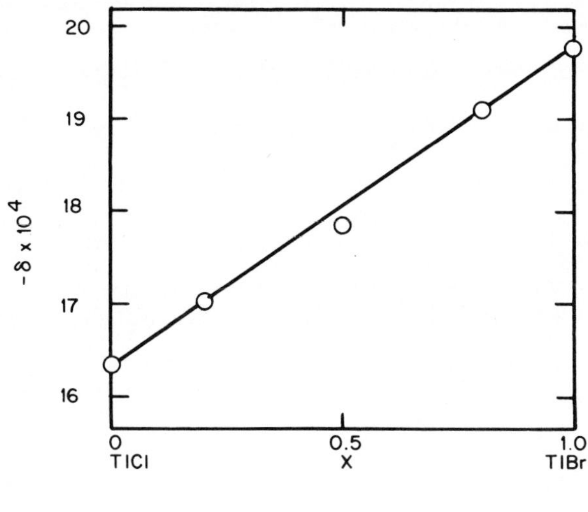

FIG. 3

Chemical shift of Tl^{205} in molten TlCl-TlBr.

The observed linearity does not imply that Tl^+ has no preference for Br^- or Cl^- as its nearest neighbors, but is a simple consequence of the complete miscibility of the system.

A more interesting case is that of a solution of two salts having a common anion and mixed cations, one species of which contains the probe nucleus. Figure 4 depicts the chemical shift of Tl^{205} in molten binary solutions with alkali halides having a common halide ion. Strong local electroneutrality in such systems would require that the foreign alkali cation lie in the 2nd coordination sphere of a given Tl^+ ion, separated from it by a shell of anions. Despite the fact that NMR effects are of short range, it is clear from Fig. 4 that a foreign cation may significantly affect $\delta_{Tl}205$. Two interesting observations are immediately apparent from Fig. 4:

1.) For a given cation, $\delta_{Tl}205$ depends linearly upon the mole fraction of the foreign salt to at least $X = 0.5$.

2.) The sign of $d\delta/dX$ may be positive or negative, and appears to be related to some property of the foreign cation.

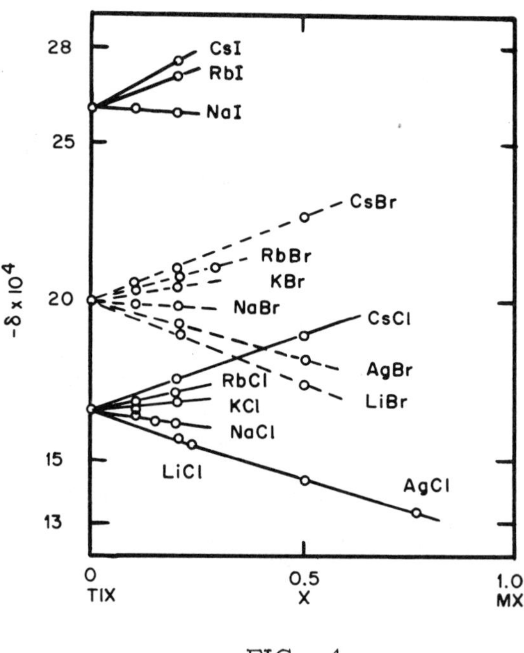

FIG. 4

Dependence of $\delta_{Tl}205$ on mole fraction in thallous halide-alkali halide mixtures. Chlorides, bromides at 520°C, iodides at 500°C

This property is closely allied to the radius of the foreign cation, as may be seen in Fig. 5, in which $\delta_{Tl}205$ is linear in the difference in radii of M^+ and Tl^+.

It is reasonable to postulate that it is the electric field of the foreign cation that is responsible for the perturbation of $\delta_{Tl}205$. Calculations indicate that direct polarization of Tl^+ by M^+ is

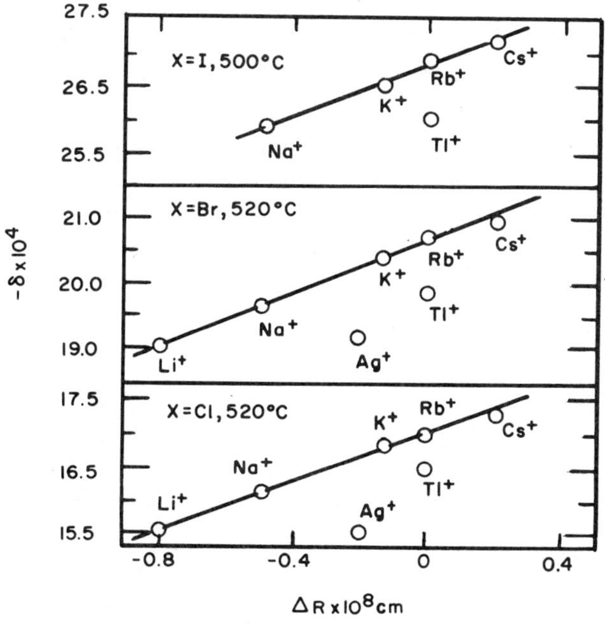

FIG. 5

Dependence of $\delta_{Tl}205$ in thallous halide-alkali halide mixtures on $\Delta R = R_{M^+} - R_{Tl^+}$.

too small at their large separation to be the origin of the effect. Moreover, Figs. 4 and 5 show that the downfield shift of $\delta_{Tl}205$ increases with increasing foreign cation radius. . . opposite to the expected direction of the shift if it were due to the direct polarization of Tl^+ by the electric field of M^+.

However, another mechanism may be envisioned that is consistent with the observed facts. On the average, an anion lies between a given Tl^+ and M^+, and the electric fields of each compete for the electron density of the intervening negative ion. The anion will experience a net field, and an electric dipole will be induced in it, whose magnitude and direction are determined by its polarizability and the relative strengths of the electric fields of the competing Tl^+ and M^+ ions. This polarization is given by

$$P_z = \alpha E_z = e\Delta r , \qquad (9)$$

where α is the anion polarizability, E_z is the net electric field component at the anion center along the direction of the laboratory magnetic field, and Δr is the length of the induced electric dipole in the anion.

Expansion of E_z in a power series, with neglect of all but the leading term, gives

$$E_z = 2e\Delta R/R_o^3 \qquad (10)$$

where

$$\Delta R = R_{M^+} - R_{Tl^+} \quad \text{and} \quad R_o = R_{Tl^+} + R_{X^-}$$

Substituting Eq. 10 into Eq. 9 yields

$$\Delta r = 2\alpha \Delta R / R_o^3 \tag{11}$$

for the anion dipole length. Anion polarization, as Sternheimer has shown[9], is principally p⟶d excitation in the outer electron shell. For present purposes, this admixture of the induced d orbital with the ground state p orbital of the anion is approximated by the simple displacement of the p orbital through the distance Δr. Thus, the anion dipole length is equated with the change in the overlap distance of the thallous ions and halide ions in this approximation, and the substitution of Eq. 11 for dr in $1/\delta_o(d\delta/dr)$ gives

$$1/\delta_o(d\delta/dr) = 1/\Lambda_o(d\Lambda/dr) = \frac{1}{\delta_o}\left(\frac{\Delta\delta}{\Delta R}\right)\frac{R_o^3}{2\alpha} \tag{12}$$

The term in parentheses in Eq. 12 is the slope of the curves of Fig. 5, so that values for the overlap dependence of $\delta_{Tl^{205}}$ may be calculated. It is important to note that the overlap dependence of the chemical shift given by Eq. 12 is based upon electric field effects of foreign cations in binary molten salt mixtures.

For comparison, a completely different calculation of the overlap dependence may be obtained from the temperature dependence of the chemical shift for the same nucleus in the corresponding pure crystalline thallium halide. This is possible because Eq. 6 relates the temperature dependence to a mean square vibrational amplitude. Since the overlap parameter is of the form, $\Lambda = \beta \exp(Cr)$, it follows that;

$$1/\Lambda_o \cdot \frac{\partial \Lambda}{\partial r} = (1/\Lambda_o \cdot \frac{\partial^2 \Lambda}{\partial r^2})^{1/2} = 1/\delta_o \cdot \frac{\partial \delta}{\partial r} \qquad (13)$$

and substitution of Eq. 6 into Eq. 13 gives;

$$1/\delta_o \cdot \frac{\partial \delta}{\partial r} = \left[1/\delta_o \cdot \frac{d\delta}{dT} \Big/ \frac{d\langle x^2 \rangle}{dT}\right]^{1/2} \qquad (14)$$

Table 3 provides a comparison of $1/\delta_o \cdot \frac{d\delta}{dr}$ based upon Eq. 12 and Eq. 14.

The agreement in $1/\delta_o \frac{d\delta}{dr}$, the fractional overlap dependence of the chemical shift, is remarkably good for the two independent methods of evaluation. It emphasizes the fact that the overlap dependence of $\delta_{Tl^{205}}$ is a property of the Tl^+X^- ion pair which is independent of physical state, and depends upon temperature and composition only to the extent that these factors affect the overlap of the Tl^+ and X^- wave functions.

ACKNOWLEDGMENTS

It is a pleasure to acknowledge the support of this research by the Air Force Office of Scientific Research under Grant No. AF-AFOSR - 1086-66 and the Advanced Research Projects Agency under Contract No. SD-89 with The University of Chicago.

TABLE 3

Overlap Dependence of Tl^{205} Chemical Shift at 500°C.

	$\alpha_{X^-} \times 10^{24}$ (cm^3)	$R_o \times 10^8$ (cm^3)	$-\frac{\Delta\delta}{\Delta R} \times 10^4$ (cm^{-1})	$-\frac{\partial\delta}{\partial T} \times 10^6$ (deg^{-1})[a]	$-\delta_o \times 10^4$ (a)	$\frac{\partial\langle x^2\rangle}{\partial T} \times 10^{20}$ ($cm^2 deg^{-1}$)	$-1/\delta_o \cdot \frac{\partial\delta}{\partial r} \times 10^8$ (cm^{-1}) (b)	(c)
TlCl	2.96	3.28	10.3	0.86	6.3	0.90	3.74	3.90
TlBr	4.16	3.42	10.4	0.60	11.0	1.13	2.53	2.20
TlI	6.43	3.66	9.8	<0.20	21.5	(d)	1.43	(d)

(a) Solid
(b) From Eq. 12
(c) From Eq. 14
(d) θ_D not available for TlI

REFERENCES

1. N. F. Ramsey, Phys. Rev., 78, 699 (1950).

2. The principle of strong local electroneutrality, which minimizes the electrostatic energy of the system, makes it statistically highly probable that the nearest neighbors of a given ion are of opposite charge.

3. K. Yosida and T. Moriya, J. Phys. Soc. Japan, 11, 33 (1956).

4. S. Hafner and N. H. Nachtrieb, Rev. Sci. Instr., 35, 680 (1964).

5. S. Hafner and N. H. Nachtrieb, J. Chem. Phys., 40, 2891 (1964).

6. S. Hafner and N. H. Nachtrieb, J. Chem. Phys., 42, 631 (1965).

7. H. A. Levy, P. A. Agron, M. A. Bredig, and M. D. Danford, Ann. N. Y. Acad. Sci., 79, 762 (1960).

8. P. O. Löwdin, Adv. Physics, 5, 1 (1956).

9. R. M. Sternheimer, Phys. Rev., 96, 951 (1954).

COORDINATION EQUILIBRIA OF NICKEL(II) IN MOLTEN CHLORIDE SALTS

G. Pedro Smith, Jorulf Brynestad, Charles R. Boston, and W. Ewen Smith

Metals and Ceramics Division
Oak Ridge National Laboratory
Oak Ridge, Tennessee 37830

INTRODUCTION

This report is a survey of recent research at Oak Ridge on the coordination properties of nickel(II) in molten chloride salts, measured by optical absorption spectroscopy.[1] Although present procedures for interpreting these data are still primitive, we are seeing much more detail than was expected a few years ago, and the emerging view of coordination chemistry in melts is one of rich diversity and unexpected regularity.

SOLID-MELT EQUILIBRIA

It is interesting to see what happens to the coordination geometry of nickel ions when a nickel compound melts. We studied this process for two compounds and found them to behave differently.[16] Fig. 1 shows the spectra of the solid

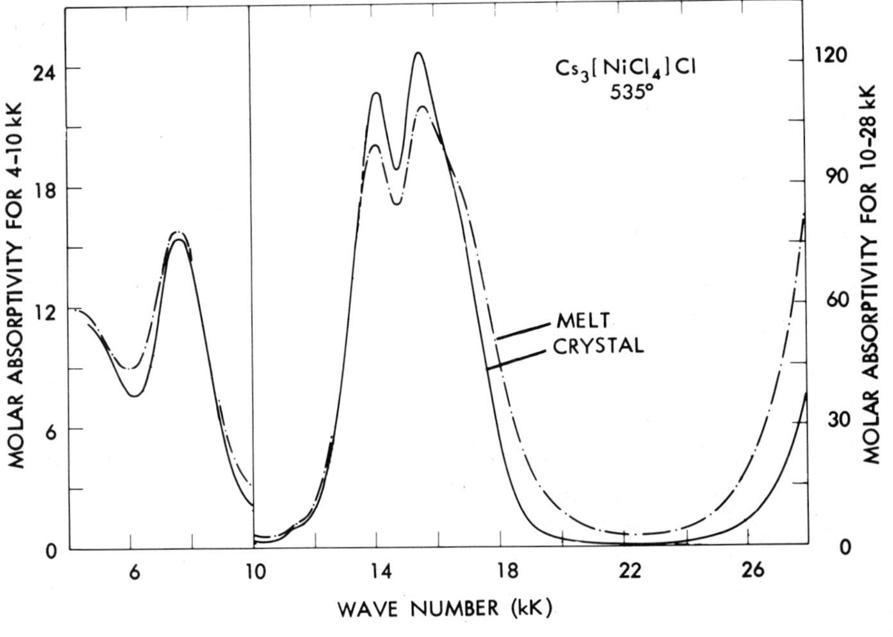

FIG. 1

Absorption spectra of the solid and liquid phases of $Cs_3[NiCl_4]Cl$ at 535°C (12° below the melting point). Adapted from Ref. 16 by courtesy of the American Institute of Physics.

and liquid phases of the compound $Cs_3(NiCl_4)Cl$ at 535°C, which is 12° below the melting point. In the solid compound x-ray diffraction measurements show that all of the nickels are present as tetrachloronickelate anions, $NiCl_4^{2-}$, which have a nearly tetrahedral coordination geometry. This complex ion has been extensively investigated in various media and its spectrum is well known. Since the spectrum of the melt is only slightly different from that of the solid, we conclude that when this compound melts, the $NiCl_4^{2-}$ anions remain largely intact. The small change in the spectrum produced by melting is further accentuated when the liquid is heated. However, additional measurements[17] show that dissociation of the $NiCl_4^{2-}$ ions is not the cause of these changes.

Hence, at the melting point of $Cs_3[NiCl_4]Cl$ we have the heterogeneous coordination equilibrium

$$Cs_3[NiCl_4]Cl(s) \rightleftharpoons NiCl_4^{2-}(l) + 3Cs^+(l) + Cl^-(l) \qquad (1)$$

in which there is no change in coordination geometry about nickel.

Fig. 2 shows similar measurements on the compound $CsNiCl_3$. These measurements were made at 743°C, which is 15° below the melting point. The very large differences

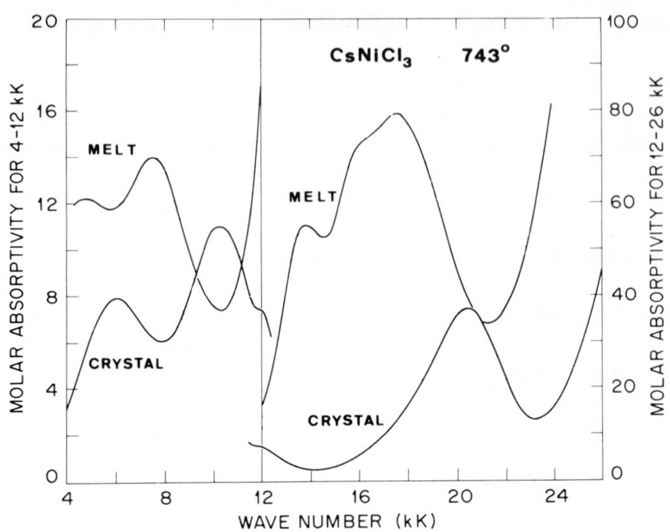

FIG. 2

Absorption spectra of the solid and liquid phases of $CsNiCl_3$ at 743°C (15° below the melting point). Adapted from Ref. 16 by courtesy of the American Institute of Physics.

between these spectra indicate that major changes in coordination geometry accompany melting. In the crystal each nickel is coordinated octahedrally to six chlorides, but the coordination octahedra share faces and are arranged in chains so that there are no molecule ions in the solid. The spectrum of the solid is characteristic of octahedral chloronickel(II) coordination, but that of the melt does not correspond to any coordination geometry known at the time the measurement was made.

In order to interpret the melt spectrum we had to carry through a series of investigations that will be described later.

ROLE OF OUTER-SHELL CATIONS IN LiCl-KCl MELTS

For most of the remainder of this report we shall be dealing with dilute solutions of nickel ions but we shall see that coordination behavior in such solutions is sometimes closely related to that in the concentrated systems

just described. When an alkali metal chloride serves as the solvent, coulombic forces impose a rough shell-like arrangement of solvent ions about a nickel ion. Each nickel ion is surrounded by a coordination shell of chloride ions beyond which is an outer shell where solvent cations predominate. When these outer-shell cations are relatively large, like Cs^+, Rb^+, or K^+, we have shown[18] that solute nickel ions form $NiCl_4^{2-}$ anions. However, when the outer-shell cations are small, like Li^+, they produce strong polarizations with complicated consequences. In order to study the effects of outer-shell cations we investigated[19] dilute solutions of $NiCl_2$ in molten mixtures of LiCl and KCl.

The LiCl-KCl system has no binary compounds, only a simple eutectic with a melting point considerably lower than those of the component salts. The spectra of nickel ions in these mixtures at a temperature well down in the eutectic trough are shown in Fig. 3. These spectra were measured at 526°C. They span solvent compositions of 20 to 55 mole % KCl, which covers most of the range that is liquid at 526°. As the solvent composition changes so do the spectra. These changes arise because nickel forms two discrete entities in these solvents at low temperatures, and the concentration ratio of these entities shifts with solvent composition.

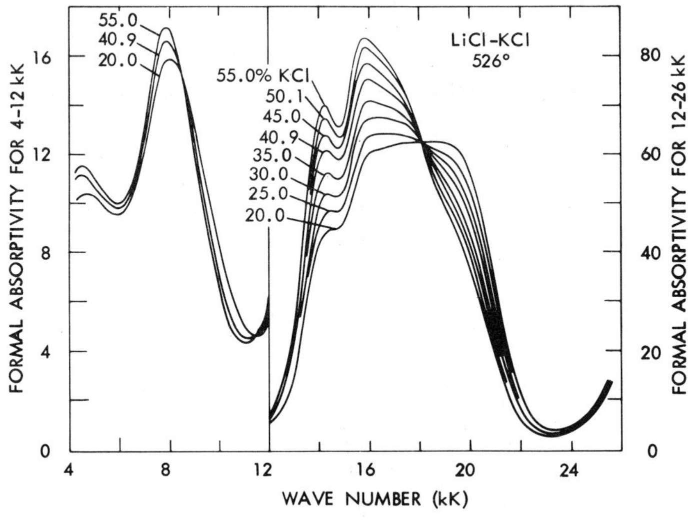

FIG. 3

Effect of solvent composition on the spectrum of nickel complexes in molten mixtures of LiCl and KCl at 526°C. Each spectrum is labeled with the solvent composition in mole percent KCl. Adapted from Ref. 19 by courtesy of the American Institute of Physics.

Even on casual inspection it is clear that these spectra behave in a highly regular way. A quantitative analysis shows that this behavior arises because of two relationships. First,

$$\epsilon_f(\lambda, X) = F_1(X)\epsilon_1(\lambda) + F_2(X)\epsilon_2(\lambda) \qquad (2)$$

where ϵ_f is the formal absorptivity of the mixture of entities, a function of wavelength λ and the mole fraction X of KCl (or LiCl) in the solvent; F_1 and ϵ_1 are, respectively, the fraction and molar absorptivity of one of the two nickel entities and F_2 and ϵ_2 are similar properties of the second entity. The sum of the two fractions $(F_1 + F_2)$ must, of course, equal unity. The unusual thing about eq. 2 is that the spectra of the individual components, ϵ_1 and ϵ_2, are independent of solvent composition. Equations of this sort are said to be internally linear with respect to the variable X, and their physical interpretation has been discussed.[20] The second regularity obeyed by the spectra in Fig. 3 is that the fractions, F_1 and F_2, of the components are linear functions of the mole fraction composition of the solvent.

As we shall see, these quantitative relations have far-reaching qualitative implications with regard to the structure of the nickel entities in the system. The development of experimental and computative procedures for determining to what degree sets of molten salt spectra obey these and other quantitative relationships is a recent accomplishment that has substantially expanded the usefulness of molten salt spectroscopy. Although we shall not take the time here to develop this subject, it is worth pointing out that such quantitative analyses underlie many of the conclusions we shall draw throughout this report.

The highly regular behavior, described above for the spectra of nickel entities in LiCl-KCl melts, permits quantitative extrapolations of the measured spectra in the directions of the two component spectra. The extrapolation in the KCl direction shows that one of the entities is $NiCl_4^{2-}$ with the same spectrum as it has in pure KCl. The spectrum of the second nickel entity cannot be determined accurately, but it is similar to the spectrum of nickel octahedrally coordinated to six chlorides, and we label it by the letter $\underset{\sim}{O}$. Thus we have an equilibrium between $NiCl_4^{2-}$ anions and $\underset{\sim}{O}$ entities that shifts in favor of $\underset{\sim}{O}$ entities upon addition of LiCl and in favor of $NiCl_4^{2-}$ ions upon addition of KCl.

One might suppose that the ratio of Li^+ to K^+ ions in the outer shells of these nickel entities would depend on the solvent composition, but the experimental facts show that at low temperatures this is certainly not the case for $NiCl_4^{2-}$ ions and probably not the case for $\underset{\sim}{O}$ entities for the following reasons. We have made measurements[18] which show that the $NiCl_4^{2-}$ spectrum is sufficiently sensitive to outer-shell cations so that if any substantial part of these cations were to change from K^+ to Li^+, we would see a measurable change in the spectrum. But we do not see such a change. Therefore, we conclude that each $NiCl_4^{2-}$ ion remains sheathed in a layer that is largely K^+ ions irrespective of the solvent composition. In the case of the $\underset{\sim}{O}$ entities we know that their spectrum is also invariant with respect to solvent composition, and their fraction changes linearly with the fraction of Li^+ ions. Therefore, it seems likely that these $\underset{\sim}{O}$ entities are surrounded predominantly by Li^+ ions.

In order to display the structural regularity of the outer shells, we propose the following notation to represent the equilibrium between $NiCl_4^{2-}$ and $\underset{\sim}{O}$ entities.

$$(NiCl_4^{2-}|K^+) + n(Cl^-|Li^+) \rightleftharpoons (O|Li^+) + K^+ \qquad (3)$$

Here the outer-shell cations about each entity are indicated to the right of the vertical bars. Tetrachloronickelate ions with K^+ ion outer shells react with chloride ions, modified by Li^+ ion polarization, and form the $\underset{\sim}{O}$ entity with a Li^+ ion outer shell.

Taken as a whole this structural behavior of nickel complexes in LiCl-KCl mixtures at 526°C is exceedingly regular, too regular to be accounted for by most theories of ion associations in molten salt mixtures. However, a model of microheterogeneities in the neighborhood of a eutectic proposed by Danilov and Radchenko for liquid metals and applied to molten salts by Delimarskii and Markov[21,22] will account for the behavior we observed. Nevertheless, there is little independent evidence for the existence of such microheterogeneities and we are skeptical of this model.

As the temperature increases above 526°C two things happen. First, the equilibrium shifts in favor of $NiCl_4^{2-}$ ions. Second, there is a progressive breakdown in the invariance of the spectra of the component entities with respect to changes in solvent composition. The first effect is shown in Fig. 4. Here we see the spectra of nickel entities in the LiCl-KCl eutectic over a wide range of temperatures. At 363°C the principal band in the $\underset{\sim}{O}$ spectrum is quite evident at about 20 kK. With increasing temperature this band fades away while the $NiCl_4^{2-}$ bands grow in.

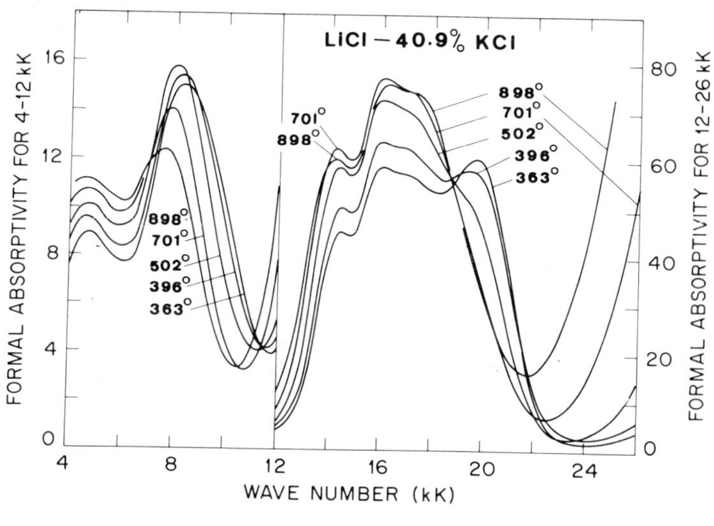

FIG. 4

Effect of temperature on the spectrum of nickel complexes in a mixture of LiCl and KCl containing 40.9 mole % KCl. Adapted from Ref. 19 by courtesy of the American Institute of Physics.

Above 700°C the $\underset{\sim}{O}$-entity bands can no longer be detected. The spectrum of what remains is quite unusual. First of all it varies with solvent composition somewhat like a mixture of two entities and even shows internal linearity over substantial ranges of solvent composition. However, its temperature dependence is quite unlike that for the spectrum of an equilibrium mixture of discrete entities. It is internally linear with respect to temperature and its temperature range. Furthermore, all of the bands in the spectrum are derived by small, continuous displacements of the bands of $NiCl_4^{2-}$. In these and other ways, the temperature dependence is like that of $NiCl_4^{2-}$ complexes in crystalline and liquid systems.[8,16,17,19]

We propose that with increasing temperature Li^+ ions in LiCl-KCl mixtures begin to invade the outer shells of $NiCl_4^{2-}$ ions in a statistical way. A $NiCl_4^{2-}$ ion so attacked is polarized and distorted away from its original geometry. The collection of these complexes with various numbers of Li^+ ions in their outer shells has a broadened distribution of coordination geometries that centers about an unknown average. This average shifts with changes in the average outer-shell composition and, hence, shifts at high temperatures with changes in solvent composition. We denote this distribution by the letter $\underset{\sim}{T}$ as a reminder that it can be converted into tetrahedral complexes in a continuous way. When we want to be more specific, we use the notation $(\underset{\sim}{T} | Li^+, K^+)$ to signify that the outer shells consist of statistical mixtures of Li^+ and K^+ ions. We shall meet further examples of such coordination distributions in other systems.

It is plausible to suppose that with increasing temperature $\underset{\sim}{O}$ entities suffer a fate like that of the $NiCl_4^{2-}$ ions, that is, they probably degenerate into (**O** | Li^+, K^+) coordination distributions. However, we do not know much about **O** entities in this system because their concentration drops rapidly with increasing temperature and their bands become lost in those of the **T** entities.

COORDINATION DISTRIBUTIONS IN $MgCl_2$-KCl and $CsNiCl_3$ MELTS

We made a similar study[23] of the behavior of nickel entities in molten mixtures of $MgCl_2$ and KCl, and again found an equilibrium between **T**-type and **O**-type entities like that which occurs at intermediate temperatures in the LiCl-KCl system. With increasing temperature the equilibrium shifted in favor of the **T**-type until at high temperatures there was only a coordination distribution.

From these measurements on $MgCl_2$-KCl melts we found a basis for inferring what happens to the coordination geometry of nickel when the compound $CsNiCl_3$ melts. The relevant results are shown in Fig. 5. The upper pair of spectra are for molten $CsNiCl_3$ at 743 and 863°C. At the beginning of this report we presented the 743° spectrum but found no basis for interpreting it. The lower pair of spectra are for dilute solutions of $NiCl_2$ in molten $KMgCl_3$ at temperatures where a coordination distribution equilbrium prevails.

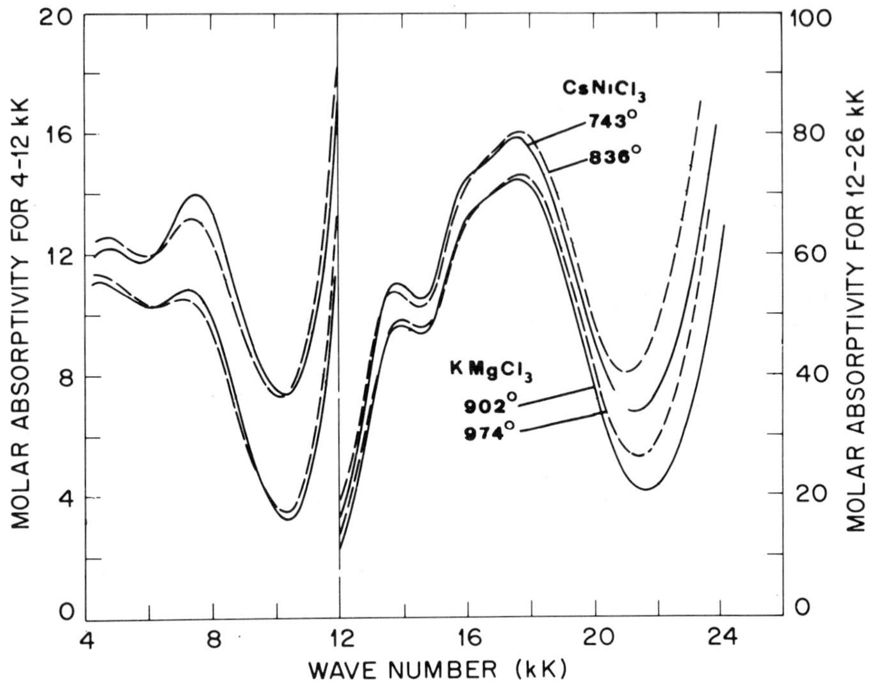

FIG. 5

Comparison of spectra of molten $CsNiCl_3$ with spectra of a dilute solution of $NiCl_2$ in molten $KMgCl_3$. Adapted from Ref. 16 by courtesy of the American Institute of Physics.

The spectra for these two systems, including their temperature dependences, are so very similar that we suppose the coordination geometry of nickel to be very similar in the two systems. In the case of the liquid nickel compound, each nickel ion is in the outer shell of some other nickel ion. We represent the heterogeneous coordination equilibrium that occurs at the melting point, thus

$$CsNiCl_3(s) \rightleftharpoons (\underset{\sim}{T}|Ni^{2+}, Cs^{+})(l) \tag{4}$$

EQUILIBRIA IN CHLOROZINCATE MELTS

One of the remarkable things about molten chloride salts is that there is such a wide variety of them with all sorts of cation-chloride ion interactions. One of the more complicated examples is zinc chloride and its mixtures with alkali metal chlorides.

Zinc chloride melts to form a viscous network liquid in which most of the zincs are bound to four other zincs by chloride bridges.[24,25] This network structure is degraded either by heating or by addition of alkali metal chlorides. Degradation by chloride additions involves breaking network bridges with the eventual formation of $ZnCl_4^{2-}$ ions, but before this stage of complete degradation is reached, various sizes of polymeric ions are formed.

We studied[26] the behavior of nickel ions in CsCl-$ZnCl_2$ mixtures, and Fig. 6 shows some spectra measured at 400°C. In 50 mole % $ZnCl_2$, there is almost entirely a single nickel complex with what appears to be a distorted tetrahedral coordination. Other measurements indicate that one or two of the ligands are $ZnCl_4^{2-}$ ions while the others are simple

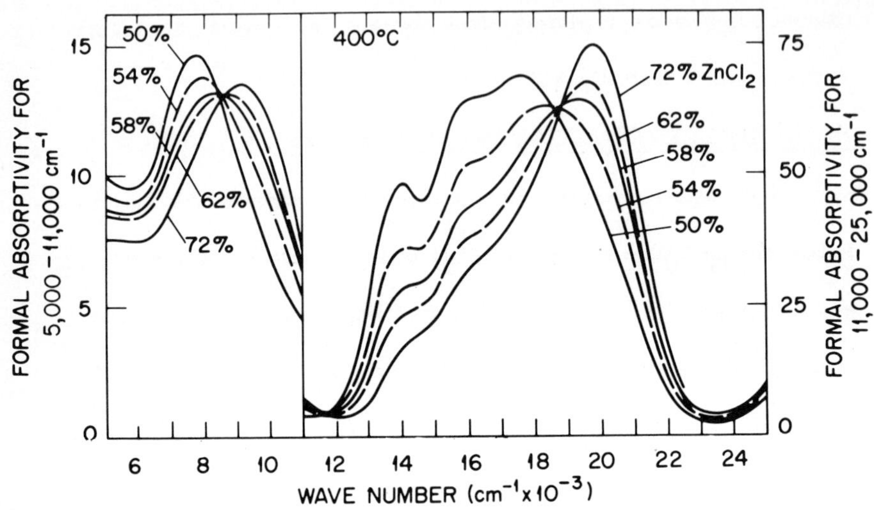

FIG. 6

Spectra of nickel complexes in molten $ZnCl_2$-CsCl mixtures containing 50-72 mole % $ZnCl_2$. Reproduced from Ref. 26 by courtesy of the American Chemical Society.

chlorides. As the $ZnCl_2$ content of the solvent is increased, the absorption bands of a second entity grow in. At 72 mole % $ZnCl_2$, conversion to the second entity has gone a long way toward completion. Virtually complete conversion can be achieved by adding a little more $ZnCl_2$ and lowering the temperature. The spectrum is found to be that of an octahedrally coordinated complex. The temperature dependence of this spectrum indicates a well-defined inversion center and, hence, a highly ordered coordination shell. Because this six-coordinate complex is favored by relatively high $ZnCl_2$ concentrations, the ligands are almost certainly chlorozincate polymers.

The equilibrium between these four-coordinate and six-coordinate complexes shifts in favor of the four-coordinate form when the temperature increases. At temperatures on the order of 700 to 800°C there is no longer a two species equilibrium but a coordination distribution.

Returning to 400°C we find that relatively small things happen to the spectrum when the solvent composition is changed from 72 to about 90 mole % $ZnCl_2$. However, as the $ZnCl_2$ content is increased beyond that point, a third complex appears and grows in at an accelerating rate as the composition of the solvent approaches pure $ZnCl_2$. This behavior is shown in Fig. 7. This figure is confusingly like the preceding one except for the crucial difference that the octahedral bands fade away, rather than grow in, as the $ZnCl_2$ concentration increases. The absorption bands of the new complex are in reasonable positions to be assigned to tetrahedral coordination. In pure $ZnCl_2$ as solvent we have a mixture of entities.

The temperature dependence of this second tetrahedral-octahedral equilibrium is very unusual. In pure $ZnCl_2$ as solvent it shifts in favor of the octahedral complex as the

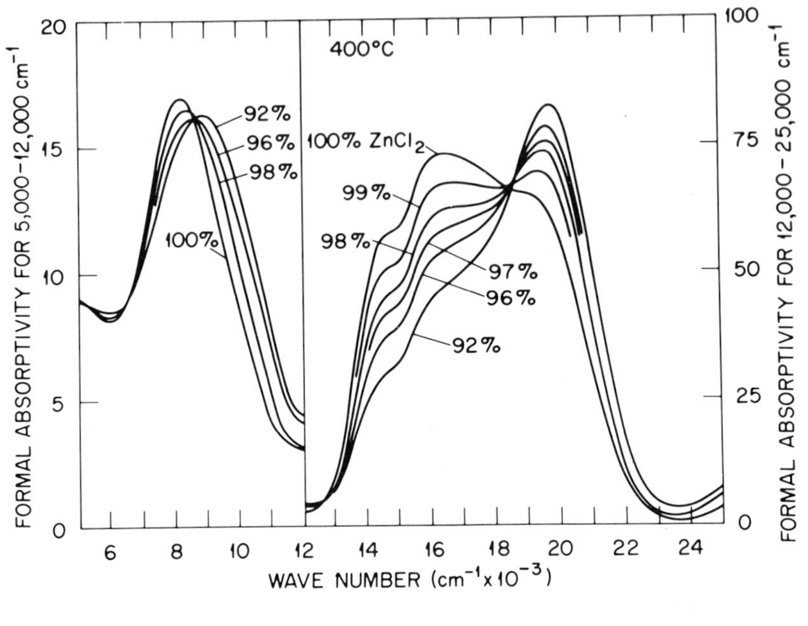

FIG. 7

Spectra of nickel complexes in molten $ZnCl_2$-CsCl mixtures containing 92-100 mole % $ZnCl_2$. Reproduced from Ref. 26 by courtesy of the American Chemical Society.

temperature increases, but in melts containing 2 mole % or more CsCl, the temperature dependence is reversed; the tetrahedral form is favored with increasing temperature.

We propose that in pure $ZnCl_2$ as solvent the tetrahedral entities are nickel ions which have substituted for zinc ions at network-forming sites, while the octahedral entities are nickel ions at network-breaking sites. However, we have not successfully rationalized the complicated sequence of events that ensues when small amounts of CsCl are added to the melt.

COORDINATION EQUILIBRIA OF NICKEL (II)

There are important relationships between our work on nickel in $ZnCl_2$-CsCl melts and that of Angell and Gruen[9] on nickel in $ZnCl_2$-KCl melts. It appears to us from examining their spectra that a basically similar tetrahedral-octahedral equilibrium occurs in both systems near the equimolar solvent composition, and that replacement of Cs^+ ions by K^+ ions shifts this equilibrium to a significant degree toward octahedral coordination.

Nickel forms three different tetrahedrally coordinated entities in $ZnCl_2$-CsCl melts in addition to an octahedrally coordinated entity. In CsCl-rich melts it forms the $NiCl_4^{2-}$ anion, in the equimolar mixture it forms a complex in which both Cl^- and $ZnCl_4^{2-}$ serve as ligands, and in $ZnCl_2$-rich melts it substitutes for Zn^{2+} at network-forming sites in the $ZnCl_2$ network. Angell and Gruen[17] found that cobalt(II) has only tetrahedral coordination in $ZnCl_2$-KCl melts, but in view of our results for nickel(II) in $ZnCl_2$-CsCl it seems likely that cobalt(II) forms several different tetrahedrally coordinated entities in this type of solvent.

Along the same vein it is interesting that Young[5,27] has studied nickel(II) in a variety of molten fluoride salts and found only octahedral coordination. Again, it seems plausible to suppose that several different octahedrally coordinated entities occur in these systems.

EQUILIBRIA IN $CsAlCl_4$ MELTS

Let us now consider the structurally simpler situation presented by the complex entities which nickel forms in $CsAlCl_4$ melts. We are currently studying[28] these complexes and finding that although the behavior is very complicated it is also very regular. In molten $CsAlCl_4$ there is a reaction between $AlCl_4^-$ ions as follows:

$$2AlCl_4^- \rightleftharpoons Cl^- + Al_2Cl_7^- \qquad (5)$$

This reaction produces chloride ions and a chloride-bridged chloroaluminate anion, $Al_2Cl_7^-$. At low temperatures, however, the extent of this reaction is very small. Hence, the chloride ion concentration can be varied over a wide range by relatively small additions of CsCl or Al_2Cl_6 which, respectively, add or remove Cl^- ions. The latter process presumably takes place by the reaction

$$Al_2Cl_6 + Cl^- \rightarrow Al_2Cl_7^- \qquad (6)$$

When the melt contains a substantial excess of chloride ions, $NiCl_2$ dissolves to form $NiCl_4^{2-}$ ions. However, as the excess chloride ion concentration is decreased to a small amount, a second chloronickel complex progressively replaces $NiCl_4^{2-}$. When the temperature is increased at constant composition, the equilibrium between these complexes shifts in favor of the new complex. It is clear from the spectrum of the new complex that its coordination is not even approximately either tetrahedral or octahedral.

The composition dependence of this equilibrium indicates that the reaction involves a loss of one or more chlorides by the $NiCl_4^{2-}$ ion, and there proves to be a simple way of determining how many are lost. Solid $CsNiCl_3$ can be equilibrated with a liquid phase containing a mixture of $NiCl_4^{2-}$ and the new complex in the presence of a small excess of CsCl. When the excess CsCl is increased, some $CsNiCl_3$ dissolves and a proportionate amount of $NiCl_4^{2-}$ is formed while the concentration of the new complex stays fixed. This shows, first, that the dissolution of $CsNiCl_3$ takes place according to the following reaction

$$CsNiCl_3(s) + Cl^-(l) \rightleftharpoons NiCl_4^{2-}(l) + Cs^+(l) \qquad (7)$$

It also shows that the Cl:Ni ratio in the new complex, apart from solvation, is the same as in $CsNiCl_3$, namely, 3:1. Hence we can formulate the equilibrium between $NiCl_4^{2-}$ and the new complex thus

$$nNiCl_4^{2-} \rightleftharpoons (NiCl_3)_n^{n-}(solv?) + nCl^- \qquad (8)$$

This experiment tells us nothing about solvation because solvent cations are present in large excess.[29]

By changing the conditions we can prepare other complexes of nickel in this solvent, and there is more to the story than we have time to tell here. The general point which we wish to make about $CsAlCl_4$ is that nickel forms well-defined molecular entities in this medium so that its coordination reactions have the simplicity envisaged by classical complex ion chemistry. The fact that the complex anions have only large Cs^+ outer-shell cations in this solvent is almost certainly essential to the observed simplicity. Thus, it might be worthwhile to write eq. 8 as

$$(NiCl_4^{2-} | Cs^+) \rightleftharpoons [(NiCl_3]_n^{n-} | Cs^+) + (Cl^- | Cs^+) \qquad (9)$$

to remind ourselves that in chloroaluminate melts containing some polarizing cations, an entity like $([NiCl_3]_n^{n-} | Li^+, Cs^+)$, if it exists, may be structurally different from $([NiCl_3]_n^{n-} | Cs^+)$.

SUMMARY

In summary, we found that at low temperatures Ni^{2+} ions participate in a wide variety of coordination equilibria among discrete entities that range from well-defined molecule ions, as in $CsAlCl_4$ melts, to strongly interacting many-atom systems, as in $ZnCl_2$. Intermediate to these extremes were equilibria, like that in LiCl-KCl melts, involving outer-shell cations that do not form well-defined chlorometallate complexes. It is not hard to imagine a wide spectrum of situations intermediate to the extremes we observed. At high temperatures some of these low-temperature equilibria degenerated into coordination distributions which might be regarded as coordination equilibria among a virtual continuum of entities.

We also found three, quite different heterogeneous coordination equilibria. One, given by eq. 1, involved no change in coordination geometry. Another, given by eq. 7, involved a change from octahedral coordination in the solid to tetrahedral coordination in the melt. The third, given by eq. 4, involved a change from a well ordered to a much less well ordered coordination geometry on passing from solid to melt.

ACKNOWLEDGMENT

This research was sponsored by the U. S. Atomic Energy Commission under contract with the Union Carbide Corporation.

REFERENCES

1. Early research and background information is given in refs. 2 and 3. More recent research, related to that reported here, is given in refs. 4-15.

2. D. M. Gruen in "Fused Salts," B. R. Sundheim, Ed., McGraw-Hill Book Co., New York, 1964, pp. 301-339.

3. G. P. Smith in "Molten Salt Chemistry," M. Blander, Ed., Interscience Publishers, Inc., New York, 1964, pp. 461-462.

4. H. A. Øye and D. M. Gruen, Inorg. Chem., 3, 836 (1964).

5. J. P. Young and G. P. Smith, J. Chem. Phys., 40, 913 (1964).

6. D. M. Gruen, Quart. Rev., 19, 349 (1965).

7. H. A. Øye and D. M. Gruen, Inorg. Chem., 4, 1173 (1965).

8. G. P. Smith and S. von Winbush, J. Am. Chem. Soc., 88, 2127 (1966).

9. C. A. Angell and D. M. Gruen, J. Phys. Chem., 70, 1601 (1966), report results for Ni^{2+} in $ZnCl_2$-KCl mixtures closely related to those reported here for Ni^{2+} in $ZnCl_2$-CsCl mixtures. The interpretation these authors give their data differs in several essential ways from that we put forward and we have given reasons (ref. 20) for supposing these authors to be wrong.

10. H. A. Øye and D. M. Gruen, Selected Topics High Temp. Chem., 1966, 61-71.

11. R. A. Bailey and J. A. McIntyre, Inorg. Chem., 5, 964 (1966).

12. R. A. Bailey and J. A. McIntyre, Inorg. Chem., 5, 1824 (1966).

13. R. A. Bailey and J. A. McIntyre, Inorg. Chem., 5, 1940 (1966).

14. C. A. Angell and D. M. Gruen, J. Inorg. Nucl. Chem., 29, 2243 (1967).

15. C. R. Boston, C. H. Liu, and G. P. Smith, Inorg. Chem., in press, studied the behavior of Ni^{2+} in LiBr-41 mole % KBr and LiI-36.8 mole % KI.

16. C. R. Boston, J. Brynestad, and G. P. Smith, J. Chem. Phys., 47, 3193 (1967).

17. G. P. Smith, C. R. Boston, and J. Brynestad, J. Chem. Phys., 45, 829 (1966).

18. G. P. Smith and C. R. Boston, J. Chem. Phys., 43, 4051 (1965); 46, 412 (1967).

19. J. Brynestad, C. R. Boston, and G. P. Smith, J. Chem. Phys., 47, 3179 (1967).

20. J. Brynestad and G. P. Smith, J. Phys. Chem., 72, 296 (1968).

21. Iu. K. Delimarskii and B. F. Markov, "Electrochemistry of Fused Salts," translated by A. Peiperl and R. E. Wood, The Sigma Press, Publishers, Washington, D. C., 1961, pp. 26-28.

22. The spectra of numerous other transition metal ions in the LiCl-KCl eutectic have been measured (refs. 2, 3, 11, 12) and some are known to correspond to multi- (maybe two) species equilibria but nothing is known about the effect of solvent composition. If the two-species equilibrium which we find for nickel(II) is associated with microheterogeneities at low temperatures, then it seems reasonable to expect the spectra of most if not all transition metal species to be composition invariant in LiCl-KCl at low temperatures.

23. J. Brynestad and G. P. Smith, J. Chem. Phys., 47, 3190 (1967).

24. R. B. Ellis, J. Electrochem. Soc., 113, 485 (1966), and references therein.

25. B. F. Markov and S. V. Volkov, Ukr. Khim. Zh., 30, 906 (1964).

26. W. E. Smith, J. Brynestad, and G. P. Smith, J. Am. Chem. Soc., 89, 5983 (1967); and unpublished results.

27. J. P. Young, reported at the Molten-Salt Chemistry Information Meeting, Oak Ridge, Tennessee, April 17, 1968.

28. J. Brynestad and G. P. Smith, unpublished research.

29. In ref. 7 there is brief mention of preliminary evidence that a tetrahedrally coordinated $NiCl_3AlCl_4^{2-}$ complex forms in molten $KAlCl_4$ containing a small excess of KCl. In pointing this out it seems worthwhile emphasizing that the $NiCl_3(solv?)^-$ complex we found in $CsAlCl_4$ melts very clearly does not have a tetrahedral-type of spectrum; it has no absorption band between 5 and 12 kK, whereas typical tetrahedral and distorted tetrahedral chloronickel complexes have an intra-term (^3F) transition at about 7-8kK.

ELECTRONIC ABSORPTION SPECTRA AND THE NATURE OF DILUTE SOLUTIONS OF ALKALI METALS IN FUSED ALKALI HALIDES

D. M. Gruen, M. Krumpelt, and I. Johnson

Argonne National Laboratory
Argonne, Illinois

INTRODUCTION

Metal-molten salt systems are of special interest because they show continuous transitions from metallic to non-metallic character. The elucidation of the nature of these solutions presents intriguing problems particularly since they represent, from a conceptual point of view, the simplest liquid systems containing "solvated" electrons.

The relative simplicity of these solutions makes a detailed study of their properties highly desirable; however, their extreme corrosivity requires the use of special containers and the high melting points of the salts usually necessitate extensive adaptations of conventional measuring apparatus. The work of M. A. Bredig and his collaborators on the phase diagrams and electrical conductivities of solutions of alkali metals in fused alkali halides furnishes a detailed overview of some of the important properties of these systems.[1-10] For each system there exists a consolute temperature above which metal and salt are miscible in all proportions. In the system involving the metals K, Rb and Cs, the consolute temperatures tend to be close to the melting points of the pure alkali halides. The electrical conductivities of the solutions are very much larger than those of the pure salts and above the consolute temperature, a smooth curve of specific conductance versus mole fraction of alkali metal connects the conductances of the pure salts with those characteristic of the pure metals. At intermediate concentrations, the solutions display a mixture of ionic and electronic conductivities, the latter exceeding the former by many orders of magnitude.

DILUTE SOLUTIONS OF ALKALI METALS

The nature of the electrons giving rise to the electronic conductivity has been investigated by means of electron spin resonance techniques,[9] magnetic susceptibility measurements[10,11] and studies of electronic absorption spectra.[12-16]

The esr study[9] and one magnetic susceptibility study[10] gave results which were inconclusive. However, a later susceptibility study concluded that solutions of K metal in molten KCl are paramagnetic.[11] Making some assumptions about the densities of the solutions and about the additivity of the ionic diamagnetism, it was possible to compute molar paramagnetic susceptibilities for the K valence electrons. Although the conductivity data of Bronstein and Bredig[7] are not suggestive of a conduction band, Wilson[17] succeeded in rationalizing their data on the basis of such a theory. Bettman also concluded that the susceptibility data can be very roughly accounted for by the equation for conduction band electrons, with the effective electron mass approximately equal to the free electron mass.[11]

The susceptibility data can equally well be accounted for, as pointed out by Bettman,[11] on the basis of a localized electron model such as the F-center like models discussed by Pitzer.[18] The electrical conductivities have been rationalized by Rice[19] assuming an atomically dispersed solution of metal atoms.

Spectrophotometric studies of fused salt solutions have been reported frequently during the last decade and the techniques primarily used to investigate transition metal ion are highly developed.[20] Earlier spectroscopic studies of solutions of alkali metals in liquid alkali halides have been carried out in pyrex, vycor or quartz containers.[12,14,15,16]

The present work was undertaken with a view to obtaining absorption spectra of such solutions in sapphire or magnesium oxide containers in order to minimize effects due to dissolved impurities. A further aim of the study was to follow the temperature dependence of the F-center absorption spectra in the solid state up to the melting point and to attempt a rationalization of both solid and liquid state spectra on the basis of a unified model.

RESULTS

The results of measurements on the Na-NaCl system are shown in Fig. 1. The measurements were made in a sapphire cell sealed by means of a sapphire plate as described in the experimental section. To obtain the absorption spectrum of the F-center at 550°C (Fig. 1C), absorption due to

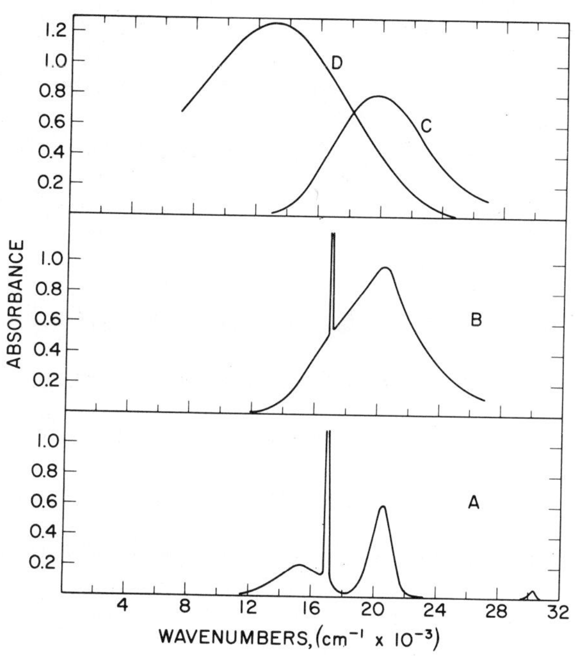

FIG. 1

Absorption spectra of (A) Na vapor; (B) Na vapor + F-center in solid NaCl at 550°C; (C) spectrum B corrected for Na-vapor absorption; (D) dilute solution of Na metal in liquid NaCl.

Na vapor (Fig. 1A) was subtracted from the measured spectrum (Fig. 1B). The correction to be applied for Na-vapor absorption in spectrum 1B was obtained by measuring the intensity of the Na-resonance line at ~17,000 cm^{-1} and calculating the Na-vapor absorption at other wavelengths from the pure Na-vapor spectrum of Fig. 1A. The spectrum of Na dissolved in liquid NaCl is show in Fig. 1D.

The maximum of the F-center band at 550°C is at 19,500 cm^{-1} (2.42 e.v.) and its half-width is 6,700 cm^{-1} (.83 e.v.). The maximum of the very broad spectrum of Na in liquid NaCl is at 13,000 cm^{-1} (1.61 e.v.) and its halfwidth is 11,400 cm^{-1} (1.41 e.v.).

The temperature dependence of the F-center band of NaCl was studied by Mollwo.[21] The (interpolated) values of the band maximum at 500°C is 2.32 e.v. and of the halfwidth .85 e.v. Mollwo[12] also obtained the absorption spectrum of Na in liquid NaCl finding a broad band with maximum at 12,650 cm^{-1} (1.57 e.v.).

DISCUSSION

The results of our measurements on the NaCl F-center absorption at 550°C and of the absorption spectrum of Na in liquid NaCl are in substantial agreement with the work of Mollwo.[12,21] This was to us a somewhat unexpected finding since the latter work had been carried out in containers of Supremax glass. Mollwo summarized his experience by stating that the glass containers became increasingly brown during the course of an experiment, due presumably to a slow attack by the alkali metals. Apparently, attack of the glass was not so serious as to prevent valid measurements of the spectra from being carried out.

Extrapolation of the shift of the F-center band maximum with temperature leads to the value 2.2 e.v. at the melting point of NaCl.[12] The large shift of the band maximum to ~1.6 e.v. on melting led Mollwo to the conclusion[21] that there is no simple relationship between the absorption bands in the melts and the absorption bands of the F-centers. We wish to point out that this conclusion is not necessarily valid if one takes into account the drastic changes in some of the physico-chemical properties such as the molar volumes which occur on melting. In the following discussion the data of Mollwo,[12,21] as well as our own results, are interpreted taking account of the large changes in molar volume that occur on melting.

The temperature dependence of the F-center band and the change in the absorption spectrum on melting is best illustrated by an examination of Fig. 2 which summarizes data on the K-KBr system taken from Ref. 12 and 21. The figure shows that the maximum of the F-center band shifts to lower energies and that the half-width increases with increasing temperature. On melting, the maximum shifts abruptly to lower energies by about 0.5 e.v. and the band broadens considerably.

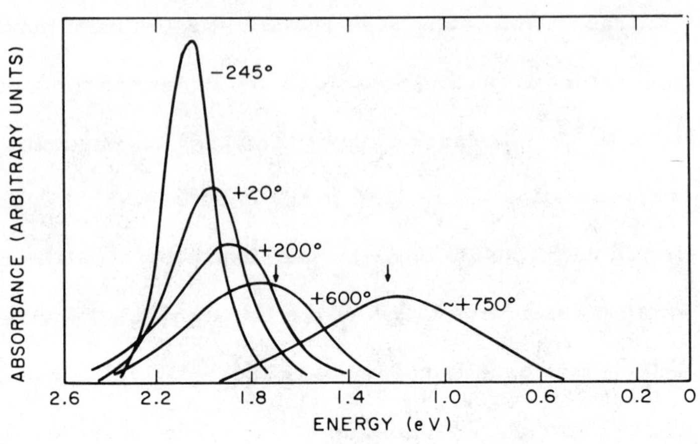

FIG. 2

Temperature variation of F-center absorption in KBr. Curve at 750°C is for a solution of K in liquid KBr. (Data from ref. 12 and 21).

A plot of the F-center band maxima for a number of alkali halides as a function of temperature is shown in Fig. 3. The data are for the melt systems: Na-NaCl; Na-NaBr; K-KCl; K-KBr; K-KI. It is clear that if the data are plotted in this way, there occurs in every case a discontinuous, abrupt change in the energy of the band maximum on melting.

It is well known that the alkali halides undergo large positive volume changes on fusion ranging from 31.7% for LiF to 9.8% for CsCl. Molar volumes of liquid alkali halides were determined by Yaffee and van Artsdalen.[22,23] Volume changes on fusion have been determined by Sauerwald and his

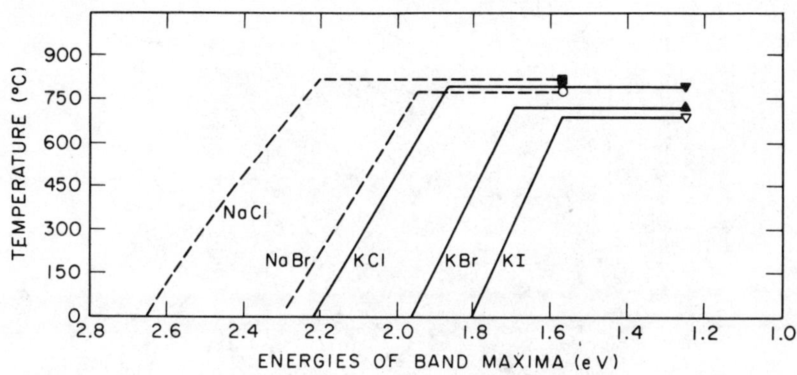

FIG. 3

Temperature variation of F-center band maxima showing discontinuous energy shift on melting. (Data for the melts are indicated by circle, square and triangles).

collaborators.[24] From these data, the molar volumes of solid alkali halides at their melting points can be calculated. The molar volumes of the alkali halides at room temperatures can be obtained from International Critical Tables.[25] The molar volumes increase about 12% between room temperature and the melting points.

Expansivities were carefully determined by Bockris et al.,[26] for solid NaCl and solid KCl. A plot of the KCl data shows that a linear interpolation of the molar volumes of the solid between room temperature and the melting point has a maximum deviation from the experimental data of 2%. In Table 1, the

TABLE 1

Temperature Variation of Electronic Absorption Band Maxima and Molar Volumes of Dilute Solutions of Alkali Metals in Solid and Molten Alkali Halides

							ΔV_{fusion}(%)
NaCl							23.0
t°C	20	550	600	700	801(MP)	809	
ν_{max}, ev	2.66	2.42	2.36	2.30	1.57*	1.61	
V_m, cc/mole	27.0	28.6	28.8	29.3	37.6	37.6	
$V_m^{-2/3}$	0.111	0.107	0.106	0.105	0.0892	0.0892	
NaBr							21.2
t°C	20	500	747(MP)				
ν_{max}, ev	2.29	2.11	1.57*				
V_m, cc/mole	32.1	34.9	43.9				
$V_m^{-2/3}$	0.099_8	0.093_7	0.0803				

TABLE 1 (Cont'd)

KCl						16.8
t^oC	20	200	400	600	771(MP)	
ν_{max}, ev	2.20	2.12	2.05	1.97	1.27*	
V_m, cc/mole	37.6	37.9	38.6	39.5	48.8	
$V_m^{-2/3}$	0.0891	0.0886	0.0876	0.0862	0.0749	
KBr						16.0
t^oC	20	200	400	600	754(MP)	
ν_{max}, ev	1.97	1.90	1.82	1.76	1.27*	
V_m, cc/mole	43.3	44.5	45.9	47.3	55.9	
$V_m^{-2/3}$	0.0811	0.0796	0.0780	0.0765	0.0684	
KI						15.0
t^oC	20 3	300	500	600	680(MP)	
ν_{max}	1.81	1.70	1.65	1.62	1.27I	
V_m	53.2	55.7	57.4	58.3	67.8	
$V_m^{-2/3}$	0.0707	0.0686	0.0672	0.0666	0.0602	

* Denotes molten halide.

molar volumes of NaCl, and KCl were taken from the data of Bockris et al.,[26] while for NaBr, KBr and KI, the densities were obtained by linear interpolation between 0^oC and the respective melting points.

In appropriately labelled rows of Table 1 are listed the molar volumes to the -2/3 power and the energies of the band maxima of those alkali halides for which data are available. A plot of the band maxima vs. molar volumes$^{-2/3}$ is shown in Fig. 4. The striking feature emerging from this plot is that the energies of the band maxima for the liquid metal-molten salt solutions lie on nearly linear lines connecting the experimental points which describe the variation of the maxima with molar volume in the solid solutions from near room temperatures to just below the melting point. The large and abrupt changes in the band maxima on melting are, therefore to be interpreted as resulting from the large changes in molar volume which accompany the fusion process. It now becomes clear that there is in fact a close and natural relationship between the spectra of the F-centers in alkali halide crystals and the melt spectra. In our view, the relationship between the solid state and melt spectra provides striking

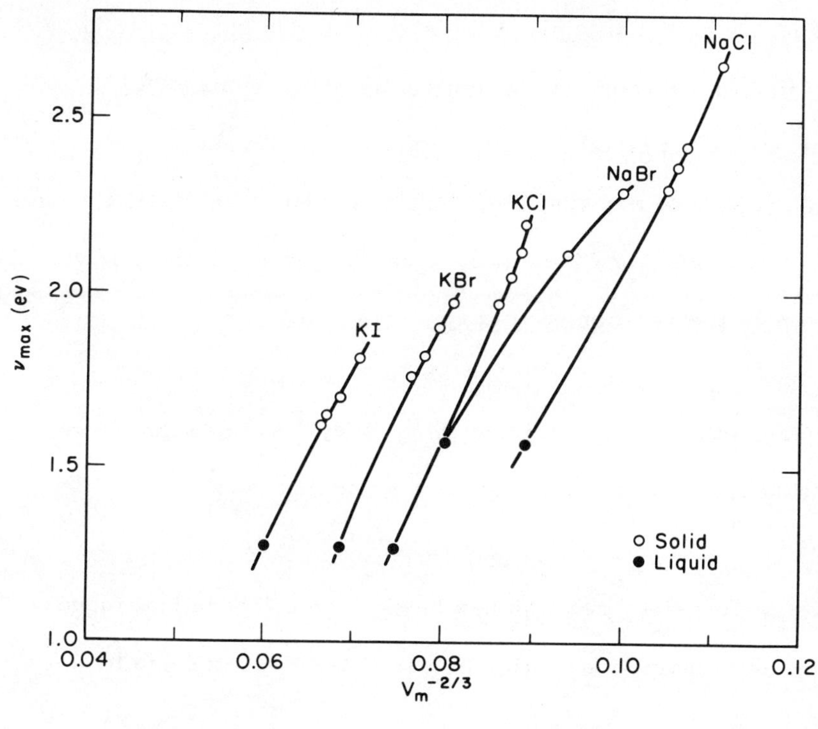

FIG. 4

Dependence of electronic absorption band maxima on molar volume for dilute solutions of alkali metals in solid and molten alkali halides.

evidence for the model of the alkali metal-alkali halide solutions which supposes that the metal dissolves to form M^+ ions and F-center like electrons.

The significance of the nearly linear relationship between the energies of the band maxima and the (molar volumes)$^{-2/3}$ power lies in the fact that for the rock salt structure common to all of the alkali halides considered here, the (molar volumes)$^{-2/3}$ are directly proportional to d^{-2}, where d is the metal-halogen internuclear distance.

Mollwo had observed[27] that the band maxima of F-centers in various alkali halides at constant temperature are fairly well given by the relation

$$\lambda_{max} = 600\ d^2$$

where both λ_{max} and d are in Å. Ivey[28] refined the curve fitting process and obtained the relation

$$\lambda_{max} = 703\ d^{1.84}$$

The Mollwo-Ivey relation has been given a theoretical foundation in the point-ion-lattice model of the F-center due to Gourary and Adrian.[29]

EXPERIMENTAL

Materials

Reactor-grade sodium metal was furnished by Dr. C. Luner of the Chemical Engineering Division. Single crystal NaCl was supplied by Harshaw Chemical Co.

Equipment

The spectra were measured in a Cary 14H spectrometer. The Na-NaCl solutions were contained in a sapphire cell, fabricated from a 1/2 x 1/2 x 1-1/2 inch boule containing a

1/8 x 1/4 x 1-1/4 well. The closure was made with a sapphire plate which was pressure sealed to the cell by means of a spring mechanism. The arrangement is schematically shown in Fig. 5. The surfaces between the cell and the cover plate were ground flat to prevent escape of sodium vapor. Best results were achieved by polishing both faces with a one micron diamond paste on a soft metal surface such as a tin plate. Flatness was checked by counting the interference fringes when both parts were pressed together. Three to five fringes were found acceptable. In another less elaborate method a 3 mil indium metal foil was used as a gasket between cell and plate. In that case, requirement of optical flatness was not as stringent because liquid indium fills any voids.

Temperatures were measured with a chromel-alumel thermocouple in contact with the cover plate.

Procedure

In most experiments the cell as well as the quartz and metal parts were outgassed at 500°C at 10^{-6} mm Hg. The furnace used for outgassing was attached to a helium dry box maintained at less than 5 ppm water. The NaCl single crystals were outgassed similarly at 200°C. The cell was loaded in the dry box with a NaCl crystal, ranging from 800-

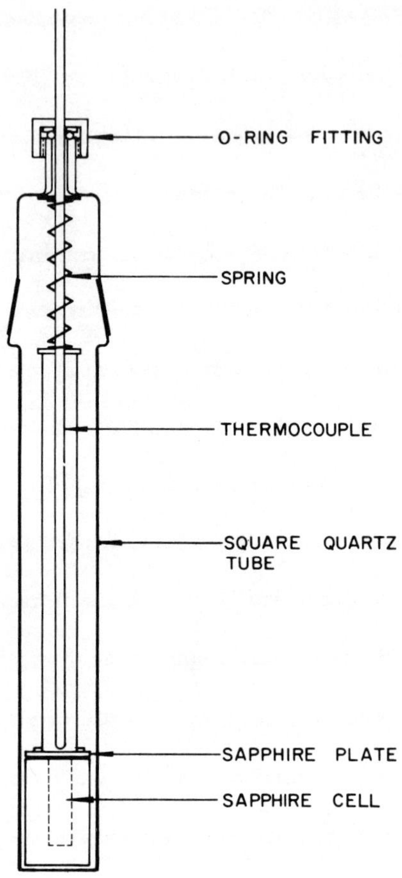

FIG. 5

Sapphire cell assembly used for measuring absorption spectra of Na metal in solid or liquid NaCl.

1000 mg and a piece of sodium metal, which was cut out of a bigger one immediately before loading the cell. The amount of sodium added to the cell was weighed on a Cahn electromicrobalance, installed in a dry box. Quantities of 40-400 micrograms Na were used. When the cell had been loaded and assembled, the cell container was removed from the dry box and spectra were taken. One cell was loaded with sodium only in order to observe the metal vapor absorption spectrum.

A common feature of all spectra was a slightly increasing background absorption with decreasing wavelength, presumably due to scattering of the quartz envelopes since it varied with different envelopes. Consequently, the background had to be established for each experiment by measuring a spectrum at room temperature and at 400°C. At 500°C, a slight coloration of the NaCl crystal became apparent. Spectra were taken of additively colored NaCl in equilibrium with Na vapor at 550°C. A spectrum of Na vapor was also taken at 550°C. At temperatures above 550°C the violet color of the crystal became too intense to measure the peak of the absorption peak. At the melting point, the color changed abruptly to blue. At first the intensity of the blue color was also too high to be measurable. However, about 20 minutes after melting, the intensity had diminished sufficiently so that spectra of the solution of Na in liquid NaCl could be obtained.

The spectra in Figure 1 were drawn as follows: Background spectra at 400°C and at room temperature were extrapolated to 550°C and 800°C and subtracted from the experimental curves. Spectrum A is the absorption of Na vapor at 550°C. The two broader peaks, corresponding to $^1\Sigma_u - {}^1\Sigma_g$ and $^1\pi_u - {}^1\Sigma_g$ transitions of the Na_2 molecule, exhibited vibrational fine structure not reproduced in the drawing.

Spectrum B represents the absorption of NaCl in equilibrium with Na vapor at 550°C. The NaCl crystal was 2 mm thick, leaving 1.1 mm of the optical path as vapor space.

Spectrum C is the result of subtracting Spectrum A reduced to one third of its absorbance, from Spectrum B. Spectrum C is taken to represent the "F center" absorption at 550°C.

Spectrum D, the absorption spectrum of a liquid Na-NaCl solution was measured with a sample originally composed of 378 μg and 970 mg NaCl. Had no sodium been lost after melting the concentration would have been 2.6 x 10^{-2} moles/liter. The concentration of Na in the solution actually measured is estimated to be less than 10^{-2} moles/liter.

ACKNOWLEDGMENT

This work was performed under the auspices of the U. S. Atomic Energy Commission.

REFERENCES

1. M. A. Bredig, J. W. Johnson, and W. T. Smith, J. Am. Chem. Soc., 77, 307 (1955).

2. M. A. Bredig, H. R. Bronstein, and W. T. Smith, J. Am. Chem. Soc., 77, 1454 (1955).

3. J. W. Johnson and M. A. Bredig, J. Phys. Chem., 62, 604 (1958).

4. M. A. Bredig and H. R. Bronstein, ibid., 64, 64 (1960).

5. M. A. Bredig and J. W. Johnson, ibid., 64, 1899 (1960).

6. A. S. Dworkin, H. R. Bronstein, and M. A. Bredig, ibid., 66, 572 (1962).

7. H. R. Bronstein and M. A. Bredig, ibid., 65, 1220 (1961); J. Am. Chem. Soc., 80, 2077 (1958).

8. H. R. Bronstein, H. S. Dworkin, and M. A. Bredig, J. Chem. Phys., 34, 1843 (1961); 37, 677 (1962).

9. J. Brown, UCRL-9944 (1961), USAEC Report.

10. N. H. Nachtrieb, J. Phys. Chem., 66, 1163 (1962).

11. M. Bettman, J. Chem. Phys., 44, 3254 (1966).

12. E. Mollwo, Nachr. Ges. Wiss. Goettingen, Math-Physik. Kl., Fachgruppe II, 1, 203 (1935).

13. J. Young, J. Phys. Chem., 67, 2507 (1963); ORNL-P-403, 1964 USAEC.

14. J. Greenberg and I. Warshawsky, J. Am. Chem. Soc., 86, 3572 (1964).

15. J. Greenberg and I. Warshawsky, J. Am. Chem. Soc., 86, 5351 (1964).

16. J. F. Rounsaville and J. J. Lagowski, J. Phys. Chem., 72, 1111 (1968).

17. E. G. Wilson, Phys. Rev. Letters, 10, 432 (1963).

18. K. S. Pitzer, J. Am. Chem. Soc., 84, 2025 (1962).

19. S. A. Rice, Disc. Far. Soc., 32, 181 (1961).

20. For pertinent references, see D. M. Gruen, Quart. Rev., 19, 349 (1965).

21. E. Mollwo, Z. Phys., 85, 56 (1933).

22. E. R. Van Artsdalen and J. S. Yaffe, J. Phys. Chem., 59, 118 (1955).

23. J. S. Yaffe and E. R. Van Artsdalen, J. Phys. Chem., 60, 1125 (1956).

24. For a compilation of volume changes of fusion of alkali halides, see H. Spindler and F. Sauerwald, Anorg. allgem. Chem., 335, 267 (1965).

25. International Critical Tables, Vol. 3, McGraw-Hill Book Co., New York, 1928, p. 43.

26. J. O'M. Bockris, A. Pilla, and J. L. Barton, J. Phys. Chem., 64, 507 (1960).

27. E. Mollwo, Goettinger Nachr., Math.-Phys. Kl., 97 (1931).

28. H. F. Ivey, Phys. Rev., 72, 341 (1947).

29. B. S. Gourary and F. J. Adrian, Phys. Rev., 105, 1180 (1957); Solid State Physics, (Seitz and Turnbull, Eds.), Vol X, Academic Press, New York, 1960.

SPECTROPHOTOMETRIC STUDIES OF SOLUTE SPECIES IN MOLTEN FLUORIDE MEDIA

J. P. Young

Oak Ridge National Laboratory
Oak Ridge, Tennessee

INTRODUCTION

Spectral studies of molten fluoride solutions have been carried out in our laboratory for some time now. As initiators of comprehensive studies in these solvents, we have been concerned with analytical applications of such spectral data to the identification and determination of solute species that could exist in molten fluoride salt mixtures of interest to the molten salt reactor program of the Oak Ridge National Laboratory. Since spectroscopy offers a direct method for the characterization of solute species which exhibit available spectra, this tool can contribute much more than analytical knowledge to the understanding of these solutions. If one had a choice of salt systems to investigate, one might not choose fluorides because of the apparent containment difficulties. We have developed or cooperated in the development of several containment techniques, however, and this problem is being minimized.

The spectra of solute species in fluoride salts yield some unique and interesting data compared with results in other salt systems. These results will be discussed later, but I might point out that a general conclusion to be drawn from the spectral studies made to date is that solute ions in fluoride melts tend to prefer a high coordination number and high symmetry. As a consequence of this behavior, transition metal ions are not highly colored in fluoride melts. The d-d transitions are relatively weak in these apparently centro-symmetric species, so that analytically sensitive determinations cannot be realized. This simplicity of structure is intriguing, however, and may be a result of the electronegativity of fluorine. An analytical consequence of the weak d-d transition spectra is that these ions offer little interference to the spectral study and analysis of various allowed transitions of other ions.

APPARATUS AND TECHNIQUES

A quick review of the experimental side of these studies is in order. Our first studies of molten fluoride salts were reported with a description of the furnace[1] and a survey of the spectra of several solute species in molten LiF-NaF-KF[2] and LiF.[3] The melts were generally contained as double concave lens-shaped drops of liquid within a horizontal platinum tube. Light from a Cary Spectrophotometer, Model 14M, passed through the melt in the tube. This is the so-called "pendent drop" technique.

A carefully controlled atmosphere is necessary for preparing and maintaining solute species in molten salts. This requirement is more easily met with melts that are compatible with silica, as the container can be flame-sealed. With the techniques developed for fluoride spectroscopy such sealing is not possible; therefore, we require a carefully controlled atmosphere surrounding the molten sample in the furnace.

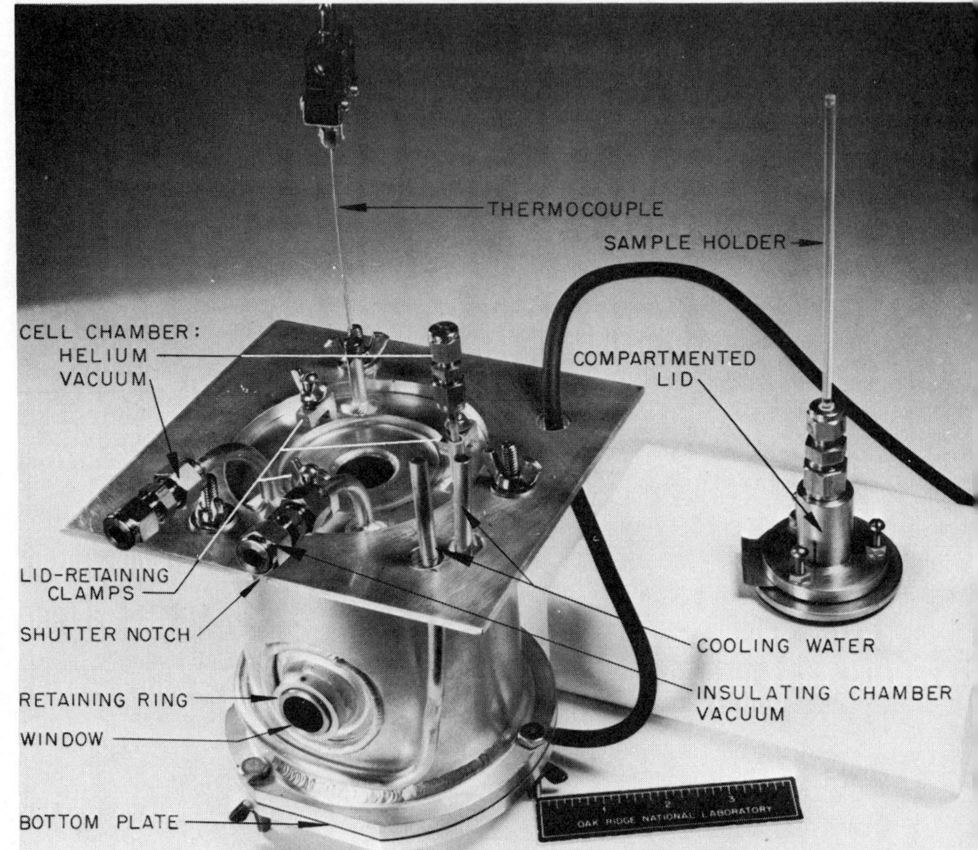

FIG. 1

Furnace Assembly used in the spectral study of molten fluoride salts.

The inert-atmosphere furnace is shown in Figure 1. Samples are held in the intersection of an inverted "T" shaped tube made of nickel. This intersection is surrounded by a mass of metal and heated. Light from the spectrophotometer passes through the arms of the "T". Light stops are located so that only the light which passes through the sample is detected. The "T" is vacuum tight, with SiO_2 windows at the end of the light tube, and a lid on the top. The exterior double walls of the furnace are water-cooled and the lower portion of the furnace can be aligned with respect to the rectangular top plate in order to maximize light transmission through the furnace. The top plate exactly fits the top of the Cary sample compartment for reproducible placement. A sample can be brought to the furnace inside the vacuum-tight compartmented lid. With the lid in place, the gas in the furnace is pumped out and filled with helium. Then, the thin plate seen in the lid is removed. This plate is held between two O-rings to seal a cavity within the lid that contains a sample salt to be melted. The furnace and inside of the lid are then re-evacuated and flushed with helium. The pumping is continued while the furnace is pre-heated to ca. 250°C; then, helium is introduced and the sample in any particular container is lowered into the light beam by means of the sample holder. The temperature of the furnace is increased to melt the sample. The purpose of these techniques is to prepare, transfer, and melt molten fluoride salt solutions without exposures to H_2O or air. The molten sample, then, is maintained in its container, of suitable material, for spectral measurements. The sample and container are, in effect, suspended in a helium atmosphere with only gas phase contact with heated nickel or gas phase contact with cooled SiO_2 windows, Viton O-rings or Teflon gaskets.

Although the largest single deterrent to the spectral study of molten fluoride salts has been the lack of a suitable spectrophotometric container, there are now in operation or almost operational, several methods of containing these melts for spectral study including both windowed and windowless containers. Depending on the specific problem, one or more of the containers may be applicable. As a general container, I use the windowless type primarily because it offers the most experimental flexibility and freedom from compatibility problems. The biggest disadvantage of windowless containers is the uncertainty with which the path length of liquids in these cells can be determined and the fact that the path length is a function of melt properties. In the way of windowed containers, use can be made of diamond,[4,5] MgO,[2] pyrolytic

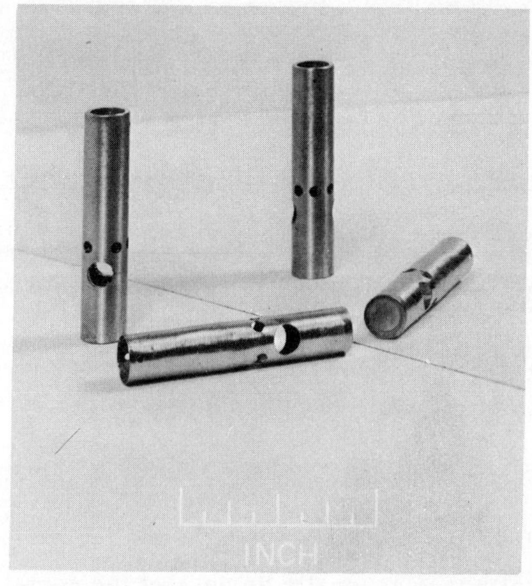

FIG. 2

Captive liquid cells.

BN,[6] and even fused SiO_2.[7] Of these, diamond windows are by far the most versatile in terms of useful wavelength range and compatibility with fluoride solutions.

After several evolutionary changes with "pendent drop" containers, a windowless container called a captive liquid cell[8] was developed for use with wetting liquids, molten salts or otherwise. The captive liquid cell is shown in Figure 2. Liquids are held in such a cell to the level of the series of small holes. Liquid, then, is also in the larger lower hole in the form of a double convex lens of reasonably reproducible path length. Methods are available for determining the pathlength of liquids in these cells.[8,9] The cells can be filled by flowing liquid or melting solid samples in them. The cells are attached to a holder which consists of a cap that fits over the cell and is held to the cell by a pin, or wire, which passes through matching holes in the cap and upper portion of the cell. The cap of the holder is attached to a rod which is seen protruding out of the top of the compartmented lid in Figure 1.

With salts that generally do not wet the container and which have a reasonably high surface tension, it has been possible to simplify the design of captive liquid cells in that no small holes are required. In Figure 3 are shown several such modified containers made of various opaque materials. In use the container is positioned with the large portion up. Pieces of solid solution or solvent and solutes are placed inside the container and then handled as described above. The loading operation can be carried out in an inert-atmosphere box, the sample and holder being loaded into the compartmented lid and sealed before removal from the dry box.

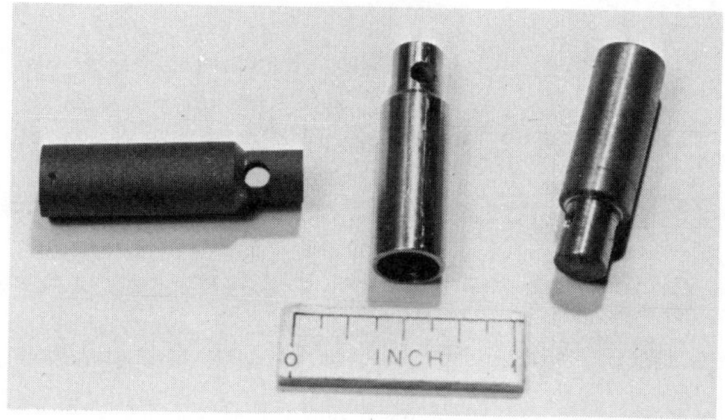

FIG. 3

Modified captive liquid cells.

RESULTS AND DISCUSSION

Solvents

As has been pointed out by others, many molten salts offer an advantage as solvents for spectral studies in that they are transparent in the near IR region of the spectrum. Transitions of a number of solute species that cannot be seen in aqueous solution, because of the absorbance of water, can be seen in a molten salt. Molten fluorides as solvents have this transparency in the near IR, but furthermore, they are transparent to much higher energies in the ultraviolet than are other salt systems. By defining the limit of transparency as the wavelength at which the relative absorbance of the solvent has increased to 1.0, a 1-cm path length of molten LiF-NaF-KF (46.5-11.5-42 mole %) transmits to ca. 210 nm at 550°C. The solvent LiF-BeF_2 (66-34 mole %) at 550°C is still transparent, relative absorbance of 0.4 at 190 nm, the practical limit of our present experimental set-up. One can assume that by proper choice of solvent and instrumentation, windowless samples of certain molten fluorides would be transparent well into the vacuum ultraviolet region. Windowless cells would be required in this application as crystalline fluorides, the presently known windows for vacuum UV work, would dissolve in the sample melt. Certainly for specific cases, excellent analytical and characterization type studies could be carried out if the interest develops.

Solutions of Transition Element Fluorides

As pointed out earlier, from the various spectral studies of transition metal ions in molten fluoride salts, a general conclusion can be made that solute species seem to prefer high symmetry and tend to achieve high coordination numbers in molten fluoride salts. This conclusion is not too surprising in alkali fluoride melts if one considers the size and electronegativity of the ligand. Experimentally, though, qualitative and quantitative evidence, within the errors of the windowless cell techniques,[8] for such a conclusion can be drawn even in melts where the fluoride ion can be thought to be tied up with the solvent, as $BeF_4^=$ in molten Li_2BeF_4 or as ZrF_n^{4-n} in NaF-ZrF_4 mixtures. A study of the spectra of several 3d transition metal ions was made using a solvent composition range of LiF-BeF_2 (72-28), (66-34), and (60-40) mole %.[10] Thus, the spectra of these ions could be observed in melts that contained an excess or deficiency of fluoride ions with respect to the stoichiometry of Li_2BeF_4 (66 2/3 - 33 1/3 mole %); the spectra in some cases could be compared with earlier work carried out in the alkali fluoride eutectic, LiF-NaF-KF (46.5-11.5-42 mole %).[2] The spectral comparison of Ni(II) in the various solvents demonstrates the nature of the results that were obtained.[10]

The spectrum of Ni(II) in LiF-BeF_2 (66-34 mole %) at 550°C consists of 3 spin-allowed absorptions with maxima at 23,100, 10,800, and <u>ca.</u> 6,000 cm^{-1}. The respective molar absorptivities are 11, 2, and <u>ca.</u> 1. The latter peak is very broad and diffuse so that both its exact position and intensity are difficult to determine. Octahedrally coordinated Ni(II), a $3d^8$ ion, would be expected to exhibit transitions in the regions seen. These transitions correspond to those reported for Ni(II) in molten LiF-NaF-KF at 525°C;[11] peaks were located at 23,000, 11,100, and _ca._ 6,500 cm^{-1}, molar absorptivities of 12, 2, and _ca._ 0.8. The similarity of the two spectra is apparent. Other than for intensity changes caused by thermal effects, which are within the errors of the windowless cell technique over the temperature range studied, the spectrum of Ni(II) was likewise unaffected if the LiF-BeF_2 solvent composition was varied over the range mentioned. Spectral measurements offer a direct method for observing the structure of amenable solute species; therefore, it can be reasoned that if the coordination of the ion being studied changes, particularly if it would be altered from a centro-symmetric coordination to any lower symmetry, this change could be observed spectrally. The initial purpose for studying the spectra of Ni(II) in these solvents was to see if the coordination would be affected, and no change was observed. The reason for the lack of coordination change is puzzling. Although possible, it does not seem likely that the $BeF_4^{=}$ grouping unsuccessfully competes with a larger Z/R transition element ion for F^-, (Z being the atomic number and R the ionic radius). It is perhaps more likely that $BeF_4^=$ can be treated as a ligand which somehow interacts with solute ions in a manner which preserves the octahedral symmetry. Another possible explanation for the observation arises from a statement Jorgensen[12] attributes to Pauling in which Pauling suggested that the fluoride ion can donate, relatively, so little electron density in the form of bonding orbitals to a central metal ion that a high coordination number (6 or 8) is preferred. Activity coefficients of LiF in various mixtures of molten $LiF-BeF_2$[13] would tend to question this possibility as the data suggest that the F^- is tied up by Be(II), at least in the solvent alone.

Exceptions to this general conclusion regarding coordination will undoubtedly be found. Iron(II), for example, does not give a spectrum which can be explained by simple O_h coordination either in molten $LiF-BeF_2$[10] or in molten alkali fluorides.[6] In both cases a single weak absorption peak would be expected in the region of 10,000 cm^{-1} for such a symmetry. Instead, Fe(II) at 550°C exhibits a weak peak and shoulder at 9800 and 5500 cm^{-1}, molar absorptivities of 4.5 and 3. At 650°C, the separate peaks appear closer together, but still definitely discernible. This ion may well be distorted by Jahn-Teller effects. The spectrum of crystalline FeF_2 at lower temperatures has been explained on this basis.[14]

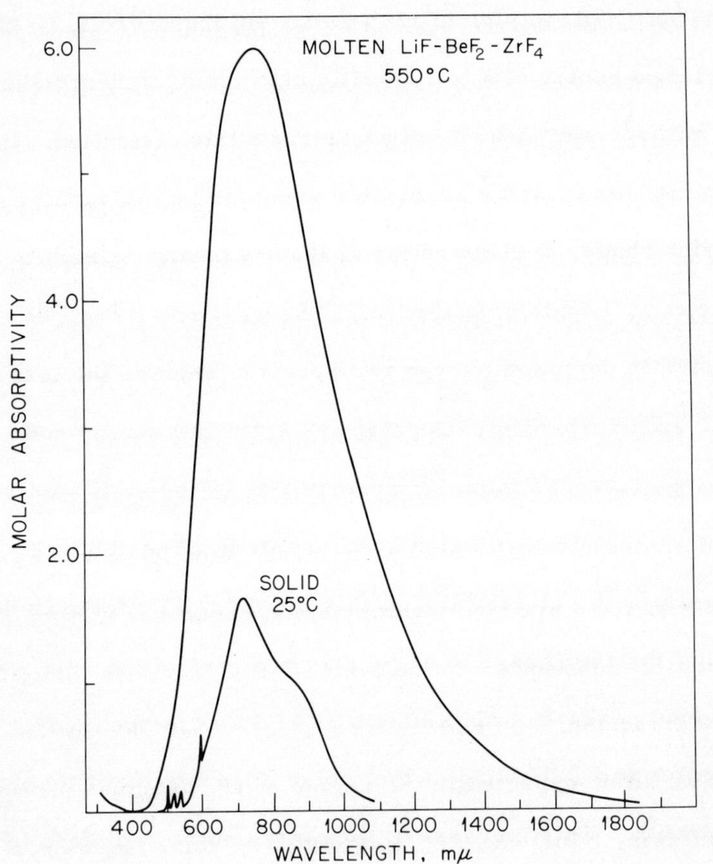

FIG. 4

Spectra of CrF_2.

SPECTROPHOTOMETRIC STUDIES OF SOLUTE SPECIES

Peripheral spectral studies of solid solute species generally augment understanding of the spectra of dissolved solute species. In Figure 4, the transmission spectrum of a small crystal of CrF_2[6] at room temperature is compared to that of Cr(II) in molten LiF-BeF_2. In the crystal, Cr(II) lies in the center of a tetragonally distorted octahedron and the spin-allowed absorption peaks show this distortion. In the melt, the complex absorption envelope is replaced by a simpler shape, demonstrating at least a change to higher symmetry, possibly octahedral.[9] Several other changes can be noted in comparing these two spectra, such as the loss of the spin-forbidden peaks that are seen in the solid and the large increase in molar absorptivity in the melt spectrum. In yet another transmission spectral study of heated CrF_2 crystals,[6] it was established that the thermal effects could account for the change in symmetry and loss of the spin forbidden peaks; solution effects at constant temperature accounted for a doubling of the molar absorptivity at the peak wavelength. This increase in intensity can be explained by assuming that the centrosymmetric cage in which Cr(II) finds itself is enlarged and/or becomes less rigid as the crystal dissolves.

Solutions of Lanthanides and Actinides

We have studied the spectra of several lanthanides in various molten fluoride solutions.[2,3] The only actinide we have investigated has been uranium in various oxidation states. In the researches concerned with molten salt nuclear reactors, the possibility of a spectrophotometric determination of U(III) in the fuel salt, a molten fluoride solution containing U(IV), is of analytical interest. A possible change of U(III) concentration in the fuel salt is indicative of chemical phenomena which should be controlled. The spectra of U(III) and U(IV) in molten LiF-BeF_2 have been reported.[9] A portion of the spectra of mixtures of U(III) and U(IV) in molten LiF-BeF_2-ZrF_4 is shown in Figure 5. A sensitive determination of U(III) can be made by absorbance measurements at 360 nm. Spectrum I is that of U(IV); spectra II and III contain both valence states. The dashed curves represent the spectrum of U(III) and were obtained by subtracting curves II and III from I. The method of preparation of these particular solutions will be discussed later. The concentration of U(IV) in these solutions is about 0.2 mole %; the concentration of U(III) represents _ca_. 100 and 250 ppm

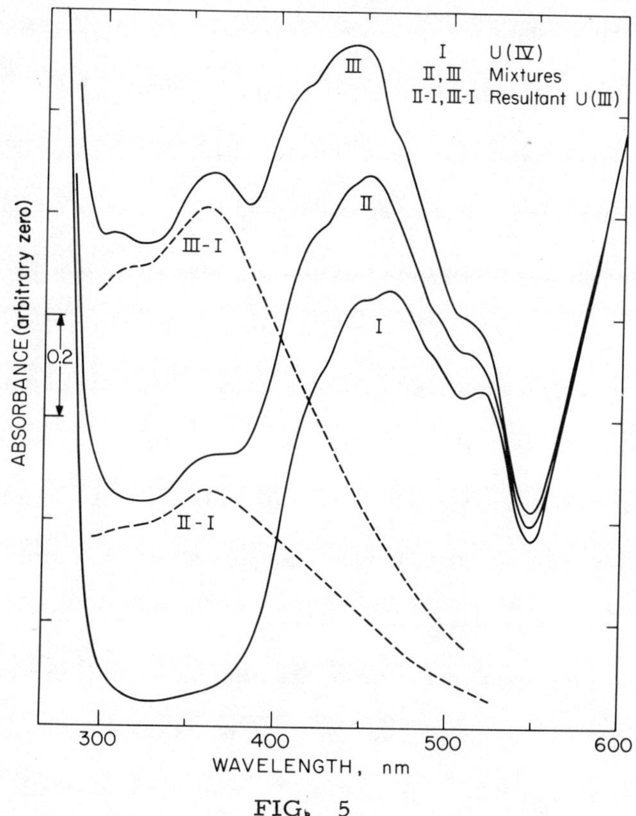

FIG. 5

Spectra of U(IV) - U(III) mixtures in molten $LiF-BeF_2-ZrF_4$. (Temperature: 540°C)

in spectra II and III, respectively. At the present time design and construction of a spectrophotometric facility in a high-radiation level cell is being pursued so that the applicability of a proposed spectrophotometric method for determination of U(III) in the highly radioactive fuel salts can be demonstrated. The facility will also be used for spectral study of other solute species in radioactive molten salts and of other actinides in molten salts.

Spectra have also been obtained for U(V)[15] and UO_2^{++}[2,6] in fluoride melts. The preparation of U(V) in molten $LiF-BeF_2$ is an example of fortuitous chemical generation of a solute species. The salt Na_2UF_8 was added to the molten fluoride solvent in a graphite windowless container and apparently the solute reacted with the graphite, not to form the expected U(IV) but to form a visually colorless melt which contained U(V). Not only can U(III), then, be identified in a molten reactor fuel salt, but U(V) can also be identified, at less than 1000 ppm concentration by its sharp absorption peak in the near IR region at 1465 nm. It is interesting to note that in reducing UO_2^{++} in $LiF-BeF_2$ melts, no conditions have yet given any spectral evidence of a UO_2^{+} species. These reductions have been carried out with metal reductants such as Fe or Cr.

ELECTROCHEMICAL-SPECTRAL STUDIES

The method of preparation of the uranium solutions, spectra of which are shown in Figure 5, demonstrates a technique that holds promise of being an excellent preparative and investigational technique for the study of most solution media and, therefore, molten fluoride salts. This technique involves the generation of a solute species at an electrode, and concurrent observation of its spectrum in the area immediately surrounding the electrode.[16] It appears that this technique offers a very clean-cut and controlled method of preparing and characterizing products of electrode reactions which are soluble solute species. This technique will be applied to the area of unusual oxidation states of solute species in fluoride melts. Because of the thermodynamic stability of such a solvent system, fluorides should be useful as solvents for this type of study, particularly in the generation of solute species of high oxidation state.

ACKNOWLEDGMENT

Research sponsored by the U. S. Atomic Energy Commission under contract with Union Carbide Corporation.

REFERENCES

1. J. P. Young and J. C. White, Anal. Chem., 31, 892 (1959).

2. J. P. Young and J. C. White, *Anal. Chem.*, *32*, 799 (1960).

3. J. P. Young and J. C. White, *ibid.*, 1658.

4. G. G. Cocks, J. B. Schroder, and C. M. Schwartz, Batelle Memorial Institute Report No. BMI-1185, April 1967, p. 13.

5. L.M. Toth, J. P. Young, and G. P. Smith, to be published.

6. J. P. Young, unpublished results.

7. C. E. Bamberger, J. P. Young, and C. F. Baes, Jr., *J. Inorg. and Nucl. Chem.*, *30*, 1979 (1968).

8. J. P. Young, *Anal. Chem.*, *36*, 390 (1964).

9. J. P. Young, *Inorg. Chem.*, *6*, 1486 (1967).

10. J. P. Young, submitted for publication.

11. J. P. Young and G. P. Smith, *J. Chem. Phys.*, *40*, 913 (1964).

12. C. K. Jorgensen, *Halogen Chemistry*, Vol. I, I. V. Gutmann, Ed., Academic Press, New York, 1967, p. 341.

13. A. L. Mathews and C. F. Baes, Jr., *Inorg. Chem.*, *7*, 373 (1968).

14. W. E. Hatfield and T. S. Piper, *Inorg. Chem.*, *3*, 1295 (1964).

15. J. P. Young and G. I. Cathers, to be published.

16. J. P. Young, Gleb Mamantov, and F. L. Whiting, *J. Phys. Chem.*, *71*, 782 (1967).

ADVANCES IN THE INFRARED SPECTROSCOPY OF MOLTEN SALTS

J. P. Devlin, P. C. Li and R. P. J. Cooney

Department of Chemistry
Oklahoma State University
Stillwater, Oklahoma

INTRODUCTION

Though research activity in the infrared spectroscopy of ionic melts has increased in the past few years, the number of recent publications is quite limited. The main contributions seem to divide, both as to content and origin into two groups: those by Hester and coworkers treating primarily nitrates and thiocyanates of doubly charged cations in solutions of either a corresponding alkali metal salt or water,[1-4] and our papers dealing with nitrate melts of singly charged cations[5,6] and mixtures thereof.[7] These studies have had the common objective of elucidating ionic melt structures, but in addition to the contrast in subject matter they have been totally lacking in any overlap in both technique and interpretation. Because of this diversity, a consideration of these two sets of papers provides an introduction to the techniques and interpretative models that have proved useful in vibrational studies of ionic melts. Thus our attention will regularly be directed to references 1-7.

INFRARED TECHNIQUES

Infrared studies of molten salts have generally been unpopular in the past because of the corrosive action of such melts. The relatively high sampling temperatures required and the unusually large intensities of many ion fundamental bands have also served as deterrents to such studies. These problems have been overcome for several systems and a brief description of the technical solutions seems in order.

TRANSMISSION SPECTROSCOPY

Greenberg and Hallgren originally sidestepped the corrosion problem, eliminating cell windows as such by supporting the molten salts on a fine mesh platinum screen which doubled as a resistance heater.[8] Hester has extended this method, using stainless steel screens etched in aqua regia to an optimum wire diameter.[1] Certain inherent limitations of the method remain, however: 1) the minimum sample pathlength exceeds by two orders of magnitude the optimum thickness (~ 1μ) for the most intense melt absorptions, 2) the sample pathlength cannot be reproduced, and 3) fluctuations in sample refractive indices with wavelength tend to distort the intense and, to a lesser extent, moderately intense bands, an effect which is greatly reduced by the salt windows in a standard infrared cell. The result is that the method is suitable for accurately measuring positions of weakly active vibrational modes as well as the observation of coarse structure for weak band complexes and the estimation of the relative intensities of such bands. However, the method has the advantage of no severe inherent limit on the maximum sampling temperature and, as Hester has shown, the positions and structures of the intense bands may sometimes be inferred from readily observed overtone and combination modes.

A rather recent emergence of infrared window materials inert to many ionic melts has made possible a second qualitative transmission technique which is popular in our laboratory. A single platelet of the window material, usually ultrapure silicon, is wetted with a thin film of a melt and mounted in the infrared beam in a suitable heating assembly. This method permits sampling with short pathlengths and substantially reduces band distortions resulting from fluctuating sample refractive indices. The latter follows because the sample is supported between two media of vastly different optical density (air and silicon) so that variations in reflectivities at the two melt interfaces caused by fluctuations in the sample refractive index as a function of wavelength are of nearly equal magnitude but opposite sign for bands up to moderate intensity.

Though the platelet method has proven very useful and is certainly preferable to silicon cavity cell transmission procedures originally considered,[9,10] limitations do exist. If sampling is to be with a vertical platelet the sample must wet the substrate and, even so, pathlengths though somewhat adjustable are not reproducible. More serious, however, are problems with distortion of very intense bands, and the absence of a generally useful substrate at temperatures >400°C.

Silicon, below ~400°C where it becomes opaque, is the most generally useful material. A 1 mm platelet transmits satisfactorily from 200 to 4000 cm^{-1} (with a 50% reflection loss), is wetted by many melts and, has been found impervious to melts except fluorides and hydroxides. However, corrosion tests have not been exhaustive. Because of a considerably higher maximum useful temperature, the Irtran materials, particularly Irtran 4 and 6, hold promise for certain studies but lack the more general inertness of silicon. It is hoped that it may soon be possible to use thin coatings, such as are commonly used on dielectric mirrors, to protect these substrates, thus generalizing their usefulness.[11] Wegdam et al. have very recently shown that a thin film of the alkali metal nitrate melts may be supported between the faces of two diamonds for direct transmission studies of particular value in the far infrared.[21] The limiting useful temperature for diamond is in the neighborhood of 600°C.

ATR SPECTROSCOPY

Though progress has been made in transmission infrared spectroscopy of melts, the fact remains that no suitable method for either absolute band intensity or Beer's Law measurements has been devised. Further, band shapes and peak positions of the intense melt absorptions are not attainable by present methods. Thus, taking a clue from Wilmshurst specular reflectance studies,[12] we prefer to augment all transmission ionic melt work by using the inert window materials such as silicon for attenuated total reflection (ATR) studies. In the ATR technique one observes the interaction of a sample with an evanescent electromagnetic field set up upon reflection of the propagating wave at the interface of an optically dense medium and the sample.[13] For example in Figure 1 depicting the assembly used in our studies the demagnified infrared beam is refracted upwards by the silicon (or Irtran) prism which has been shaped to give a resultant 54° angle of incidence (θ) at the upper prism surface. This angle is such as to guarantee that $\theta \gg \theta_{critical}$ at all wavelengths so that total internal reflection occurs. If this surface is coated with a melt the reflected wave will be attenuated at frequencies where the sample normally absorbs because of interaction of the standing wave with the sample. Analysis of the reflected wave thus gives an ATR spectrum which will closely resemble a transmission spectrum provided the parameters such as incident angle and prism refractive index have been optimized for that objective.[13,14]

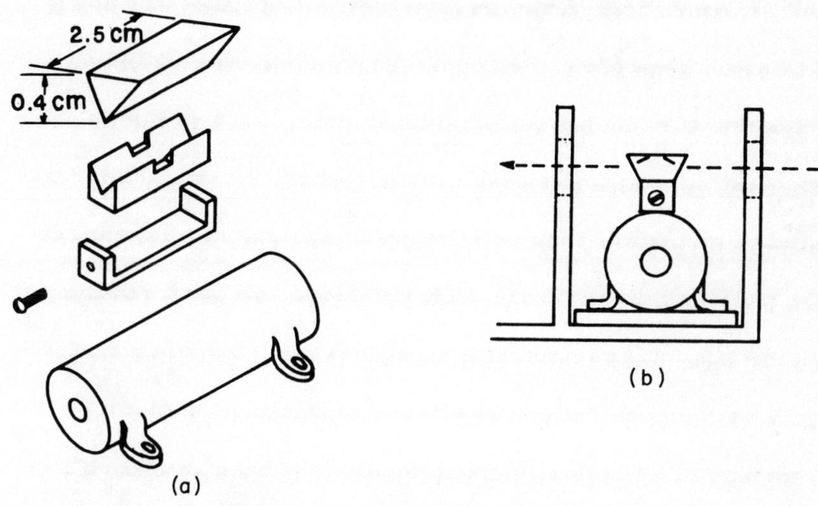

FIG. 1

ATR assembly for molten salts: (a) exploded view of the prism, holder and heater; (b) assembly mounted on ceramic base in the beam condenser.

At temperatures $< 400^{\circ}C$ silicon is a satisfactory ATR element material, particularly for observation of intense and moderately intense absorption bands. Again the Irtran materials find some application since their lower refractive indices give greater effective ATR pathlengths[13] and thus permit observation of the less intense bands particularly when used in a multiple reflection mode. When it becomes technologically possible to protectively coat the Irtran materials, the ATR method will be generalized for molten salts.

More elaborate foreoptics than those pictured in Fig. 1 are required for reflection measurements at a variety of carefully defined angles of incidence, a necessity for precise absolute intensity and band profile studies.[15] Even with our simple foreoptics, however, ATR offers notable advantages over all transmission methods by making possible 1) reproducible and adjustable pathlengths, 2) short effective pathlengths in the 1 μ range and 3) the virtual elimination of band-distorting reflection problems.[6] In brief the ATR and single plate transmission methods compliment each other so that many systems can be closely studied using the combined techniques.

RESULTS AND DISCUSSION

A brief consideration of the more significant data obtained for the nitrate melts of monovalent and divalent cations best exemplifies the value of the methods described above as well as the general utility of vibrational data in understanding melt structures. As has been reiterated numerous times, the unperturbed nitrate ion is of D_{3h} symmetry which implies that ν_1 (~ 1050 cm^{-1}) is Raman active only, ν_2 (~ 820 cm^{-1}) is infrared active only, and both ν_3 and ν_4 are doubly degenerate and Raman as well as infrared active. Since any observed splitting of these degeneracies or deviations from the predicted selection rules for optical activity can be interpreted in terms of the symmetry and strength of forces perturbing the ion, the spectrum of the nitrate ion becomes a useful probe of the ion environment.

MONOVALENT NITRATES

Thus, the critical infrared observations for the pure alkali metal nitrates and silver nitrate are 1) ν_1 is not only activated but the 1050 cm^{-1} band is structured in certain cases and differs by as much as 10 cm^{-1} from the Raman value, 2) ν_3 is split into two or more components in all cases with the magnitude of the splitting related to the cation polarizing power (Figure 2), 3) the infrared ν_3 component frequencies seem to differ significantly from the corresponding Raman component values, and 4) ν_2 the out-of-plane bending mode shows some shouldering, particularly for $LiNO_3$.[6] A comparison of the infrared and Raman frequencies for $LiNO_3$ are presented in Table 1 so as to exemplify points 1) and 3) above.

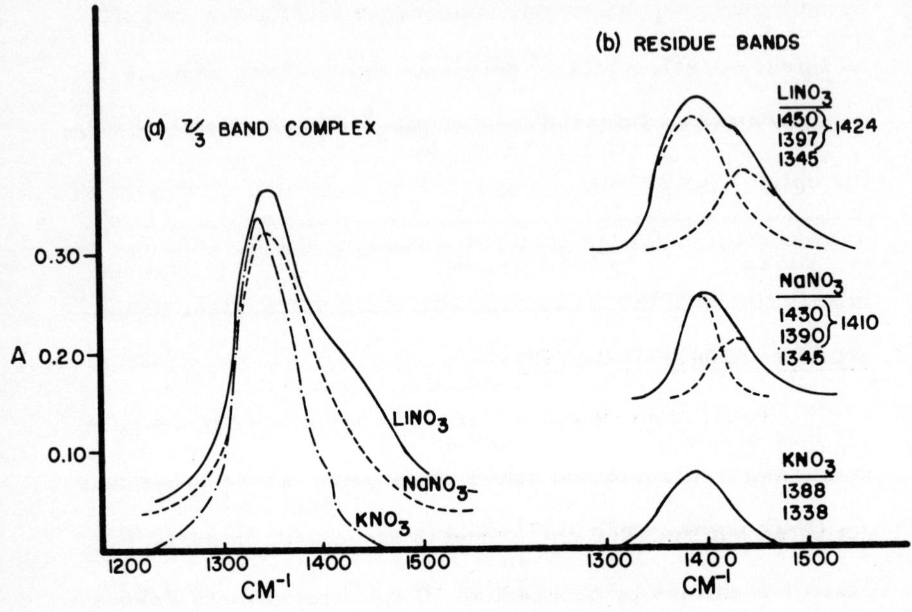

FIG. 2

The 7-μ ν_2 band system for alkali metal nitrate melts: (a) unresolved band complex; (b) residue bands left after resolving out the principal band component.

The evidence is substantial that in addition to cation perturbation, which is largely responsible for the splitting of ν_3 and the activation of ν_1, a coupling of the vibrational modes of neighboring anions is important. Such coupling to give "in-phase" and "out-of-phase" modes is the only logical explanation of the discrepancies observed between infrared

TABLE 1

Fundamental Infrared[a] and Raman[b] Frequencies for Molten $LiNO_3$ (cm^{-1})

Mode	Infrared	Raman
ν_1 (N-O sym. str.)	1053 s	1061
	1030 sh	
ν_2 (out-of-plane)	827 sh	
	818 m	
ν_3 (N-O asym. str.)	1450 m	1478
	1397 ms	
	1345 s	1362
ν_4 (in-plane bend)	742 w	747
		716

[a]From ref. 6.
[b]Private communication from Professor David James.

and Raman frequencies and the structuring of the bands for ν_1 and ν_2. The inference is that these melt structures are best described in terms of a perturbed lattice, with the above observations the outgrowth of combined site (cation-anion) and correlation field (anion-anion) effects. This conclusion has gained invaluable affirmation from recent low-frequency Raman studies by James and Leong (private communication) and the far infrared data of Wegdam et al.,[21] both of which studies clearly show the preservation of lattice supported phonon spectra in certain alkali metal nitrate melts.

Additional insight into these structures is gained from data presented in Figure 3 for the ν_2 band in the KNO_3-$AgNO_3$ melt mixture. A statistical analysis has shown that the behavior of this band with varying melt composition is most consistent with random distribution of the silver and potassium cations in roughly four lattice sites about the nitrate ions.[7] The low coordination number is consistent with earlier diffraction results[16] and is supported by recent observations made on solid ionic surfaces.[17]

DIVALENT NITRATES

Data for the fundamental region for 10% and 30% solutions of $Mg(NO_3)_2$ in KNO_3 (infrared)[2] and $NaNO_3$ (Raman)[18] are presented in Table 2 as an example of results reported for melt solutions of divalent metal nitrates. Primarily because of the large splitting of the ν_3 degeneracy as judged from overtone bands, the infrared activity of ν_1, the splitting of ν_4, the observation of a low frequency band assignable to the M-O stretching mode, and the composition

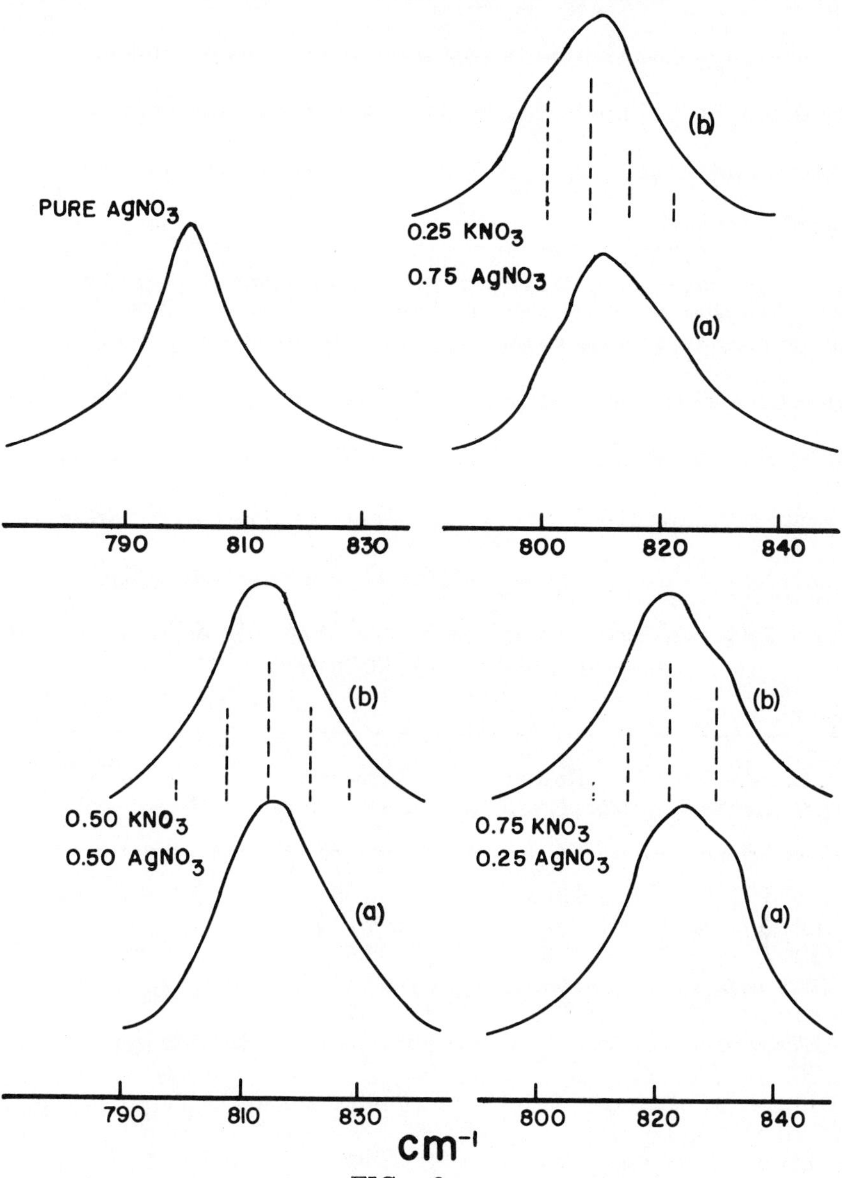

FIG. 3

Experimental band contours for ν_2 in pure $AgNO_3$ and molten mixtures of KNO_3 in $AgNO_3$ (a); and theoretical band contours for ν_2 of KNO_3-$AgNO_3$ mixtures assuming four coordination (i. e., five subbands separated by 7 cm^{-1}) (b).

dependence of the infrared spectra, Hester and Krishnan have identified Mg-O-NO_2 as the dominant species in these systems.

However, one may question whether or not this complex plus the uncomplexed (solvent) nitrate would yield the observed vibrational spectra. For example, it seems unlikely that the two forms of nitrate would give rise to the 1038 cm^{-1} infrared and the 1053 cm^{-1} Raman without producing corresponding Raman (1038 cm^{-1}) and infrared (1053 cm^{-1}) features.

TABLE 2

Infrared[a] and Raman[b] Data (cm^{-1}) for $Mg(NO_3)_2$ in Alkali Metal Nitrate Solutions

Infrared 30% in KNO_3	Raman 30% in $NaNO_3$	Infrared 10% in KNO_3	Raman 10% in $NaNO_3$
1550 to 1200 vs	1526 to 1285	1500 to 1200 vs	1580 1355
1038 s, b, as	1053	1042 s	1053
830 s		832 s	
815 s			
755 s		750 m	
745 s, sh		740 m, sh	
710 w		715 w, sh	705
380 s		380 m	
340 s		330 m	

[a]From ref. 2.
[b]From ref. 18.

Rather the coupling of the vibrational modes of two or more nitrate ions associated with a common cation, or possibly a perturbed lattice structure, may be suggested.

In view of possible inadequacies in the M-O-NO_2 complex species model, the structures of the divalent nitrate melts deserve further study. Two observations from our laboratory may be relevant. The splitting of the ν_3 components in a 45% $Ca(NO_3)_2$ glass forming mixture with KNO_3 increases from 85 cm^{-1} at room temperature to 105 cm^{-1} at 280°C. This might possibly result from an increase in the number of lattice vacancies with temperature accompanied by stronger anion-distorting forces between residual neighboring ions. Secondly, we find that the ν_2 mode for this system produces a relatively sharp symmetric band at 820 cm^{-1}, approximately halfway between the value for pure KNO_3 (829 cm^{-1})[6] and that projected for pure molten $Ca(NO_3)_2$ (810 cm^{-1}).[1] Neither observation seems consistent with a structure based on a well defined 1:1 complex but each tends to support consideration of a perturbed lattice structure. Precise Raman data for this system should clarify the situation.

SURFACES OF NITRATE CRYSTALS

Surprisingly, a very useful source of data for elucidating molten salt structures may exist in infrared ATR studies of crystal surfaces. The ATR and transmission spectra of ionic melts have invariably been found to be in excellent agreement in the many instances compared to date. This is not true for ionic crystals.[17] To cite a particular case, although excellent agreement is found for the nitrate in-plane modes in solid $LiNO_3$, ν_2 appears as a sharp band at 844 cm^{-1} by transmission while the ATR value, using a silicon prism, is 834 cm^{-1}. Further, as is obvious in Figure 4, the 834 cm^{-1} ATR band is oddly shaped, tailing off in an unnatural fashion toward the 844 cm^{-1} transmission value. A similar phenomenon has been found for $NaNO_3$, but probably more informative is a shift toward the transmission value produced by substitution of Irtran 2 (greater penetration depth) for the silicon prism (Figure 4). Of the many explanations considered, such as epitaxial growth and the possibility of solid state dichroic effects, the only possibility consistent with all observations is that ionic solid surfaces have a unique structure which converges on the bulk structure, probably exponentially with penetration.

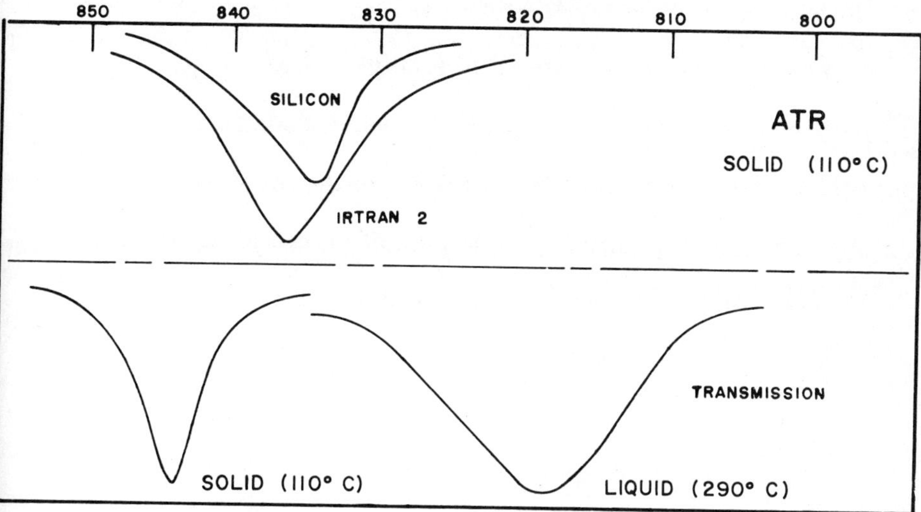

FIG. 4

Position and profile of the ν_2 absorption as observed by ATR and transmission studies of molten and solid $LiNO_3$.

The surprising feature is that the surface structure must have a "depth" in the micron range rather than the five or so ion layers commonly believed.[19] The relevance to salt melt structures is that, in all cases studied ($LiNO_3$, $NaNO_3$ and KNO_3), the ν_2 surface value is between the bulk solid and melt values (844(s), 834 (surface) and 818 (liq.) for $LiNO_3$). This fact, together with surface models that have evolved from theoretical[19] and low energy electron diffraction studies[20] indicating a tendency for pairing of ion layers at the surface, seems to identify reduced coordination numbers, accompanied by stronger nearest neighbor interactions, as a common characteristic of the melt and crystal surface. In other words, the pairing of layers pictured in Figure 5 may be viewed as a relaxation of the bulk structure

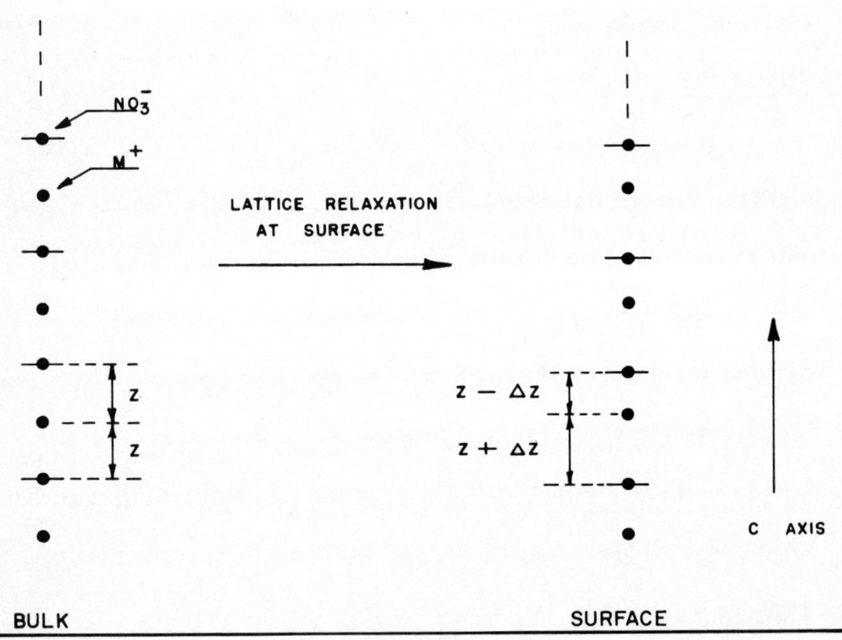

FIG. 5

Representation of the change in anion-cation coordination as a result of surface relaxation.

so as to ease the coordination insufficiency at the solid surface. The effect is to propagate a reduced coordination number some depth into the surface. By analogy the numerous holes (lattice vacancies) in a melt spread this coordination insufficiency throughout the liquid phase. This view is supported by the identical nature of the bulk and surface melt spectra mentioned earlier (i.e., the feature that distinguishes the crystal surface from the bulk pervades the entire liquid phase).

SUMMARY

It has been shown that existing ATR and transmission methods permit detailed infrared studies of many molten salt systems. The results of such studies when compared with Raman data seem to lead to quite firm conclusions regarding melt structures. For example, the vibrational spectra for nitrate melts of several monovalent cations have been found most consistent with a perturbed lattice structure in which coordination forces exceed those of the corresponding solid as a result of reduced coordination numbers. The situation for the divalent nitrates seems less settled, but it is suggested that the 1:1 complex model has been overworked.

ACKNOWLEDGMENTS

We are grateful to the United States Atomic Energy Commission for support of our research efforts in this area under Contract No. AT (11-1)-1615.

REFERENCES

1. R. E. Hester and K. Krishnan, J. Chem. Phys., 46, 3405 (1967).

2. R. E. Hester and K. Krishnan, J. Chem. Phys., 47, 1747 (1967).

3. R. E. Hester and C. W. J. Scaife, J. Chem. Phys., 47, 5253 (1967).

4. R. E. Hester and K. Krishnan, J. Chem. Phys., 48, 825 (1968).

5. J. P. Devlin, K. Williamson and G. Austin, J. Chem. Phys., 44, 2203 (1966).

6. K. Williamson, P. Li and J. P. Devlin, J. Chem. Phys., 48, 3891 (1968).

7. P. C. Li and J. P. Devlin, J. Chem. Phys., scheduled for publication August, 1968.

8. J. Greenberg and L. J. Hallgren, J. Chem. Phys., 33, 900 (1960).

9. A. Bandy, J. P. Devlin, R. Burger and B. McCoy, Rev. Sci. Instr., 35, 1206 (1964).

10. The cavity cell approach does permit controlled path-lengths in the mm range and is thus probably the most useful for quantitative studies of dilute solutions.

11. The hard coating presently used by Spectra Physics is inert to nearly all molten salts but, unfortunately, they have not learned to coat the Irtran materials without pinholes and flaking.

12. See for example, J. K. Wilmshurst and S. Senderoff, J. Chem. Phys., 35, 1078 (1961).

13. N. J. Harrick, Internal Reflection Spectroscopy, Interscience Publishers, 1967.

14. J. Fahrenfort, Spectrochim. Acta, 17, 698 (1961).

15. A. C. Gilby, J. Burr, Jr., W. Krueger, and B. Crawford, Jr., J. Phys. Chem., 70, 1525 (1966).

16. H. A. Levy and M. D. Danford in Molten Salt Chemistry, Interscience Publishers, 1964, pp. 109-125.

17. R. P. J. Cooney, P. C. Li and J. Paul Devlin, submitted J. Chem. Phys., August, 1968. Also paper Dl, Molecular Structure and Spectroscopy Symposium, Columbus, Ohio, September 1968.

18. G. J. Janz and T. R. Kozlowski, J. Chem. Phys., 40, 1699 (1964).

19. G. C. Benson and T. A. Claxton, J. Chem. Phys., 48, 1356 (1968).

20. A. A. Lucas, Surface Sci., 11, 19 (1968).

21. G. H. Wegdam, R. Bonn, and J. Van Der Elsken, Chem. Phys. Let., 2, 182 (1968).

A REVIEW OF RAMAN SPECTROSCOPY OF FUSED SALTS AND STUDIES OF SOME HALIDE-CONTAINING SYSTEMS

Victor A. Maroni and Elton J. Cairns

Argonne National Laboratory
Argonne, Illinois

INTRODUCTION

During the last decade or two, there has been an increasing effort to determine the structure of molten salts, and to understand the principles which explain the structure and behavior of molten salts. Spectroscopic methods have been among the most powerful used in this field, which is still in its formative years. The most successful spectroscopic techniques for elucidation of the structure of polyatomic molecules and ions in fused salts have been those which depend upon changes in vibrational energy levels (in contrast to rotational and electronic). These are Raman spectroscopy and infrared spectroscopy. The Raman and infrared spectra are generally complementary, and both are usually required to obtain a complete set of vibrational frequencies used for assignment of the molecule (or ion) to a certain point group, and for determination of the force constants.

Vibrational spectroscopy has been used in two ways in the study of fused salts: a) to determine the structure of ions and molecules; b) to study environmental effects (e.g. from cations, other salts or solvents) on certain ions and molecules. The latter area has been studied systematically only in the last few years, and has been limited primarily to the nitrate anion.[1,2,3]

Part of the reason for the relative paucity of vibrational spectra, especially Raman spectra, has been the experimental difficulty associated with preparing, handling, and containing fused salts in optically suitable sample containers at elevated temperatures. Because of this, the less-corrosive and low-melting salts have been studied most thoroughly. Most conspicuous by their absence have been the fluorides, carbonates, and highly colored salts.

Since the infrared studies of fused salts are being treated in another chapter of this volume, the subsequent discussion in this chapter is limited to Raman studies.

Until the introduction of the laser to Raman spectroscopy in 1962,[4] and the use of continuous gas lasers[5-7] with photoelectric recording of the spectra,[6,7] the mercury arc was the standard source of radiation. Since the mercury arc was of only marginal intensity for Raman work, a great deal of effort was expended in optimizing the optical arrangement of sample containers with respect to the mercury tube. This introduced an additional burden when the samples also required heating to temperatures of several hundred degrees. The use of lasers as light sources greatly simplifies the experimental designs and procedures, since very high-intensity radiation in the form of a collimated beam is available and can be transmitted into remote, heated sample containers with great ease. This development has resulted in an increase of activity in the Raman study of fused-salt systems, and a considerable improvement in the quality of spectra obtained.

REVIEW OF RAMAN SPECTROSCOPY OF FUSED SALTS

Several reviews of Raman spectroscopy of fused salts have appeared since 1962[1-3] covering the literature to 1966. This review will attempt to assimilate in a systematic way the bulk of work which has appeared primarily in the last decade up to the middle of 1968.

Nitrates

The Group I nitrates have been the most extensively studied fused salts, both as pure salts[1-3] and as mixtures.[3,8] The pure-salt spectra and their interpretation in terms of D_{3h} point group symmetry for the nitrate ion, which may be reduced to C_{2v} by cation environmental effects ("ion-pairing") for alkali nitrates, and silver and thallium nitrates have been discussed by Janz and Wait.[3] Correlation of the symmetrical stretching frequency ν_1 (D_{3h}) of the nitrate ion with the polarizing power ($P = [5z^{1.27}/r^{1/2}I][z/r]$ where z = ionic charge, r = ionic radius in Angstroms, I = ionization potential in volts) of the cation has been proposed. The plot of frequency against polarizing power is a smooth curve.[2,3] James[2] also reports correlations of stretching force constants for N-O stretch with polarizing power of the cation, and a correlation of the ν_1 frequency with free volume. Most recently, Wait, Ward and Janz[9] have found a correlation between the splitting of the $\nu_3(E', D_{3h})$ band and the ratio of the force constant for the N-O_I stretch to that of the N-O_{II} stretch where O_I is the oxygen atom interacting most strongly with the cation. The splitting of the degeneracy is measurable for Tl^+, Li^+, and Ag^+ in increasing order. It should be noted here, however, that the vibrational analysis of Brintzinger and Hester[9a] is in disagreement with that of Wait, Ward and Janz, and indicates that the splitting of the ν_3 band of NO_3^- caused by M-O bonding should be so small as to be unobservable, and that polarization effects on the NO_3^- ion by the cation should account for an appreciable splitting of the ν_3 band. The matters of the detailed vibrational analysis and the cause for the splitting of the ν_3 band seem unsettled at present.

In considering all of the data available, it seems most reasonable that the D_{3h} symmetry for NO_3^- is reduced to C_{2v} or C_s by anion-cation interactions in the cases of molten $LiNO_3$, $TlNO_3$, and $AgNO_3$. The other alkali nitrates may be regarded as retaining essentially D_{3h} symmetry in the melt. These conclusions are consistent with the results of normal coordinate analyses of the spectral data for all of the Group I nitrates.[9,9a]

The only Group II nitrates which have been investigated in the molten state by Raman spectroscopy are $Zn(NO_3)_2 \cdot xH_2O$,[10] $Ba(NO_3)_2$,[11] and $Ca(NO_3)_2$.[12] Hester and Scaife[10] found that removal of water from $Zn(NO_3)_2 \cdot xH_2O$ (down to x = 1.42) brings about a change from D_{3h} symmetry of the NO_3^- to C_{2v} symmetry, caused primarily by the cationic environment, creating ion-associations. There was also evidence for hydrogen-bonding and hydration effects, enhancing the tendency toward C_{2v} symmetry in the melts of higher water content. In keeping with the results for Group I nitrates, the splitting of $E'(D_{3h})$ to form A_1 and B_1 (C_{2v}) bands increased with decreasing water content, corresponding to larger anion-cation interaction. A shift in ν_1 (A'_1, D_{3h}) to lower frequencies becoming ν_2 (A_1, C_{2v}) was also observed. The effect of temperature on the spectra was used to distinguish qualitatively between the cation and water effects on the anion spectra. The conclusion was drawn that anhydrous $Zn(NO_3)_2$ would be a molecular melt with almost no ionic character.

The spectrum of fused $Ba(NO_3)_2$ was marred by the presence of large amounts of nitrite and other decomposition products, and was not resolved well enough to decide the extent of perturbation on the nitrate ion.[11] The data on $Ca(NO_3)_2$ also do not allow this decision to be made. The frequencies for $Zn(NO_3)_2 \cdot 1.42\ H_2O$, $Ba(NO_3)_2$ and $Ca(NO_3)_2$ are shown in Table 1.

There is need for additional Raman studies of Group II nitrates with the objective of determining their character (ionic vs. molecular melts) and the effect of environment on the symmetry of the nitrate group.

<u>Halides</u>

The halides of Group II B have received much more attention by Raman investigators than those of any other group with the chlorides being more thoroughly investigated than bromides or iodides (see Table 2). In general, the Group II B halides have a tendency to form polymers which retain the structure of the corresponding solid well above the melting point.

In the case of zinc chloride, a polymer formed by the linking of the basic $ZnCl_4^{-2}$ tetrahedra is the predominant species.[13-16] This polymeric species has a symmetrical

TABLE 1

Vibrational Frequencies of some Group II Nitrates According to D_{3h} Symmetry

Compound	Temp, °C	ν_1	ν_2	ν_3	ν_4	Ref.
$Zn(NO_3)_2 \cdot 1.42\ H_2O$	90	1023	806	1495, 1302	-	10
$Ca(NO_3)_2$	?	1050	-	1415	742, 720	12
$Ba(NO_3)_2$	600	1047	-	1383	720	11

TABLE 2

Recent Literature References for Raman Studies of Molten Halides

Cations	Anions		
	Cl^-	Br^-	I^-
Group II B			
Zn^{+2}	13 - 16	15	-
Cd^{+2}	16, 17	-	-
Hg^{+2}	18 - 22	18, 20	23, 24
Group II A			
Mg^{+2}	25	-	-
Ba^{+2}	32	-	-
Group III A			
Al^{+3}	28	-	-
Ga^{+3}	29	30	-
Group IV A			
Sn^{+2}	26	-	-
Pb^{+2}	27	-	-

Zn-Cl bond stretch near 230 cm^{-1} and a low-lying band (probably a bending mode) below 100 cm^{-1}. As the temperature is raised,[13] or as additional Cl^- is added to the melt[14] (in the form of KCl, for instance), the polymer of tetrahedra breaks into smaller units. $ZnCl_4^{-2}$ predominates at the higher Cl^-/Zn^{+2} ratios and has been assigned a stretching frequency of 280 cm^{-1}.[15] The $ZnCl_2$ molecule was assigned the symmetrical stretch frequency of 305 cm^{-1} in keeping with the assignments for aqueous solutions of $ZnCl_2$.[15] The bands observed in the region 275-290 cm^{-1} have been assigned to $ZnCl_3^-$ and $ZnCl_4^{-2}$. Analogous results were found for $ZnBr_2$ and its mixtures with KBr.[15] The general picture is one of a melt which retains its crystalline structure in polymer-like form near the melting point and for X^-/Zn^{+2} ratios near 2. As the temperature is raised or X^-/Zn^{+2} is increased, ZnX_2 molecules and ZnX_3^- and/or ZnX_4^{-2} ions become the predominant species. The work of Ellis[15] provides the most complete review and discussion of these systems.

The structure of molten $CdCl_2$, both pure and mixed with KCl, has been studied by Bues[16] and by Tanaka, Balasubrahmanyam and Bockris.[17] The spectra were not as well-defined as those for $ZnCl_2$, and there has been a difference of opinion concerning the stoichiometry and structure of the complex ion. Bues[16] proposed $CdCl_3^-$ of D_{3h} symmetry ($\nu_1(A'_1)$ = 259), while Tanaka et al. proposed pyramidal (C_{3v}) $CdCl_3^-$. In considering Bues' data, Bredig and Van Artsdalen[33] proposed $CdCl_4^{-2}$, which would be tetrahedral. Unfortunately, only the polarized symmetrical stretch bands have been observed, so it is not possible to make a choice with any degree of certainty. The major portion of the melt just above the fusion point seems to retain its layer-like crystal structure in the form of a polymer of $CdCl_6$ coordination with a symmetrical stretch frequency at 212 cm^{-1}. Since the questions concerning the stoichiometry and structure of the complex ions in molten $CdCl_2$ and $CdCl_2$-MCl mixtures are still largely unanswered, some further, more definitive work seems desirable.

The halides of the last member of Group II B, mercury, are the most thoroughly investigated of all, with recent investigations having been performed on mercuric iodide.[23,24] Molten mercuric halides differ significantly from the other Group II B halides in that they do not seem to form polymeric liquids near the melting point. The Raman spectrum of molten $HgCl_2$ is easily interpreted in terms of the single molecular species $HgCl_2$ of $D_{\infty h}$ symmetry having ν_1 at 313 cm^{-1} and ν_3 at 376 cm^{-1}.[20] As chloride ion is added (in the form of KCl), additional polarized Raman bands appear at 282 and 267 cm^{-1}, accompanied by a weakening of the band at 313 cm^{-1}. Janz and James[21] interpreted these results to mean that $HgCl_3^-$ and $HgCl_4^{-2}$ were formed as Cl^- was added to $HgCl_2$. A cross-plot of peak intensities against melt composition showed the band assignments to be reasonable. The structures favored by Janz and James were D_{3h} for $HgCl_3^-$ and T_d for $HgCl_4^{-2}$. Other weaker bands observed in the spectra seemed consistent with these assignments. Further work with $HgCl_2$-KCl mixtures in KNO_3 as a solvent lent support to the above assignments.[22]

Mercuric bromide shows a relatively intense polarized band at 195 cm^{-1} assigned to $\nu_1(D_{\infty h})$ of $HgBr_2$, and a weak band at 271 cm^{-1} assigned to ν_3.[20] No evidence was seen for any other species; therefore $HgBr_2$ is essentially a molecular melt. The spectra for HgClBr are also consistent with the idea of a molecular melt, but with some disproportionation to form $HgCl_2$ and $HgBr_2$. The structure of HgClBr is $C_{\infty v}$ with the stretching frequencies, ν_1 at 335 and ν_3 at 236 cm^{-1}, and the bending frequency ν_2 at 111 cm^{-1}.[20]

The first spectra for melts containing HgI_2 were reported earlier this year by Clarke and Solomons[23] for HgI_2 in $HgCl_2$ and in $HgBr_2$. The HgI_2 in $HgCl_2$ was found to be a molecular species of $D_{\infty h}$ symmetry with ν_1 at 146 cm^{-1} (30% HgI_2 in $HgCl_2$). They found it impossible to study pure fused HgI_2 due to its intense color (dark red to black) in the molten state. The formation of HgICl and HgIBr was also observed; the bands appeared at 330 cm^{-1} (ν_1, Hg-Cl stretch), 179 cm^{-1} (ν_2, Hg-I stretch for HgICl), 229 cm^{-1} (ν_1, Hg-Br stretch) and 164 cm^{-1} (ν_2, Hg-I stretch for HgIBr). Both compounds were assigned $C_{\infty v}$ symmetry. No bands which could be assigned to complex ions were observed.

Melveger, Khanna, Guscott and Lippincott[24] have reported the first spectra of pure molten HgI_2, as well as spectra of the solid in both the red (low temperature) and yellow (high temperature > 137°C) forms. In their work on the melt, $\nu_1(D_{\infty h})$ was observed at 138 cm^{-1}, and a new band at 41 cm^{-1} was found and assigned to ν_2, the bending mode. The Raman spectrum for the yellow solid was the same as that for the (deep red) melt and is characteristic of a molecular crystal (orthorhombic). The red modification of the solid gave a spectrum more analogous to those for the other Group II B halides, that is, a structure in which the anions are shared by several cations, forming a network. The crystal structure of the red form is tetragonal.

There is a great paucity of Raman spectra for Group IIA halides; $MgCl_2$ and $BaCl_2$ are the only compounds that have been studied. Balasubrahmanyam[25] reported Raman spectra for $MgCl_2$ at 730°C, a 50 mole % mixture of $MgCl_2$

with KCl at 525°C, and a mixture of 33 mole % $MgCl_2$ with KCl at 475°C. The spectrum for pure $MgCl_2$ showed four bands, at 269, 241, 214, and 142 cm^{-1}, which were assigned to the solid-like polymer structure of octahedral symmetry (241 cm^{-1}) and $MgCl_6^{-4}$ of octahedral symmetry (O_h, 269 $cm^{-1} = \nu_1$, 214 $cm^{-1} = \nu_2$, 142 $cm^{-1} = \nu_3$). In the case of $MgCl_2$ mixed with KCl, different lines appeared (451, 250, 232, ∿ 200 and ∿ 158 cm^{-1}), and were assigned to a pyramidal $MgCl_3^-$ (C_{3v}, $\nu_1(A_1) = 250$ cm^{-1}, $\nu_2(A_1) = 158$ cm^{-1}, $\nu_3(E) \simeq 451$ cm^{-1}, $\nu_4(E) \simeq 200$ cm^{-1}), with the remaining weak line at 232 cm^{-1} assigned to tetrahedral $MgCl_4^{-2}$. Additional Raman studies of the $MgCl_2$-KCl system, yielding somewhat different frequencies and different band assignments, are discussed in a later section of this chapter.

The only study of $BaCl_2$ which has resulted in a Raman band being assigned to a barium-containing species is that of Vallier and Lira,[32] who studied mixtures of $BaCl_2$ with $AgNO_3$, and with KCl. The spectra for the nitrate-containing melts showed only characteristic nitrate frequencies, but three weak bands appeared for the $BaCl_2$-KCl mixtures at 700°C. The frequencies, ∿ 380, ∿ 472, and ∿ 541 cm^{-1}, were independent of $BaCl_2$ concentration over the range 25 to 40 mole percent. The bands were assigned to $BaCl_3^-$ (D_{3h}, $\nu_1 = 472$, $\nu_3 = 541$, and $\nu_4 = 380\ cm^{-1}$). These frequencies are much higher than those of all other Group II chlorides. Furthermore, estimation of the Ba-Cl force constant for such a species with the frequencies indicated yields a value of greater than 4.0 md/Å, much higher than could reasonably be expected. This would seem to indicate that further investigation of $BaCl_2$-containing melts is in order.

The molten halides of two Group III A metals have been studied by Raman spectroscopy: aluminum chloride,[28, 40, 41] bromide,[40, 41] and iodide,[40] and gallium chloride[29] and bromide.[30]

The Raman spectra of pure fused $AlCl_3$, $AlBr_3$ and AlI_3 were all consistent with an Al_2X_6 bridged dimer structure. Nine Raman lines were reported by Balasubrahmanyam and Nanis[28] for an $AlCl_3 \cdot KCl$ melt at 300°C, four of which were said to be polarized. This spectrum was interpreted in terms of an $AlCl_4^-$ tetrahedron distorted to C_{2v} symmetry by strong cationic association. The observation of tetrahedral symmetry for $AlCl_4^-$ in an $AlCl_3 \cdot NaCl$ melt which showed four bands at the same frequencies as four of the nine bands reported for $AlCl_3 \cdot KCl$ was used to support the assumption of a distorted tetrahedral arrangement in $AlCl_3 \cdot KCl$. The published spectra for both melts ($AlCl_3 \cdot KCl$ and $AlCl_3 \cdot NaCl$) show that a number of the reported bands are weak and poorly defined and may in fact be spectral artifacts.

Woodward, Garton and Roberts[29] found that the spectrum of molten gallium dichloride is identical to that for $GaCl_4^-$ ion in aqueous solution. The melt is therefore composed of tetrahedral (T_d) $GaCl_4^-$ ions and Ga^+ ions. No other frequencies were observed. The assignments are $\nu_1(A_1) = 356\ cm^{-1}$, $\nu_2(E) = 115\ cm^{-1}$, $\nu_3(F_2) = 380\ cm^{-1}$, and $\nu_4(F_2) = 153\ cm^{-1}$. The gallium dibromide melt was studied by Woodward et al.[30] with similar results: $GaBr_4^-$ of T_d symmetry, $\nu_1(A_1) = 209$, $\nu_2(E) = 79$, $\nu_3(F_2) = 288$, $\nu_4(F_2) = 107\ cm^{-1}$.

The assignments that have been made for the melts of Group III A halides are the least ambiguous of all the halides except possibly for some of the mercuric halides.

The only Group IV molten halides that have been studied by Raman spectroscopy are $SnCl_2$ and $PbCl_2$. Clarke and Solomons[26] found that pure molten $SnCl_2$ exhibited two Raman bands similar to those observed for the solid. The structure was therefore assumed to be $(SnCl_2)_n$ polymer formed by single Cl-Sn bridging bonds, giving Sn a coordination of three. The symmetric stretching mode was assumed to yield the polarized band at 228 cm^{-1}, and the Cl-Sn-Cl deformation was held responsible for the polarized band at 112 cm^{-1}. When KCl was added to $SnCl_2$, three new bands appeared at the expense of the old ones. These three bands were assigned to $SnCl_3^-$ of C_{3v}(pyramidal) symmetry, analogous to the structure of $SnCl_3^-$ in ether, reported by Woodward and Taylor.[31]

Lead(II) chloride and its mixtures with KCl were investigated by Balasubrahmanyam and Nanis.[27] No Raman spectrum was found for pure $PbCl_2$ but the mixtures containing 33 to 67 mole % $PbCl_2$ in KCl exhibited several bands. For 67 and 50 mole % $PbCl_2$, the four bands observed were assigned to pyramidal (C_{3v}) $PbCl_3^-$: $\nu_1(A_1)$ = 217 cm^{-1}, $\nu_2(A_1)$ = 172 cm^{-1}, $\nu_3(E)$ = 245 cm^{-1} and $\nu_4(E)$ = 205 cm^{-1}. It seems unlikely, however, that $\nu_3(E)$ and $\nu_4(E)$ would be separated by only 40 cm^{-1}. Normally $\nu_3/\nu_4 \approx 2$ for pyramidal XY_3 species. In the case of 33 mole % $PbCl_2$, the authors report nine Raman bands which were assigned to a distorted tetrahedral structure (C_{2v}) for $PbCl_4^{-2}$. From an inspection of the published spectra, however, it appears that a number of these bands may be spectral artifacts and not genuine Raman lines.

In general, there is a need for Raman studies of the fluorides, bromides, and iodides of all groups (Hg^{+2} being the only exception), and all halides of the missing Group II A members, the higher members of Group III A, and all of Group IV B. A reinvestigation of $CdCl_2$ and $BaCl_2$ melts seems appropriate.

<u>Oxygen-Containing Anions</u>

Raman spectra of melts containing various oxyanions have been studied during the last several years,[34-38] as indicated in Table 3. The anions are those containing members of Groups III A (borates), V A (nitrates, phosphates, arsenates), VI A (sulfates), and VI B (molybdates and tungstates). Nitrates have been discussed earlier in this chapter because of the extensive studies that have been performed on them. Noticeable because of their absence are the carbonates, which are very corrosive toward the usual materials used for Raman cells (quartz, Vycor, pyrex).

Molten lithium orthoborate and metaborate have been studied by Bues, Förster and Schmitt,[34] who found BO_3^{-3}

TABLE 3

Literature References for Raman Studies of Molten Salts Containing Oxyanions

Anions	Cations		
	Li	Na	K
$B_xO_y^{(3x-2y)}$	34	-	-
$P_xO_y^{(3x-2y)}$	35	35	-
$As_xO_y^{(5x-2y)}$	35	-	-
SO_4^{-2}	36	36	-
HSO_4^-	-	-	37
$S_2O_7^{-2}$	-	-	-
MoO_4^{-2}	-	38	-
WO_4^{-2}	-	38	-

ions of D_{3h} symmetry in a $3Li_2O \cdot BO_3$ melt. Only three Raman bands were found, and all were assigned to BO_3^{-3}. When the composition was changed to $2Li_2O \cdot B_2O_3$, additional bands appeared (a total of 10) which were assigned to $B_2O_5^{-4}$ and BO_2^-. The melts $5Li_2O \cdot B_2O_3$, $3Li_2O \cdot 2B_2O_3$, and $LiKO \cdot B_2O_3$, had 7, 6, and 4 Raman lines, respectively. The frequency for the orthoborate anion, BO_3^{-3}, was found in the first two of these melts, in addition to bands assigned to polyborates. In the last melt, the BO_3^{-3} symmetrical stretch band at 910 cm^{-1} was absent, and the melts were concluded to be polymeric in nature.

Similar Raman studies of phosphates, arsenates, and their mixtures were reported by Bues, Bühler and Kuhnle.[35] The melt of equal parts $Na_4P_2O_7$ and $Li_4P_2O_7$ at 850°C showed seven Raman bands, six of which corresponded to IR bands found in $Na_4P_2O_7$ crystals; thus the viscous liquid retained the network structure characteristic of the solid. The same was true for the mixture of $Li_4As_2O_7$ and $Na_4As_2O_7$ at 800°C. In mixtures of diarsenate and diphosphate, two new Raman bands appeared which were assigned to a POAs bend (276 cm^{-1}) and a POAs stretch (649 cm^{-1}), characteristic of arsenophosphate, $AsPO_7^{-4}$. Otherwise the network structure observed for $P_2O_7^{-4}$ and $As_2O_7^{-4}$ was maintained. No bands were found for small fragments or single complex ions.

Walfrafen[36] reported Raman spectra for Li_2SO_4 and Na_2SO_4 at 950°C. These spectra contained three bands, assigned to T_d symmetry of SO_4^{-2}, at $\sim$ 980 $cm^{-1} = \nu_1(A_1)$, 460 $cm^{-1} = \nu_2(E)$, and 625 $cm^{-1} = \nu_4(F_2)$. There was a moderate cation effect on ν_1 similar to that found by Janz and James[1-3] for the nitrates. A more complex situation was found by Walrafen, Irish, and Young[37] for potassium bisulfate melts. The formation of $S_2O_7^{-2}$ and SO_4^{-2} from HSO_4^- was evident in these Raman spectra. The HSO_4^- was reported to have C_s symmetry, the SO_4^{-2} possessed the expected T_d symmetry, and the $S_2O_7^{-2}$ was proposed to be C_2 (17 lines appeared in the Raman spectrum). The intensity of the HSO_4^- bands decreased with temperature in the range 300 - 600°C, while $S_2O_7^{-2}$ went through a maximum at 500°C and SO_4^{-2} increased with temperature although always being a very minor component.

Vallier[38] studied the melts of Na_2MoO_4 and Na_2WO_4 at about 710°C, and found that in alkaline melts, the simple tetrahedral MoO_4^{-2} and WO_4^{-2} were the only polyatomic species present. With excess MoO_3 and WO_3 respectively, the melts showed three additional Raman lines, which were not assigned to any specific structure. The suggestion was made that perhaps bimolybdate and bitungstate ions had formed.

Some alkali metal thiocyanates have been studied by Baddiel and Janz, who found the expected $C_{\infty v}$ symmetry for the SCN^- ion.[39] These results are reviewed and discussed by Janz and Wait.[1]

EXPERIMENTAL METHODS

Review of Some Recent Experimental Advances

The well-established methods of sample containment and heating, designed for use with various types of mercury arcs, which have been reported by Janz et al., T. F. Young et al., and Bues et al., have been adequately reviewed by James[2] and by Janz and Wait.[1] The recent developments involving the use of laser excitation will be discussed here.

The room-temperature laser-excitation methods developed at Bell Laboratories[4-7] were extended to fused-salt systems by Clarke, Solomons, and Balasubrahmanyam,[23a] who took advantage of the great optical flexibility offered by the intense, well-collimated laser beam. The samples were held in cylindrical quartz tubes which had optical flats at their lower ends. The laser beam (pulsed ruby, 6943 Å) was passed upward through the sample, and the scattered light was viewed perpendicular to the direction of the exciting radiation. The samples were heated in an ordinary electrical furnace with small (8 mm dia) viewing ports. At temperatures near 950°C, the thermal radiation was blocked from the spectrometer by a shutter, except during the flashes of the laser. Although they worked with a pulsed laser, Clarke et al. recognized the advantages of a continuous laser source.

The use of a continuous-wave laser (He-Ne, 50 mW at 6328 Å), a double spectrometer and continuous photoelectric recording for the study of fused salts was reported by Gasner and Young.[42] They used a metal boat sample container similar to that shown in Figure 3 (see later), which performed adequately with melts of low volatility. Condensation of salt vapors on the quartz envelope presented a problem which was minimized by use of a shutter arrangement above the boat.[43]

The use of lasers is particularly valuable where highly colored melts are involved,[39,44] and sample containers of unusual geometry are used.[24] They allow a maximum of flexibility in heating arrangements, which have taken the form of tubular electric furnaces,[39,42-44] or simple wire coils.[24]

An interesting sample cell for use with certain highly corrosive melts at temperatures below 300°C has been reported by Arighi and Evans.[45] The cell is made from

Monel and has sapphire windows seated to polytetrafluoroethylene gaskets. One window is 12 x 50 x 1.6 mm and is located on the side of the cell, while the other, at the bottom of the cell, is 12 mm dia x 1.6 mm. Volatile samples are distilled from a monel sidearm into the window zone of about 3 cm^3 volume. This cell has evidently been used with Hg arc excitation, but it has obvious advantages for use in laser excitation of Raman spectra at moderate temperatures.

A simple nichrome wire-heated Raman cell for fused salts was reported by Beattie.[46] The cell was intended for use with a horizontal laser beam with back-scattered Raman radiation passing to the spectrometer. The cell consisted of glass or quartz tubing having a diameter of a few millimeters with a 1-cm solid end to act as a light-pipe and thermal insulator. The tube was bent at the other end, and could be sealed off. The light-pipe end can be placed near the spectrometer optics, but the 180° scattering makes it difficult to obtain good depolarization ratios. Methods for extending the usable temperature range of quartz in corrosive media by the application of metal coatings have been suggested by Toth.[47]

REVIEW OF RAMAN SPECTROSCOPY

Solomons et al.[47a] have recently reported on the usefulness of a windowless cell for containing corrosive melts of high surface tension. The cell, similar to one used by Young for absorption spectrophotometry,[47b] was made of boron nitride, and contained two mutually perpendicular intersecting horizontal holes, one for the passage of the laser beam, the other for the exit of the scattered light. A filling chamber was located immediately above the intersection of the two horizontal holes. The solid sample was placed in the filling chamber, and on melting ran into the sample chamber formed by the intersecting holes, where it was held by surface tension forces. This cell was used with cryolite at about 1000°C.

The photomultiplier tube best suited for use with a He-Ne laser (6328 Å) is that with S-20 response. The S-20 cathodes have been used with both dc and chopped detection methods. For low-level signals, photon counting techniques have been employed. In general, chopping the laser beam has proved most successful for high-temperature work because it eliminates the dc component of the black-body thermal radiation from the sample, which is of significant amplitude in the region of interest when 6328 Å exciting radiation is used. Of course, the (ac) noise components of the black-body radiation are not eliminated by chopping, but a large fraction of the noise can be eliminated by careful use of active band-pass filters prior to amplification in a phase-sensitive lock-in amplifier. Another suitable approach is to use green or blue exciting radiation (as from an argon ion laser) with colorless samples. Red and yellow samples such as HgI_2 require red exciting light.

At least four commercial laser-Raman spectrometers are available, all with He-Ne Lasers and S-20 photomultipliers for continuous recording of spectra. The best results are obtained from those instruments which use two spectrometers in series (double monochromator) to minimize the stray light.

Experimental Methods for This Work

A laser-Raman spectrophotometer for studies of fused-salt systems was constructed using a Spectra-Physics Model 125 continuous helium-neon gas laser (nominal power,

50 mW at 6328 Å), and a Spex Industries 1400 double spectrometer, as shown in Figure 1. The light exiting from the spectrometer was focused on the cathodeof a Hamamatsu Ltd. R136 photomultiplier tube cryostated at -50°C to reduce tube shot noise. Thermal radiation from the molten sample was removed from the spectral background by mechanically chopping the exciting radiation at 400 cps and measuring the

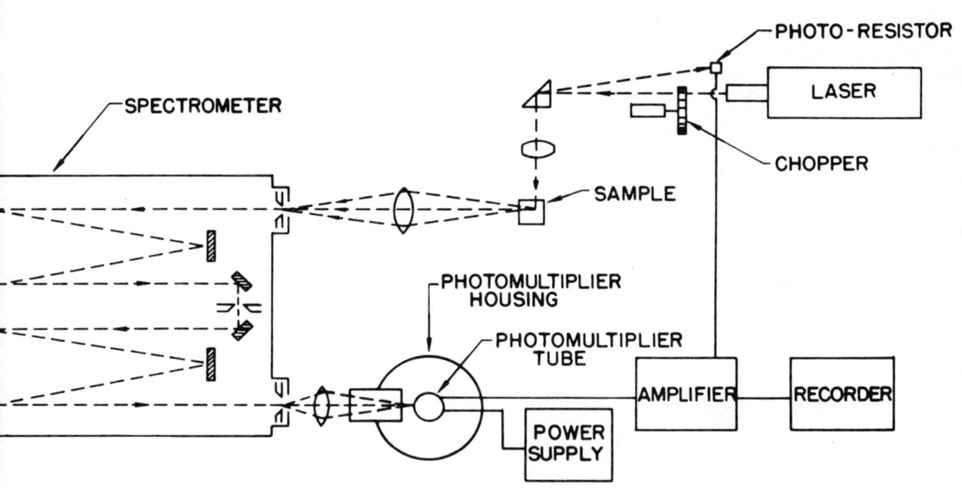

FIG. 1

Laser-excited Raman spectrophotometer (schematic).

photomultiplier output with an ac lock-in amplifier coupled to a phase-angle voltmeter. A small fraction of the laser radiation that was reflected from the front surface of the prism to a photo-resistive detector provided a reference signal for the ac amplifier. The spectra were plotted by a strip-chart recorder.

Scanning conditions of 2 cm^{-1}/min with a 50 second time constant or 5 cm^{-1}/min with a 10 second time constant were normally used in the recording of spectra. Spectral slit widths were in the range of 5 to 10 cm^{-1}.

Quartz sample cells in various configurations, depending on the relative volatility of the melt components, have proven suitable for obtaining the Raman spectra of most molten salts below 600°C. The furnace assembly and three types of quartz cells are shown in Figure 2. The laser beam enters along the axis of the cylindrical furnace and the scattered light is observed perpendicular to the furnace axis through a 4 x 2 cm slot cut through the heating element and insulation. A curved quartz window covers the slot inside the furnace to reduce the thermal loss through the opening. By focusing

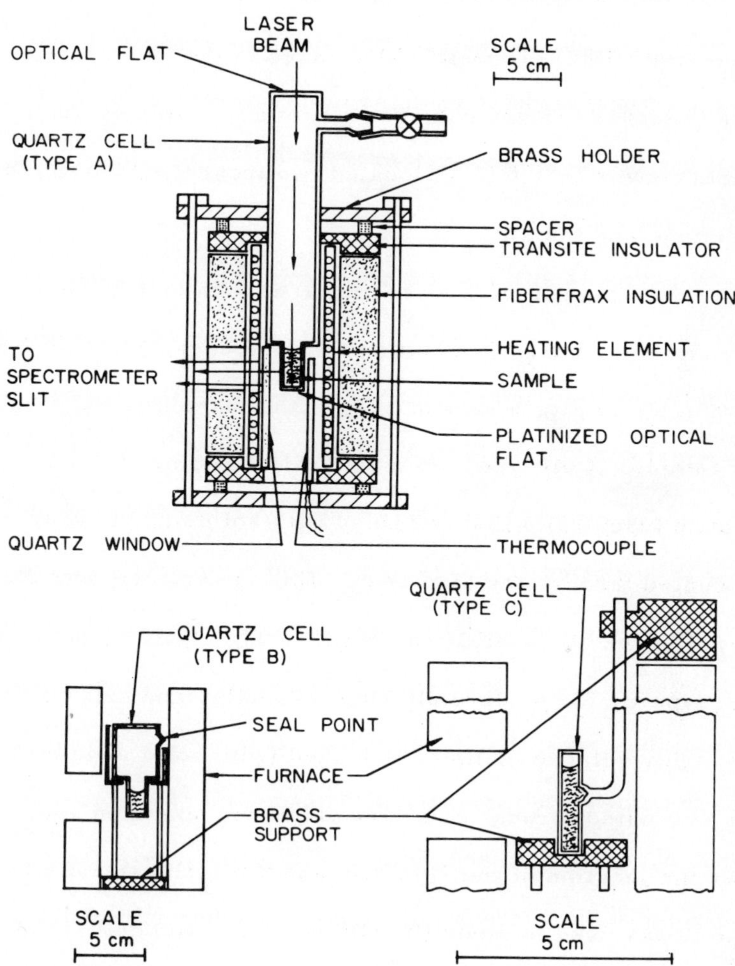

FIG. 2

Furnace assembly and three quartz cells for Raman studies of fused salts.

the laser beam at the center of the sample meniscus, a bright "rod" of illumination approximately 1 mm in diameter is created within the melt. The image of the illuminated zone is focused on the entrance slit of the spectrometer by a double convex lens (f/2.0) located two focal lengths from both the image and the slit.

The quartz cells were prepared with optical flats fused to both ends. The bottom flat is then platinized[59] on the outside to give a second pass of the laser beam through the sample. Cell type A is a reusable container for non-volatile fused salts that is loaded with solid sample and evacuated through the side arm. This type of cell was employed in the $MgCl_2$-KCl and $MgBr_2$-KBr studies (see below). Cell type B, for moderately volatile melts, and cell type C, for highly volatile melts, are loaded (with solid samples) and evacuated through a narrow side arm, and then sealed near the junction of the side arm and the cell wall. Cell type B was used to study UO_2Cl_2 solutions and cell type C was used for HgI_2-LiI-KI melts.

A cell assembly for use with highly corrosive melts (e.g., those containing fluorides) has also been designed and tested. A cross-section of this cell assembly is shown in Figure 3. The sample is contained in a tungsten boat with a polished tungsten mirror immersed in the sample at a 45°

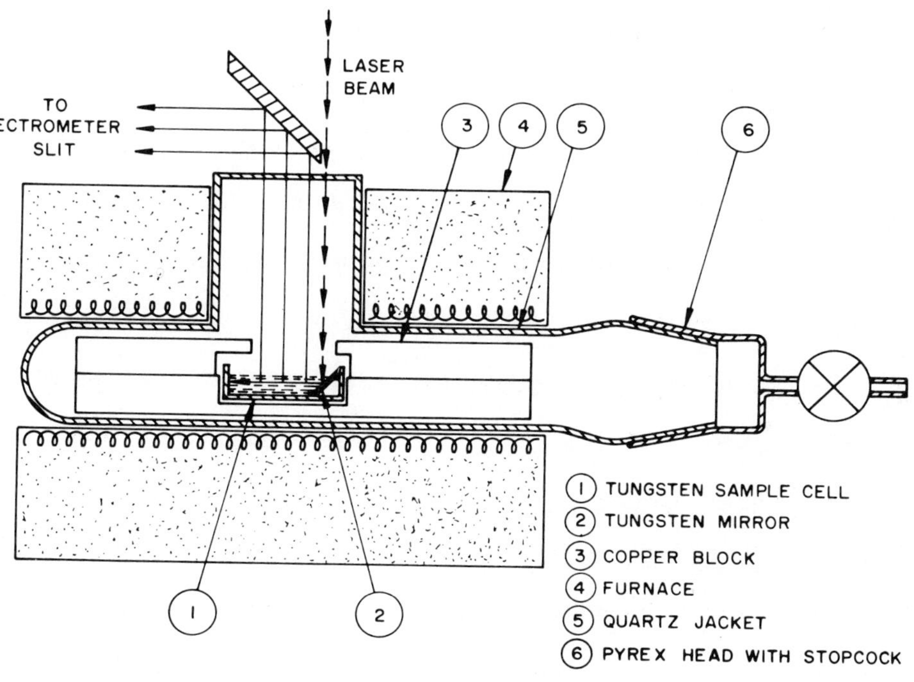

FIG. 3

Raman cell for corrosive fused salts.

angle so that the laser beam is reflected horizontally through the melt. The boat is held in a two piece cylindrical copper block with a narrow slot cut through the upper portion to permit the entrance of the laser beam and the exit of scattered light. A larger 45° mirror outside the furnace permits the scattered light originating from the horizontal "rod" of illuminated sample to be focused on the slit of the spectrometer. The copper block is enclosed in a quartz jacket to which a rectangular side-arm has been attached. The slot in the copper block is aligned so that the entering laser beam and exiting scattered light pass through the optical flat fused to the top of the rectangular side-arm. With this configuration, small amounts of salt vapor from the melt condense on the walls of the rectangular side-arm before reaching the optical flat.

The preparation of clear uncontaminated melts is essential to high-quality Raman spectroscopic studies of fused salts. Most halide melts are corrosive toward quartz in the presence of oxygen, and traces of organic materials cause the melts to become clouded by small carbon particles. Therefore, the cells described above were pretreated first with aqua regia, then dilute hydrofluoric acid, and were heated under vacuum at 500-600°C to remove all traces of organic materials, oxygen, and other impurities.

High purity $MgCl_2$, $MgBr_2$, and HgI_2 were obtained by sublimation of the commercially available anhydrous materials. Prior to being loaded into sample cells, the salt mixtures were first equilibrated and then filtered through the fine quartz fritted discs under helium pressure. All sample handling was carried out in inert (helium) atmosphere boxes having a moisture content of less than 1 ppm.

A sample of UO_2Cl_2 in $MgCl_2$-KCl-NaCl eutectic was prepared by bubbling chlorine gas through fused $MgCl_2$-KCl-NaCl (50 mol %, 20 mol %, 30 mol %) to which anhydrous UO_2 had been added. The reaction was continued at 550°C until all of the UO_2 had reacted. The resultant solution was filtered through a fine quartz fritted disc under helium pressure.

RAMAN STUDIES OF SOME MOLTEN HALIDE-CONTAINING SYSTEMS

The Magnesium Chloride-Potassium Chloride System

Studies of electrical conductance,[48] molar volume,[48] vapor pressure,[49] viscosity,[50] surface tension,[51] and activity[52] in $MgCl_2$-KCl melts have been interpreted in terms of the existence of complex species. Solid $MgCl_2$, like $ZnCl_2$, crystallizes in the rhombohedral space group, $R\bar{3}m$ (D_{3d}^5) and has the $CdCl_2$ layer structure. Raman studies of molten $ZnCl_2$[13-15] and $CdCl_2$[16,17] have supported the persistence to some extent of this layer structure into the melt phase. A previous Raman investigation of molten $MgCl_2$ and $MgCl_2$-KCl mixtures[25] concluded that the layer structure in the solid was at least partly broken up in the liquid to form octahedral $MgCl_6^{-4}$ units for pure $MgCl_2$, and $MgCl_3^-$ and possibly $MgCl_4^{-2}$ for molten $MgCl_2$-KCl. A reinvestigation of the Raman spectra for $MgCl_2$-KCl melts at ratios of Cl^-/Mg^{+2} from 2.5 to 4.0 has been conducted to more thoroughly assess the nature of the species present.

The Raman spectrum for Cl^-/Mg^{+2} = 4.0, shown in Figure 4, is typical of the results obtained for the $MgCl_2$-KCl system. The strong Raman band between 200 and 300 cm^{-1} is completely polarized and the broad band centered at about 160 cm^{-1} is depolarized. As the Cl^-/Mg^{+2} ratio is reduced, the position of the band maximum in the 200-300 cm^{-1} region shifts continuously to lower frequency, as illustrated by the circled data points in Figure 5, all of which were obtained at 550°C.

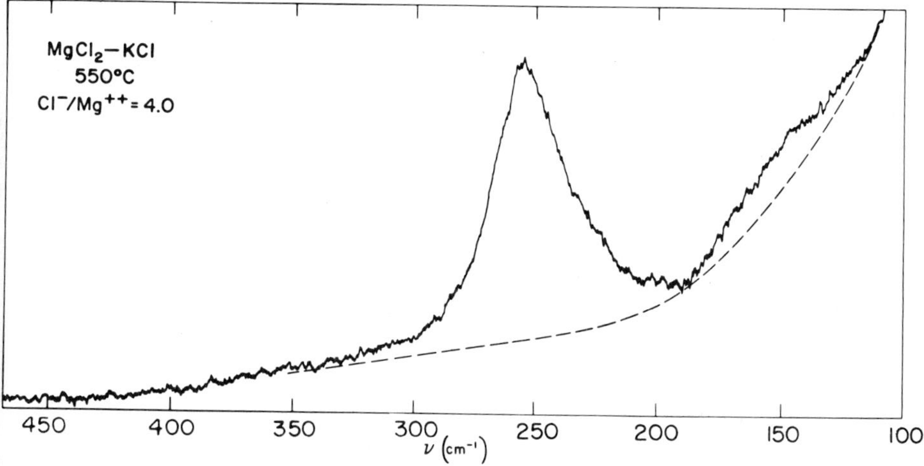

FIG. 4

Raman spectrum of an $MgCl_2$-KCl melt, Cl^-/Mg^{+2} = 4.0.

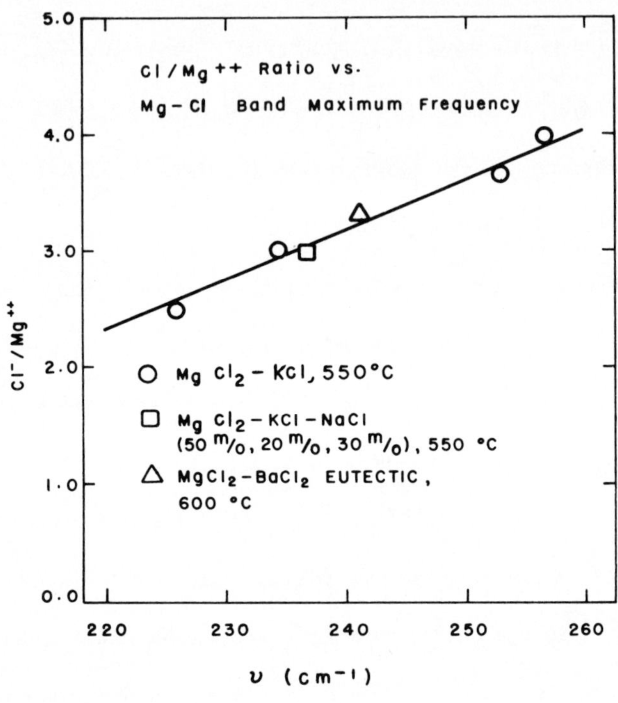

FIG. 5

Plot of the Cl^{-}/Mg^{+2} mole ratio against band maximum frequency in the 200-300 cm^{-1} region for $MgCl_2$-containing melts.

Spectra of the $MgCl_2$-$BaCl_2$ and $MgCl_2$-KCl-NaCl eutectics have also been obtained and these data points are also plotted in Figure 5. The point for each eutectic falls very close to the line determined by the $MgCl_2$-KCl data, indicating that the substitution of other alkali and alkaline earth cations for potassium ion has relatively little effect on the frequency of the band maximum in the 200-300 cm^{-1} region.

These results may be explained by assuming a magnesium(II)-chloride complex equilibrium involving several structurally different species. In order to study this equilibrium in more detail, a series of $MgCl_2$-KCl spectra of sufficient quality (for Cl^-/Mg^{+2} = 4.0, 3.6 and 3.0) were subjected to a computerized Gaussian curve resolution analysis. The results of this analysis, shown in Figures 6, 7, and 8, are summarized in Table 4. In the absence of an internal standard, it is not possible to obtain a quantitative relationship among the results for the three Cl^-/Mg^{+2} ratios; however, an increase in the intensities of the 282, 255, and 221 cm^{-1} bands relative to the intensity of the 233 cm^{-1} band with increasing Cl^-/Mg^{+2} ratio is apparent in Figures 6, 7, and 8, where the squares represent the recorded spectrum, the lines are the computer-resolved Gaussian peaks, and the x's are $(I_{obs}-I_{calc})$.

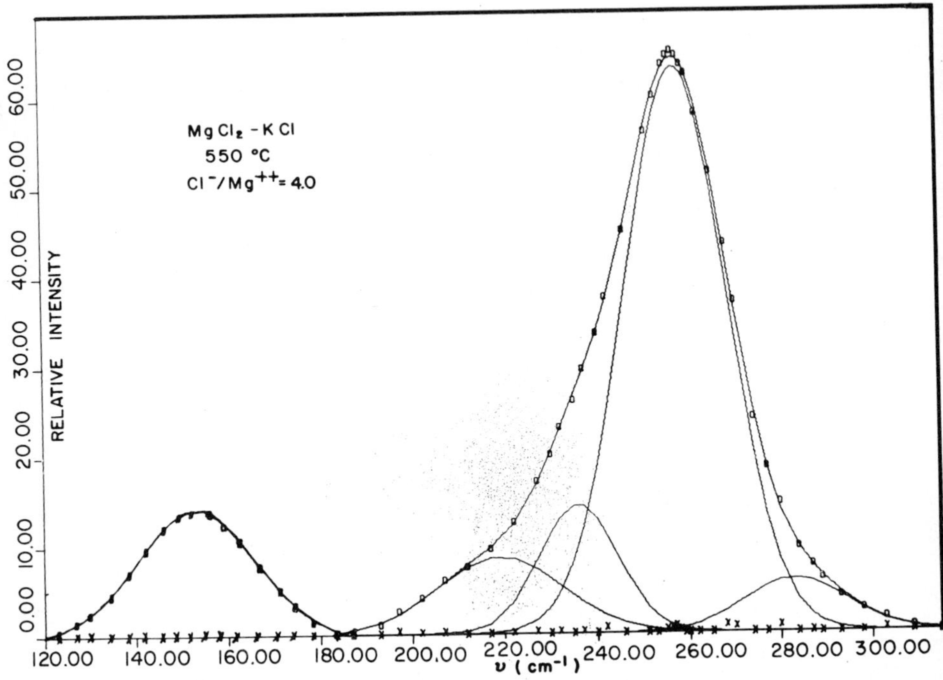

FIG. 6

Resolved spectrum of molten $MgCl_2$-KCl, Cl^-/Mg^{+2} = 4.0.

Since the four bands, 282, 255, 233, and 221 cm^{-1}, compose the completely polarized 200-300 cm^{-1} envelope, they must all arise from totally symmetric vibrations of the lattice structure or of species of the type $MgCl_n^{(2-n)}$. The

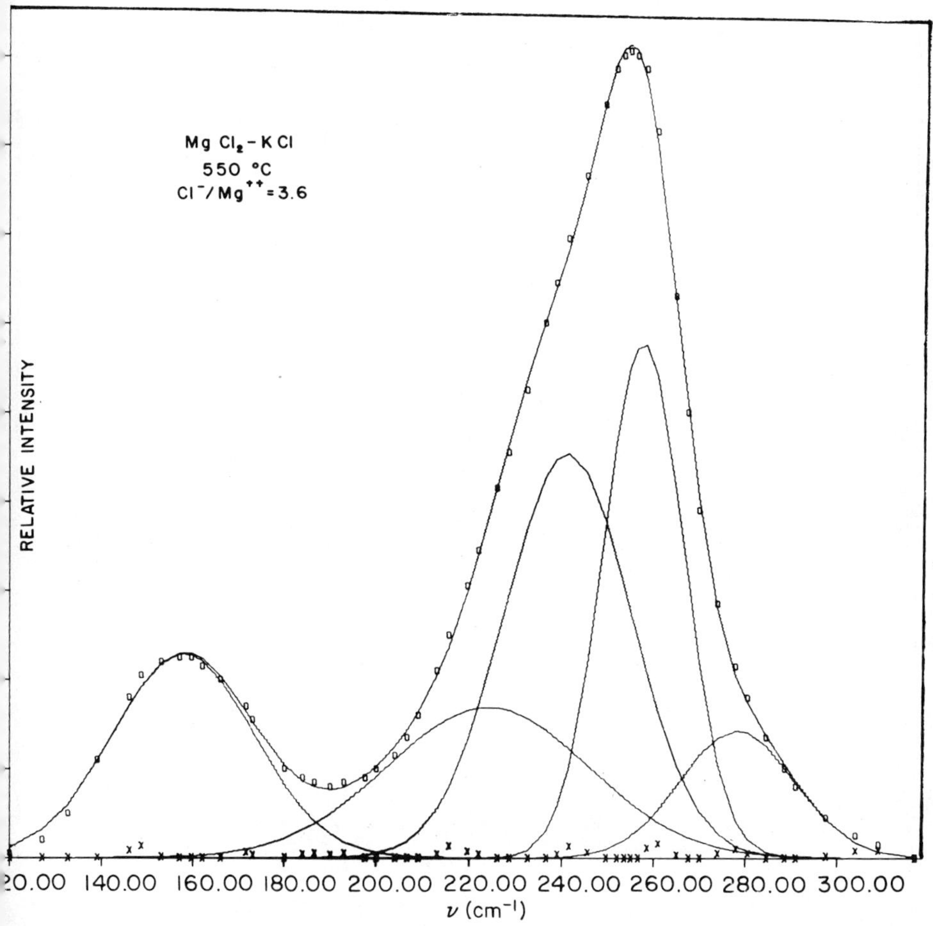

FIG. 7

Resolved spectrum of molten $MgCl_2$-KCl, $Cl^-/Mg^{+2} = 3.6$.

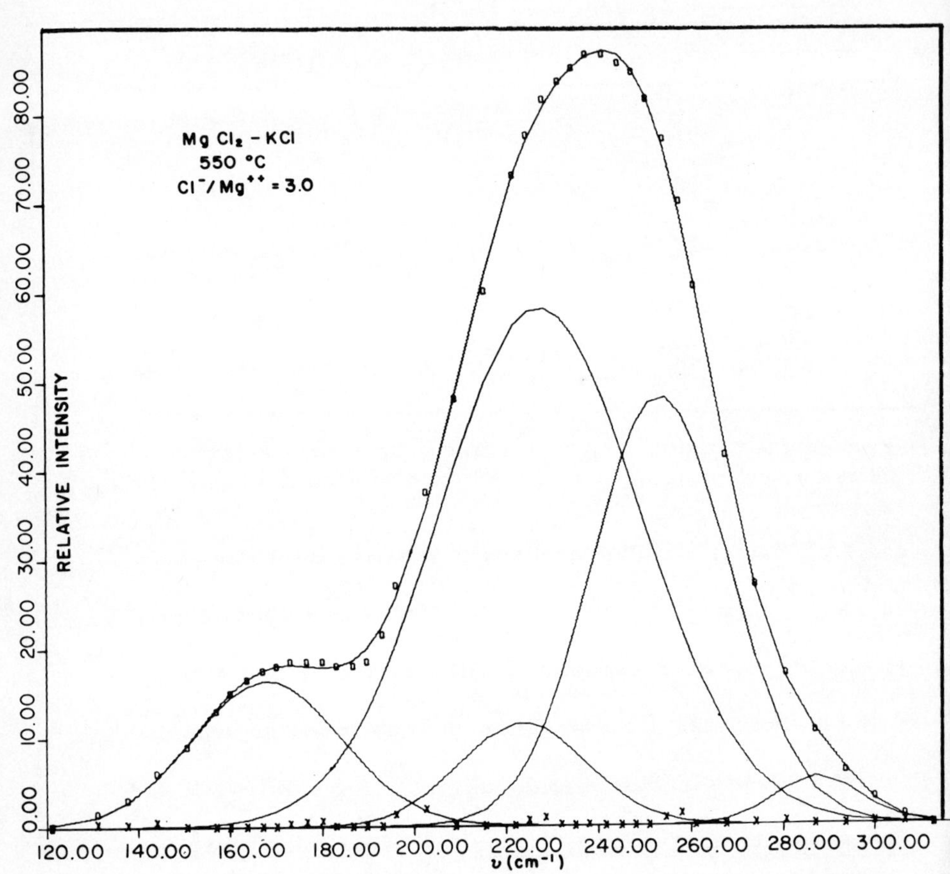

FIG. 8

Resolved spectrum of molten $MgCl_2$-KCl, Cl^-/Mg^{+2} = 3.0.

TABLE 4

Summary of Curve Resolution Data for the $MgCl_2$-KCl System[a]

Cl^-/Mg^{+2} Ratio			Average[b]	Assignment
4.0	3.6	3.0		
283	277	287	282 ± 5, p	ν_1 $MgCl_n^{(2-n)}$
257	257	253	255 ± 2, p	$\nu_1(A_1)$ $MgCl_4^{-2}$
236	239	227	233 ± 6, p	$(MgCl_2)_n$ polymer
218	220	224	221 ± 3, p	$(MgCl_2)_n$ polymer
158	158	168	163 ± 5, dp	Cl-Mg-Cl bending modes

[a] Frequencies in cm^{-1}, p = polarized, dp = depolarized.
[b] These are the numbers referred to in the text.

233 cm^{-1} band, which decreases in relative intensity as the Cl^-/Mg^{+2} ratio is increased, is attributed to a symmetric stretching vibration associated with residual lattice structure in the melt phase. This assignment is in good agreement with the results of Balasubrahmanyam[25] for solid $MgCl_2$. The strong band at 255 cm^{-1} which increases markedly in relative intensity as the Cl^-/Mg^{+2} ratio is increased is tentatively assigned to $\nu_1(A_1)$ for $MgCl_4^{-2}$. The weak band at 282 cm^{-1} which also appears to increase in relative intensity with increasing Cl^-/Mg^{+2} ratio is attributed to a second species of the type $MgCl_n^{(2-n)}$, $n \neq 4$.

For $MgCl_4^{-2}$ or for any species of the type $MgCl_n^{(2-n)}$ with n > 1 (except linear Cl-Mg-Cl) higher frequency (> 300 cm^{-1}) antisymmetric stretching modes should be obtained in the Raman spectrum, but no depolarized bands above 300 cm^{-1} were observed in this investigation. These antisymmetric vibrations are normally much weaker in intensity than the totally symmetric mode and for this reason may have escaped detection. The predominantly bond-bending Raman-active vibrations also expected for these species may be contained in the broad depolarized band centered at about 163 cm^{-1}.

The 221 cm^{-1} band appears to decrease slightly in intensity relative to the 282 and 255 cm^{-1} bands as the Cl^-/Mg^{+2} ratio is increased. This band may arise from a totally symmetric vibration of briding chlorine atoms in the lattice structure other than the ones associated with the 233 cm^{-1} band. The assignments of the frequencies based on this work are summarized in the last column of Table 4.

The results of this investigation differ from those of the previous Raman study of the $MgCl_2$-KCl system by

Balasubrahmanyam[25] in a number of areas. His spectra show a depolarized band at 450 cm^{-1} of medium intensity relative to the 255 cm^{-1} band; the 450 cm^{-1} band was not observed in this investigation for melts of the same composition. Also, he suggests a pyramidal species $MgCl_3^-$ based on the observation of a second depolarized band at 198-208 cm^{-1} for Cl^-/Mg^{+2} = 4.0 and 3.0. Depolarization studies of the spectrum in Figure 4 and of the other spectra obtained in this investigation do not indicate the presence of a depolarized band near 200 cm^{-1}. Furthermore, he finds the band at 152-158 cm^{-1} to be polarized, but results of the present investigation indicate that this band (163 cm^{-1}) is depolarized.

An attempt was made in this investigation to obtain the Raman spectrum of pure $MgCl_2$ reported by Balsubrahmanyam. The molten $MgCl_2$ sample attacked the quartz container so rapidly at 710°C that it was not possible to obtain a spectrum, and the existence of $MgCl_6^{-4}$ as suggested by Balsubrahmanyam could not be corroborated. It seems unlikely, however, that $MgCl_6^{-4}$ would exist in detectable amounts in pure $MgCl_2$, Cl^-/Mg^{+2} = 2.0, and not be detectable at higher Cl^-/Mg^{+2}

ratios. It is the suggestion of this investigation that the symmetric stretch for $MgCl_4^{-2}$ is located at 255 cm^{-1}; therefore, the frequency of 269 cm^{-1} for the symmetric stretch of $MgCl_6^{-4}$ reported by Balasubrahmanyam seems rather high. The totally symmetric stretching frequencies for a series of complexes, $MX_n^{(q-n)}$, normally decrease in value as n is increased.

Studies of UO_2Cl_2 in $MgCl_2$-KCl-NaCl Eutectic

The Raman spectrum of UO_2^{+2} in $MgCl_2$-KCl-NaCl eutectic and the effect of dissolved UO_2Cl_2 on the magnesium (II)-chloride complex equilibrium have been investigated. The spectrum of 1.5 M UO_2Cl_2 in $MgCl_2$-KCl-NaCl eutectic (uranium/magnesium = 0.18) is shown in Figure 9. The band at 850 cm^{-1} is the totally symmetric stretching vibration characteristic of UO_2^{+2}.[58] The broad band centered near 205 cm^{-1} is assigned to vibrations of the various magnesium(II)-chloride species, but this band is displaced to a significantly lower frequency relative to that for the $MgCl_2$-KCl-NaCl eutectic alone (238 cm^{-1}, see Figure 5). A plot of the magnitude of the displacement from the expected peak

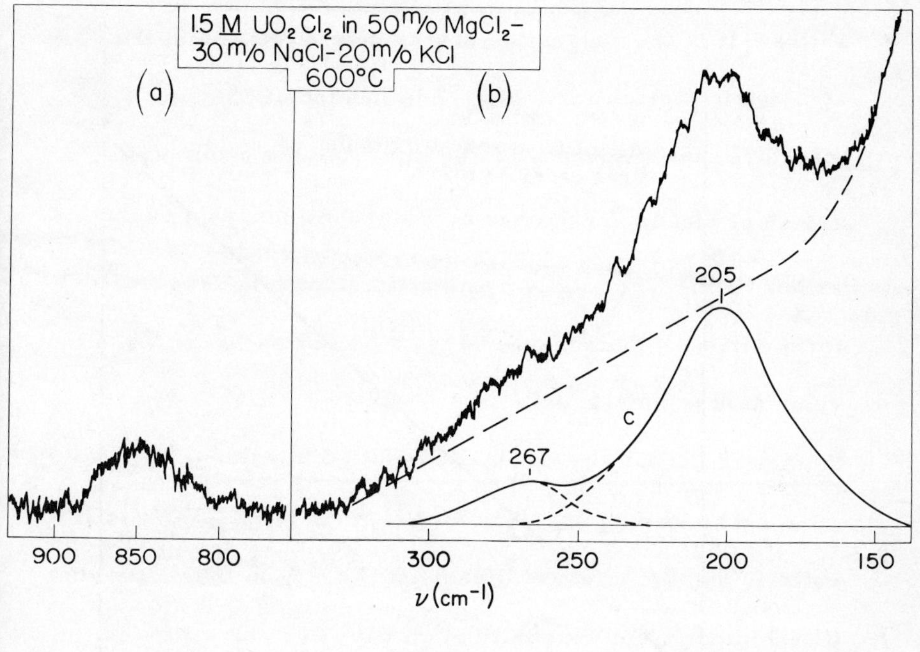

FIG. 9

Raman spectrum of UO_2Cl_2 dissolved in $MgCl_2$-KCl-NaCl eutectic.

position in the absence of UO_2^{+2} for four different uranium/magnesium mole ratios is given in Figure 10. The extent of shift increases linearly as the amount of uranium relative to magnesium is increased. From the direction of the shift (to lower frequency), it appears that UO_2^{+2} is complexing some of those chlorine atoms which give rise to the higher

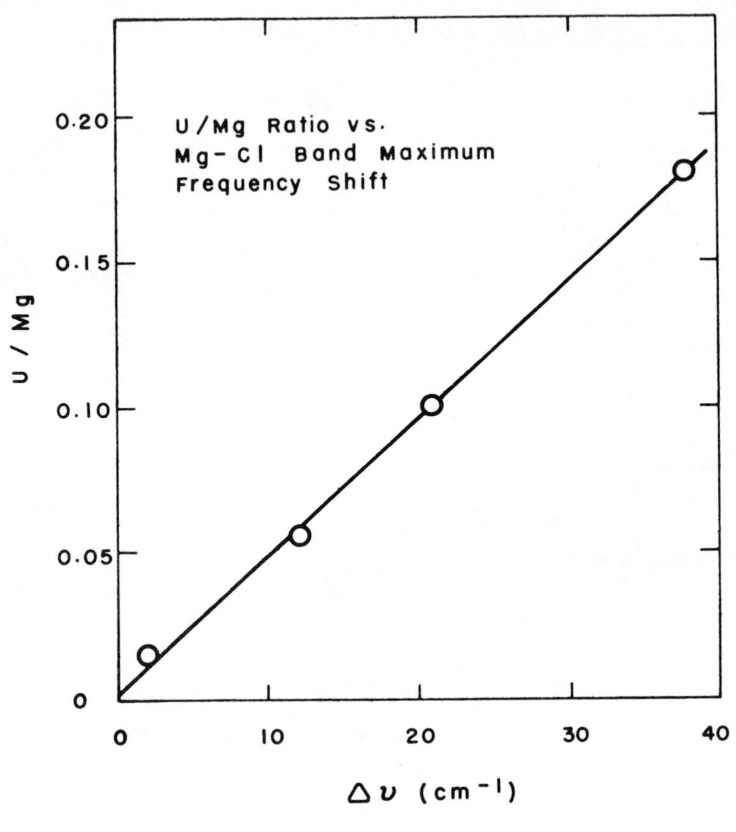

FIG. 10

Plot of the uranium/magnesium mole ratio against the band maximum frequency shift in the 200-300 cm^{-1} region.

frequency vibrations in the 200-300 cm^{-1} region. The band at 267 cm^{-1} is, therefore, assigned to vibrations associated with a structure having uranium-chlorine-magnesium bridges.

THE MAGNESIUM BROMIDE-POTASSIUM BROMIDE SYSTEM

A Raman study of complex formation in $MgBr_2$-KBr melts as a function of the Br^-/Mg^{+2} ratio has been made for compositions in the range $Br^-/Mg^{+2} = 2.5$-5.0. All of the spectra show a strong polarized band in the 125-175 cm^{-1} region and a broad weak band centered near 90 cm^{-1}. The spectrum for an intermediate composition, $Br^-/Mg^{+2} = 3.48$, is reproduced in Figure 11. Curves A and B, taken at 530°C and 410°C, respectively, illustrate the negligible effect of temperature on the band shapes and peak positions. Curve C shows the results of the depolarization experiment. A plot of the peak position of the polarized band for five Br^-/Mg^{+2} ratios is shown in Figure 12. As in the case of the $MgCl_2$-KCl system, the position of this maximum shifts steadily to higher frequency when the halide/magnesium(II) ratio is increased toward 4.0. However, because of the lower melting temperatures in the $MgBr_2$-KBr system, spectra could be taken at compositions up to a Br^-/Mg^{+2} ratio of 5.0.

The data in Figure 12 are interpreted in terms of a magnesium(II)-bromide complex equilibrium involving at least two species. There appear to be two components in the 125-175 cm^{-1} envelope, one at 156 cm^{-1} and a second at a

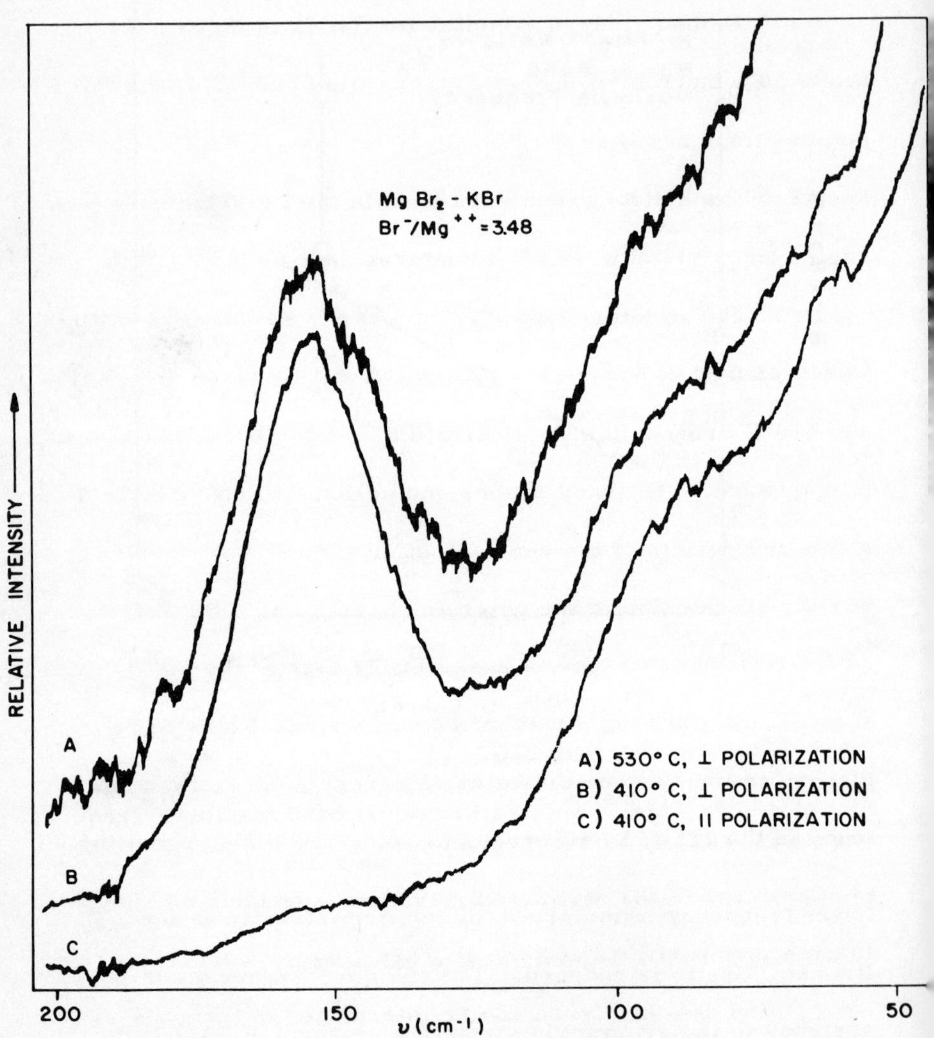

FIG. 11

Raman spectrum of an $MgBr_2$-KBr melt, Br^-/Mg^{+2} = 3.48.

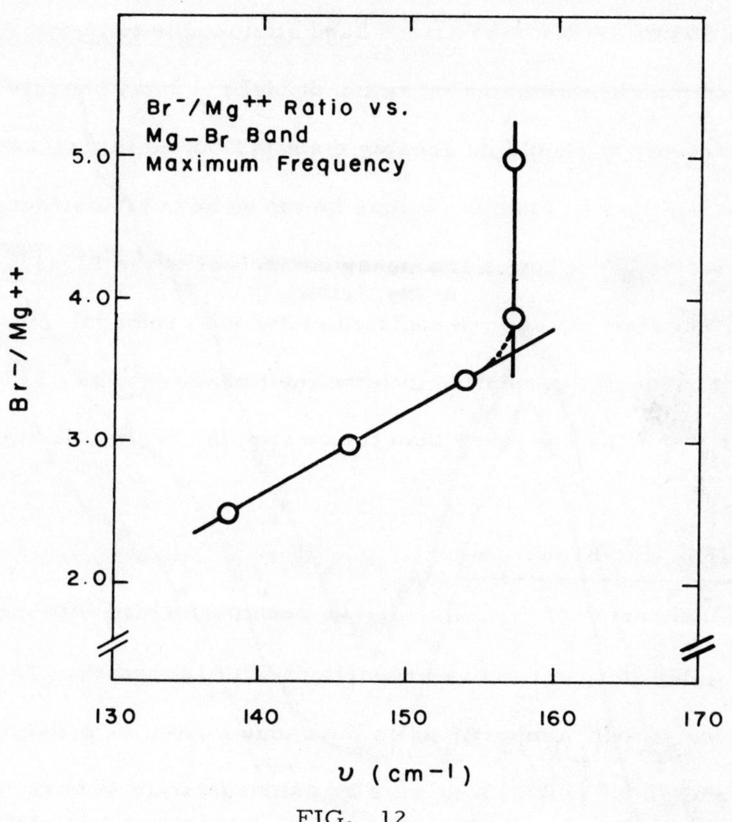

FIG. 12

Plot of the Br^-/Mg^{+2} mole ratio against band maximum frequency in the 125-175 cm^{-1} region.

lower frequency which grows in relative intensity as the Br^-/Mg^{+2} ratio is reduced. The 156 cm^{-1} frequency is assigned to the symmetric stretching motion, $\nu_1(A_1)$, for tetrahedral $MgBr_4^{-2}$. Of the remaining Raman active modes expected for this species, the predominantly bond-bending vibrations could be contained in the depolarized band at about 90 cm^{-1}. No depolarized band attributable to the antisymmetric stretching vibration of $MgBr_4^{-2}$ was observed. This frequency should be greater than ν_1, but as in the case of the $MgCl_2$-KCl studies, it may be too weak in intensity to be detected. The lower frequency component of the 125-175 cm^{-1} envelope could be associated either with residual lattice structure persisting into the melt phase or with $MgBr_n^{(2-n)}$ species other than the one producing the 156 cm^{-1} band.

The HgI_2-LiI-KI System

A series of experiments has been performed with melts containing HgI_2 in various proportions with LiI and KI. The spectra of four compositions in the range I^-/Hg^{+2} = 2.0-7.0 are shown in Figure 13. Curve a is the spectrum of pure HgI_2, and curves b, c, and d are the spectra of melts with various I^-/Hg^{+2} ratios. The proportions of LiI, KI, and HgI_2 in these melts were adjusted to give samples of the desired I^-/Hg^{+2} ratio which melted below the boiling point of pure HgI_2, 354°C.

The spectrum of pure HgI_2 contains a single polarized band at 143 cm^{-1} which is assigned to ν_1 for the linear triatomic HgI_2 molecule with $D_{\infty h}$ symmetry. This band has been observed at 155 cm^{-1} in the gas phase Raman spectrum of HgI_2,[53] at 146 cm^{-1} in Raman spectra of melts containing mixed mercuric halides,[23] and at 138 cm^{-1} in a previous study of molten HgI_2.[24] At I^-/Hg^{+2} = 2.7, the maximum position of this band is at 136 cm^{-1}, and at I^-/Hg^{+2} = 4.5, the maximum has shifted to 131 cm^{-1}, but no new bands are observed in either spectrum. At I^-/Hg^{+2} = 7.0, the maximum position is 126 cm^{-1}, and a second weak band appears at 185 cm^{-1}.

The above results are explained by assuming gradual formation of planar HgI_3^- as the I^-/Hg^{+2} ratio is increased. The totally symmetric vibration, $\nu_1(A_1')$, of the HgI_3^- species begins to grow in on the low frequency side of the 143 cm^{-1} HgI_2 fundamental causing this band to broaden and shift to lower frequency. At I^-/Hg^{+2} = 7.0, there is sufficient HgI_3^- formed so that the antisymmetric stretching motion, $\nu_3(E')$, at 185 cm^{-1} becomes detectable. The $\nu_4(E')$ vibration, also expected in the Raman spectrum of planar HgI_3^-, is probably buried in the Rayleigh scattering.

The formation of HgI_3^- in other media containing HgI_2 and added iodide has been demonstrated by previous investigators. A study of the ultraviolet spectra of HgI_2 in a variety of solvents[54] concluded that HgI_3^- forms readily in the presence of excess iodide ion. Conductance measurements using the method of continuous variations[55] have determined that the stoichiometry for the species present in a tributyl phosphate (TBP) solution of HgI_2 and LiI in a 1:1 mole ratio is HgI_3^-. Raman spectra of TBP solutions of HgI_2 have also been obtained.[56] A TBP solution of pure HgI_2 contained one strong polarized band at 148 cm^{-1} which was attributed to ν_1 for the HgI_2 molecule. For a TBP solution of HgI_2 with LiI added in a 1:1 mole ratio, however, a single band was observed at 125 cm^{-1}, and this band was assigned to the ν_1 fundamental of HgI_3^-.

An alternative assignment of the 185 cm^{-1} band to the antisymmetric stretching vibration, ν_3, for linear HgI_2 is not likely, since this vibration has been observed[57] in the gas phase infrared spectrum of HgI_2 at 237 cm^{-1}. Also, both ν_2 and ν_3 for an unperturbed linear triatomic species with $D_{\infty h}$ symmetry are strictly Raman inactive. Melveger, et al.[24] claim the observation of ν_2 in the Raman spectrum of pure molten HgI_2 at 41 cm^{-1}, and attribute the selection rule violation to environmental effects, but this band is not apparent in curve a of Figure 13.

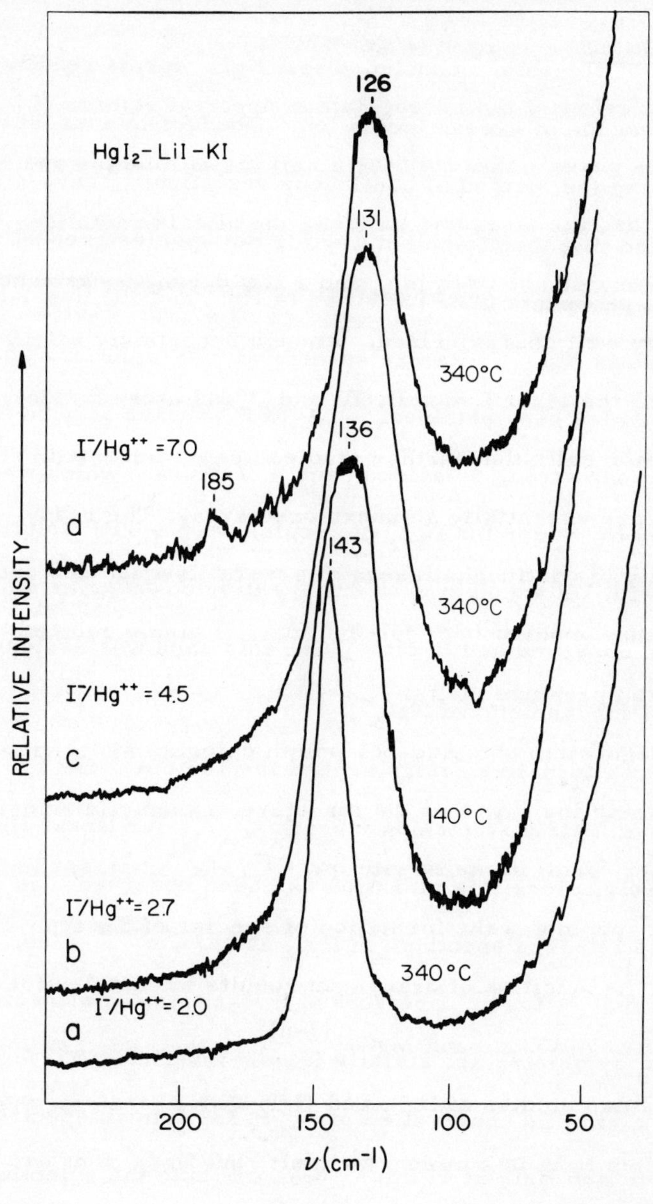

FIG. 13

Raman spectra of HgI_2 containing melts at four I^-/Hg^{+2} ratios.

CONCLUSIONS

A review of the recent Raman spectral studies of fused salts reveals that only the alkali metal nitrates and thiocyanates, the mercuric halides, the aluminum halides, gallium chloride and bromide, and a few oxyanions have been reasonably well characterized. Enough uncertainty exists concerning the other Group II, III and IV nitrates, halides, and oxyanion salts that further studies seem warranted. The fluorides are essentially an unexplored area. The ready availability of continuous lasers of several frequencies opens up many new opportunities for the study of highly colored and high temperature melts.

Magnesium chloride-potassium chloride melts yield spectra which indicate that the structure of the solid remains (in polymer form) in melts with low Cl^-/Mg^{+2} ratios. Addition of Cl^- promotes the formation of species of the type $MgCl_n^{(2-n)}$. Addition of uranyl ion results in complex formation between UO_2^{+2} and $MgCl_n^{(2-n)}$.

Raman studies of HgI_2 and HgI_2-LiI-KI indicate that pure molten HgI_2 is a molecular melt, but HgI_3^- ions are formed at high I^-/Hg^{+2} ratios.

ACKNOWLEDGMENTS

We thank T. F. Young and E. L. Gasner for their help with the design and construction of the apparatus. The support and encouragement of R. C. Vogel and A. D. Tevebaugh, and the expert technical and editorial assistance of Mrs. Ellen J. Hathaway are gratefully acknowledged.

This work was performed under auspices of the U. S. Atomic Energy Commission.

REFERENCES

1. G. J. Janz and S. C. Wait, Jr., in Raman Spectroscopy, Theory and Practice, H. A. Szymanski, Ed., Plenum Press, New York, 1967.

2. D. W. James, in Fused Salt Chemistry, M. Blander, Ed., Interscience, New York, 1964.

3. G. J. Janz and S. C. Wait, Quart. Rev. (London), 17, 225 (1963).

4. S. P. S. Porto and D. L. Wood, J. Opt. Soc. Amer., 52, 251 (1962).

5. H. Kogelnik and S. P. S. Porto, J. Opt. Soc. Amer., 53, 1446 (1963).

6. S. P. S. Porto and R. C. C. Leite, J. Opt. Soc. Amer., 53, 1503 (1963).

7. J. A. Koningsteinand R. G. Smith, J. Opt. Soc. Amer., 54, 1061 (1964).

8. G. J. Janz, Molten Salts Handbook, Academic Press, New York, 1967.

9. S. C. Wait, A. T. Ward, and G. J. Janz, J. Chem. Phys., 45, 133 (1966).

9a. H. Brintzinger and R. E. Hester, Inorg. Chem., 5, 980 (1966).

10. R. E. Hester and C. W. J. Scaife, J. Chem. Phys., 47, 5253 (1967).

11. Y. Doucet and J. Vallier, C. R. Acad. Sci., Paris, 259, 1315 (1964).

12. E. F. Gross and V. A. Koksova, Akad. Nauk SSSR Pamyati S.U. Vavukiva, 231 (1952).

13. D. E. Irish and T. F. Young, J. Chem. Phys., 43, 1765 (1965).

14. J. R. Moyer, J. C. Evans, and G. Y-S. Lo, J. Electrochem. Soc., 113, 158 (1966).

15. R. B. Ellis, J. Electrochem. Soc., 113, 485 (1966).

16. W. Bues, Z. Anorg. Allg. Chem., 279, 104 (1955).

17. M. Tanaka, K. Balasubrahmanyam and J. O'M. Bockris, Electrochim. Acta, 8, 621 (1963).

18. G. J. Janz and J. D. E. McIntyre, J. Electrochem. Soc., 109, 842 (1962).

19. G. J. Janz and T. R. Kozlowski, J. Chem. Phys., 39, 843 (1963).

20. G. J. Janz and D. W. James, J. Chem. Phys., 38, 902 (1963).

21. G. J. Janz and D. W. James, J. Chem. Phys., 38, 905 (1963).

22. G. J. Janz, C. Baddiel and T. R. Kozlowski, J. Chem. Phys., 40, 2055 (1964).

23. J. H. R. Clarke and C. Solomons, J. Chem. Phys., 48, 528 (1968).

23a. J. H. R. Clarke, C. Solomons, and K. Balasubrahmanyam, Rev. Sci. Instrumen., 38, 655 (1967).

24. A. J. Melveger, R. K. Khanna, B. R. Guscott, and E. R. Lippincott, Inorg. Chem., 7, 1630 (1968).

25. K. Balasubrahmanyam, J. Chem. Phys., 44, 3270 (1966).

26. J. H. R. Clarke and C. Solomons, J. Chem. Phys., 47, 1823 (1967).

27. K. Balasubrahmanyam and L. Nanis, J. Chem. Phys., 40, 2657 (1964).

28. K. Balasubrahmanyam and L. Nanis, J. Chem. Phys., 42, 676 (1965).

29. L. A. Woodward, G. Garton and H. L. Roberts, J. Chem. Soc., 1956, 3723.

30. L. A. Woodward, N. N. Greenwood, J. R. Hall, and I. J. Worrall, J. Chem. Soc., 1505 (1958).

31. L. A. Woodward and M. J. Taylor, J. Chem. Soc., 407 (1962).

32. J. Vallier and R. Lira, C. R. Acad. Sci., Paris, 259, 4579 (1964).

33. M. A. Bredig and E. R. VanArtsdalen, J. Chem. Phys., 24, 478 (1956).

34. W. Bues, G. Förster and R. Schmitt, Z. Anorg. Allg. Chem., 344, 148 (1956).

35. W. Bues, K. Bühler and P. Kuhnle, Z. Anorg. Allg. Chem., 325, 8 (1963).

36. G. E. Walrafen, J. Chem. Phys., 43, 479 (1965).

37. G. E. Walrafen, D. E. Irish and T. F. Young, J. Chem. Phys., 37, 662 (1962).

38. J. Vallier, J. Chim. Phys., 63, 1530 (1966).

39. C. B. Baddiel and G. J. Janz, Trans. Faraday Soc., 60, 2009 (1964).

40. H. Gerding and E. Smit, Z. Phys. Chem. (Leipzig), B50. 171 (1941).

41. E. V. Pershina and Sh. Sh. Raskin, Optika i Spectroskopiya, 13, 488 (1962).

42. E. L. Gasner and T. F. Young, in Argonne National Laboratory Chemical Engineering Division Semiannual Report, July-Dec., 1966, ANL-7325, p. 184.

43. E. L. Gasner, E. J. Cairns, and E. J. Hathaway, in Argonne National Laboratory, Chemical Engineering Division Semiannual Report, Jan.-June, 1967, ANL-7375, p. 168.

44. E. J. Cairns, V. A. Maroni, E. J. Hathaway and K. A. Davis, in Argonne National Laboratory, Chemical Engineering Division Semiannual Report, July-Dec., 1967, ANL-7425, p. 181.

45. L. S. Arighi and M. V. Evans, Appl. Spectrosc., 21, 43 (1967).

46. I. R. Beattie, Chem. Brit., 3, 347 (1967).

47. L. M. Toth, Oak Ridge National Laboratory Report No. ORNL-TM-2047, November, 1967.

47a. C. Solomons, J. H. R. Clarke, and J. O'M. Bockris, J. Chem. Phys., 49, 445 (1968).

47b. J. P. Young, Anal. Chem., 36, 390 (1964).

48. P. W. Huber, E. V. Potter, and H. W. St. Clair, U. S. Bureau of Mines, Report of Investigation No. 4858, 1952.

49. E. G. Schrier and H. M. Clark, J. Phys. Chem., 67, 1259 (1963).

50. S. Karpachev and A. Stromberg, Z. Anorg. Allg. Chem., 222, 78 (1935).

51. O. G. Desyatnikov, J. Appl. Chem. USSR (English Transl.), 29, 945 (1956).

52. H. Flood and S. Urnes, Z. Elektrochem., 59, 834 (1955).

53. H. Braune and G. Engelbrecht, Z. Phys. Chem. (Leipzig), B19, 303 (1932).

54. T. R. Griffiths and M. C. R. Symons, Trans. Faraday Soc., 56, 1752 (1960).

55. H. Specker, E. Jackwerth, and H. G. Kloppenburg, Z. Anal. Chem., 183, 81 (1961).

56. E. L. Short, D. N. Waters, and D. F. C. Morris, J. Inorg. Nucl. Chem., 26, 902 (1964).

57. W. Klemperer, J. Chem. Phys., 25, 1066 (1956).

58. G. K. T. Conn and C. K. Wu, Trans. Faraday Soc., 34, 1483 (1938).

59. Engelhard Industries, Inc., (Newark, N. J.) No. 5 Liquid Bright Platinum Can Be Used For Platinizing Quartz.

THE INVESTIGATION OF FUSED SALTS BY THE INELASTIC SCATTERING OF COLD NEUTRONS

J. K. Wilmshurst and J. M. Bracker

Nuclear Engineering Department
Brookhaven National Laboratory
Upton, New York 11973

INTRODUCTION

Considerable progress has been made in the past decade in the investigation of the vibrational spectra of fused salts.[1-6] However, due to the rather formidable experimental difficulties encountered in optically sampling these often very corrosive liquids at the high temperatures necessarily involved, all investigators to date have chosen to study either the infrared spectra[3-6] only or alternatively the Raman effect.[1-2] Thus, for most salts, only either infrared or Raman data are available and accordingly it is impossible to ascertain whether each technique individually gives the complete vibrational spectrum or whether there might in fact be some selection rules operative. Only in the case of the nitrates,[3,7] zinc chloride,[1,8-11] and a few zinc chloride—potassium chloride[1,8,10,11] and zinc chloride–lithium chloride[8,10] mixtures have both infrared (to 200 cm^{-1}) and Raman data been reported. For the internal modes of the nitrates, the selection rules based on D_{3h} point symmetry for a free nitrate ion appear to have been relaxed; thus, the a_1' symmetrical stretching mode, normally inactive in the infrared, is observed in the infrared spectra of all nitrates[3] studied, whilst the a_2'' bending mode, normally Raman inactive, is observed in the Raman spectrum of silver nitrate.[7] In the case of zinc chloride and its mixtures with potassium and lithium chloride there does not appear to be any obvious agreement between the infrared[10] and Raman data[1,8,9,11] and one must therefore conclude that some type of selection rule is indeed operative in these systems.

The inelastic scattering of cold neutrons presents an attractive approach to the study of the vibrational spectra of fused salts for three reasons: (1) no selection rules are operative and the complete vibrational spectrum should therefore be obtained; (2) sampling is much simpler than in the infrared experiment and compares favorably with that in the Raman experiment, and (3) the technique is especially useful for the study of low frequency vibrations ($<800\ cm^{-1}$) and represents a welcome adjunct to both the Raman effect, where background scattering is so bad in the vicinity of the exciting line, and to the infrared data which at present extend only as far as 200 cm^{-1}. In this paper therefore, we will discuss some of the experimental aspects of neutron scattering and illustrate its usefulness to the study of fused salts by presenting data on the inelastic scattering of cold neutrons from molten zinc chloride.

EXPERIMENTAL

The cold neutron facility at the Brookhaven Graphite Research Reactor utilizing the filter-chopper technique coupled with a time-of-flight measurement was used for the present investigation. Although this technique is well known to all neutron workers, and the Brookhaven equipment has been described previously,[12] a short description is given here for completeness.

The experimental apparatus is shown schematically in Figure 1. Neutrons, generated by thermal fission of U^{235}, are moderated or "thermalized" by the graphite lattice and assume a Maxwellian velocity distribution corresponding to the temperature of the moderator. This thermal beam, taken

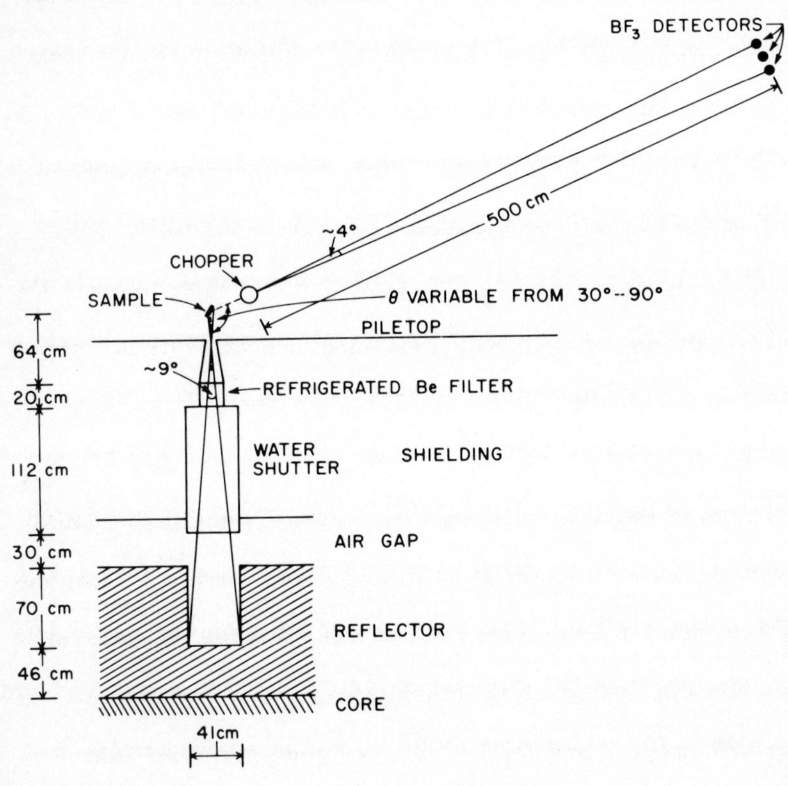

FIG. 1

Schematic diagram of the cold neutron facility at the Brookhaven Graphite Research Reactor.

from a hole in the reflector, is passed through a polycrystalline block of beryllium which acts as a low pass filter, those neutrons having a DeBroglie wavelength ($\lambda = h/mv$) shorter than a certain value given by

$$\lambda = 2d_{max}$$

where d_{max} depends upon the crystal structure, being Bragg scattered and absorbed in the cadmium shielding on the vertical faces of the beryllium block. Neutrons having a wavelength longer than this critical value, which for beryllium at liquid nitrogen temperature is 3.95Å, are transmitted with virtually no loss. The filtered neutron beam is now scattered from the sample, passes through a rotating chopper consisting of a set of 34 cadmium-plated sheets spaced 0.06 in. apart and then travels five meters to a set of three banks of boron trifluoride detectors. A magnetic pickup on the chopper provides a trigger for the time base on a multichannel analyzer, actuating the analyzer time sweep each time the chopper opens. Thus, we measure the time taken for the neutron to travel from the chopper to the detector (500 cm) and hence determine its velocity. The three banks of detectors are each provided with time delay circuits thereby allowing some "time focusing" and consequent improvement in the collimation of the flight path from $\approx 4^{\circ}$ to $\approx 1^{\circ}$.

The time resolution, δt, of the equipment is determined by the burst time of the chopper, $t_B = d/R\omega$, where d is the height of the chopper slit, 2R its length, and ω is its angular speed of rotation; the time broadening $t_d = \phi/\omega$, due to the angle ϕ subtended by the detector at the chopper, and the channel width Δt of the multichannel analyzer, as[13]

$$\delta t = \left\{ \left[t_B^2 + t_\phi^2 + (2\Delta t)^2 \right] / 24 \right\}^{\frac{1}{2}}$$

For the present experimental conditions the time resolution is calculated to be $\approx 30\ \mu$sec.

The action of the beryllium filter is shown in Figure 2. Besides the sharp cut-off at 3.95Å additional cut-offs also occur at 3.58Å due to beryllium and at 4.05Å due to aluminum in the neutron beam. These cut-offs, shown in Figure 2, were obtained by elastic scattering of the neutron beam from vanadium and are somewhat "rounded-off" due to the finite resolution of the equipment. Effectively then, we may consider the filtered neutron beam as being nearly monochromatic having a mean energy of ≈ 0.0048 eV corresponding to a wavelength of ≈ 4.11Å and a velocity of 0.9625×10^5 cm sec^{-1} with a half-band width of ≈ 0.56Å (≈ 10 cm^{-1}). Obviously only inelastic scattering processes that provide energy gain for the neutron can be observed with this experimental arrangement.

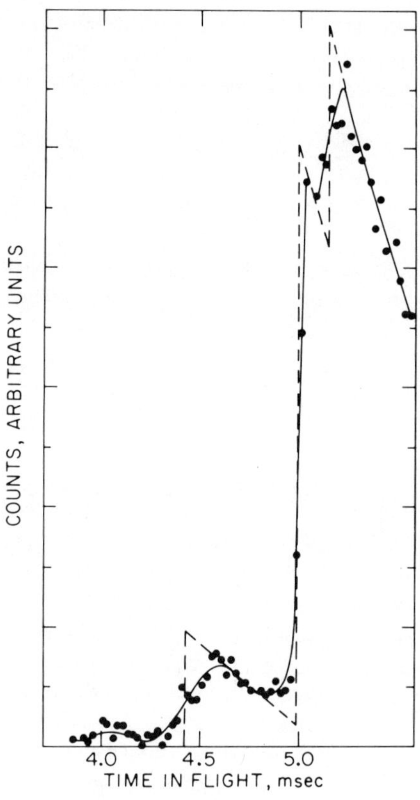

FIG. 2

Incident beryllium filtered neutron spectrum as observed by elastically scattering the incident beam through 90° from a vanadium sample.

Anhydrous zinc chloride obtained from Mallinckrodt Chemical Company was melted under vacuum, treated with dry hydrogen chloride for 24 hours and filtered. The solid material was transferred to the fused silica sample cell shown in Figure 3 using a glove box, and then melted and kept molten under

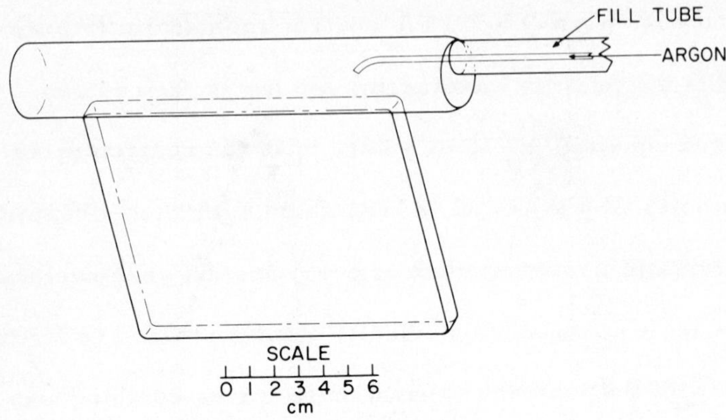

FIG. 3

The fused quartz sample cell used to contain the molten salt under an atmosphere of dry argon.

a flow of dry argon. The sample cell was contained in a conventional furnace provided with ports at the correct angles to allow passage of neutrons without scattering or capture by the furnace insulating material. To further prevent any unwanted background scattering the walls of these ports were lined with boron or boron nitride. The sample measured ≈12x10x0.5 cm and was studied in an essentially reflective arrangement.

THEORY

A neutron may be considered as having a wavelike nature and accordingly we may write a wavefunction such as

$$\chi_o = \exp[i(\underline{k_o} \cdot \underline{r} - \omega_o t)]$$

to describe it, where $\hbar k_o$ is the neutron momentum (= $2\pi/\lambda$, where λ is the neutron wavelength) and $\hbar\omega$ is the neutron energy and equals $\hbar^2 k_o^2/2m$, where m is the neutron mass. If now a collection of nuclei is irradiated with such a beam of monochromatic neutrons, then at every instant each nucleus becomes the source of a spherically scattered wave coherent in phase with the incident wave. These waves combine and interfere producing, at large distances from the scattering sample, a total wave field representing the outgoing neutrons of various energies and momenta. If the nuclei are fixed, the frequency of every scattered wave is the same as that of the incident wave and elastic scattering results. If, however, the nuclei are moving, Doppler shifts change the frequency and scattered neutrons will be found having momenta changed by $\hbar\underline{Q}$ and energy changed by $\hbar\omega$ from the incident neutrons, these changes corresponding to changes in the dynamical state of the scattering system.

In the present experiment, the incident neutrons have a wavelength of ≈ 4Å corresponding to a velocity of $\approx 10^5$ cm sec^{-1} and accordingly the time taken for a neutron to travel between neighboring nuclei spaced ≈ 3Å apart is $\approx 3 \times 10^{-13}$ sec which is comparable to the period of oscillation of the nuclei. Again since these incident neutrons have an energy of ≈0.005 eV, interactions with the vibrating system will produce relatively large charges of energy and momenta. (These results may be compared with those for x rays where the similar observation time would be $\approx 10^{-18}$ sec and the vibrational energy is extremely small compared to the photon energy.)

The scattering cross section for thermal neutrons per unit angle and unit energy range may be written as[14]

$$\frac{d^2\sigma}{d\Omega dE} = \frac{k}{k_o} \sum_{\substack{m,n \\ i,f}} P_i b_m b_n \langle i | \exp[i\underline{Q} \cdot \underline{r} \delta(\underline{r} - \underline{R}_n)] | f \rangle$$

$$\langle f | \exp[-i\underline{Q} \cdot \underline{r} \delta(\underline{r} - \underline{R}_m)] | i \rangle \, \delta(\Delta E + \hbar\omega)$$

where $\hbar k$ is the momentum of the scattered neutron, $\Delta E = (E_i - E_f)$ and $\langle i|$, $\langle f|$ are the wavefunctions of the initial and final states of the scattering system. P_i is the thermal occupation probability of the initial states and b_m and b_n are the scattering lengths of the m and n nuclei. For a regular array of nuclei the states i and f may be associated with wavevectors $\underline{q}$ and $\underline{q}'$ and it is easy to show that for $m \neq n$ (where m and n refer to nuclei in different primitive cells) the scattering cross section is now zero if and only if

$$\underline{q}' - \underline{q} = \underline{Q} + 2\pi\underline{\tau}$$

where $\underline{\tau}$ is a reciprocal lattice vector. This may be considered as an equation expressing momentum conservation. For m = n such momentum conservation does not apply and accordingly we write (considering, for simplicity, a single element only) the differential scattering cross section in two parts, a coherent part

$$\frac{d^2\sigma^{coh}}{d\Omega dE} = \frac{k}{k_o}\,\bar{b}^2 \sum_{\substack{m,n\\ i,f}} P_i \langle i|\exp[i\underline{Q}\cdot\underline{r}\delta(\underline{r}-R_n)]|f\rangle\cdot$$

$$\langle f|\exp[-i\underline{Q}\cdot\underline{r}\delta\ \underline{r}-\underline{R}_m\ ]|i\rangle\ \delta(\Delta E + \hbar\omega)$$

for which both momentum and energy are conserved, on scattering, and an incoherent part

$$\frac{d^2\sigma^{incoh}}{d\Omega dE} = \frac{k}{k_o}(\overline{b^2} - \bar{b}^2) \sum_{\substack{n\\ i,f}} P_i \langle i|\exp[i\underline{Q}\cdot\underline{R}_n\delta(\underline{r}-R_n)|f\rangle\cdot$$

$$\langle f|\exp[-i\underline{Q}\cdot\underline{R}_n\delta(\underline{r}-\underline{R}_n)|i\rangle\ \delta(\Delta E + \hbar\omega)$$

for which only energy is conserved. In these equations

$$\bar{b}^2 = \frac{1}{N^2}\sum_{m,n}^{m\neq n} b_m b_n \quad \text{and} \quad \overline{b^2} = \frac{1}{N}\sum_{m,n} b_m b_n \delta_{mn}$$

The energy conservation on scattering is given by

$$k^2 - k_o^2 = (2m/\hbar)\omega_j(\underline{q})$$

Thus, for incoherent scattering we see that $|k|$ is limited to the range k_o to $k_o + \sqrt{(2m/\hbar)\omega_{max}}$, where ω_{max} is the maximum value of $\omega_j(\underline{q})$ for all j and $\underline{q}$, and the direction of $\underline{k}$ is unrestricted. Thus neutrons are scattered in all directions with energies ranging continuously from 0 to $\hbar\omega_{max}$. For coherent scattering the same energy conservation must be obeyed but in addition momentum must also be conserved as given by the expression

$$\underline{k} - \underline{k}_o = \underline{Q} = 2\pi\underline{\tau} + \underline{q} - \underline{q}'$$

and this places severe restrictions on the direction and magnitude of $\underline{k}$.

For a liquid or polycrystalline material the sample has no preferred orientation and hence the momentum conservation must be written as

$$|Q| = 2\pi|\tau| + |q - q'|$$

Further, the differential scattering cross section for incoherent scattering may be expressed in terms of the distribution of vibrational modes $g(\omega)$ in the sample as

$$\frac{d^2\sigma^{incoh}}{d\Omega d\omega} = (\overline{b^2} - \overline{b}^2)N\frac{k}{k_o}\frac{Q^2}{\omega}\frac{\hbar}{2M}\frac{e^{-2W}}{\exp(\hbar\omega/k_BT) - 1}g(\omega)$$

where W is the Debye-Waller factor, M the mass of the scattering nuclei, and k_B is Boltzmann's constant.

RESULTS

The raw data observed for liquid zinc chloride at 400°C at two different angles of scattering, 90° and 45°, are shown in Figures 4 and 5, whilst a plot of the energy of the scattering intensity maxima [ignoring the apparent fine structure in the lower energy (<130 cm^{-1}) portions of the data] against the corresponding momentum changes is given in Figure 6.

Both zinc and chlorine have approximately equal coherent and incoherent scattering lengths[15] and hence it is extremely difficult to analyze the data other than in a qualitative manner. Some dispersion is obvious in Figure 6 and, if the highest energy component of the pair of broad low energy bands is taken as the longitudinal acoustic mode and we assume a linear dispersion for this mode, the velocity of sound propagation in molten zinc chloride may be calculated to be 535 meters sec^{-1}, the edge of the zone boundary being found at 3.32 $Å^{-1}$. This edge must correspond to an integral multiple of $2\pi\underline{\tau}$, where $\underline{\tau}$ is a reciprocal lattice vector, and thus, depending on whether this is the edge of the second, fourth, or sixth zone the Cl-Cl distance in the melt must be either 1.89Å, 3.79Å, or 5.68Å.

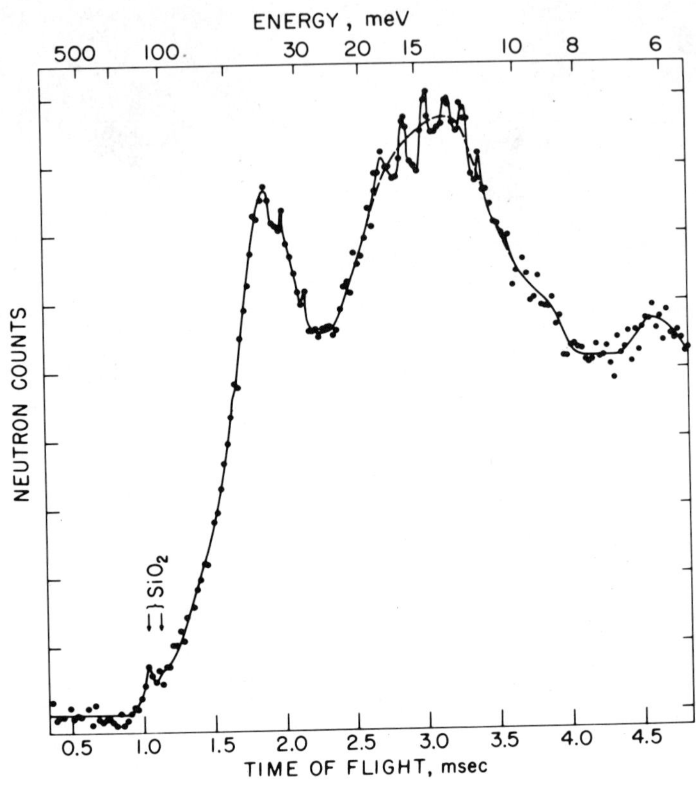

FIG. 4

Time-of-flight distribution of neutrons scattered at an angle of 90° from molten zinc chloride contained in a quartz cell at 400°C.

The Cl-Cl distance[16] in lithium chloride is 3.80Å and in potassium chloride 4.7Å, thus we conclude that the edge above corresponds to the fourth Brillouin zone and the Cl-Cl distance in zinc chloride is 3.8Å.

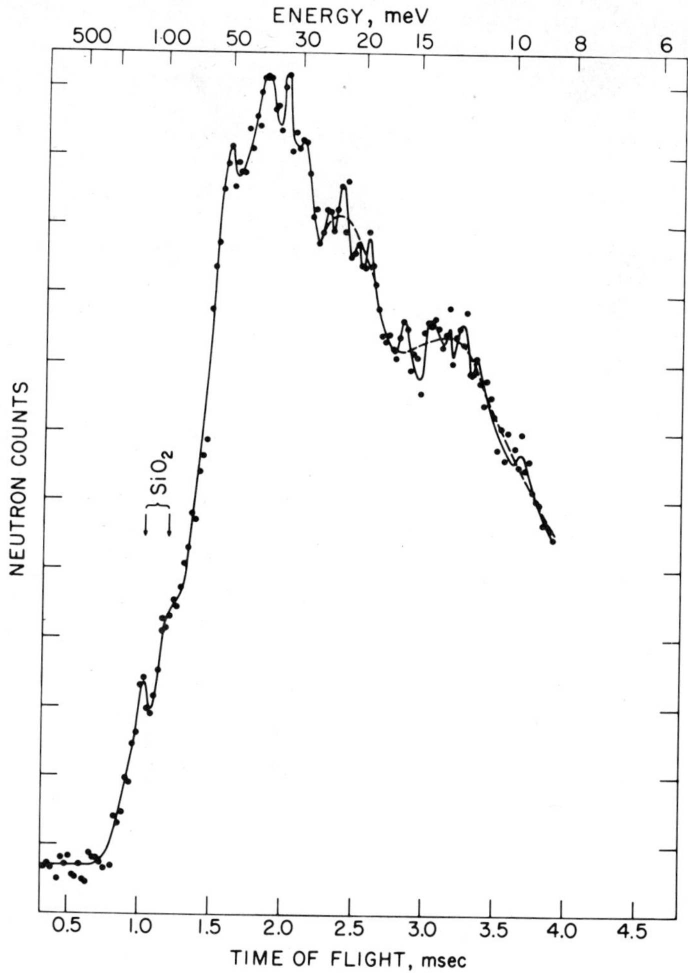

FIG. 5

Time-of-flight distribution of neutrons scattered at an angle of 45° from molten zinc chloride contained in a quartz cell at 400°C.

With both infrared absorption or Raman scattering the momentum change is virtually zero and hence we may include this optical data[10, 11] on Figure 6 by plotting it on a vertical axis at Q = 0, or what is equivalent, at Q = 1.66, 3.32, or 4.98Å^{-1}. If it is assumed that the dispersion curves for all modes other than the acoustical mode have zero gradients at the zone boundaries, simple dispersion curves may be drawn

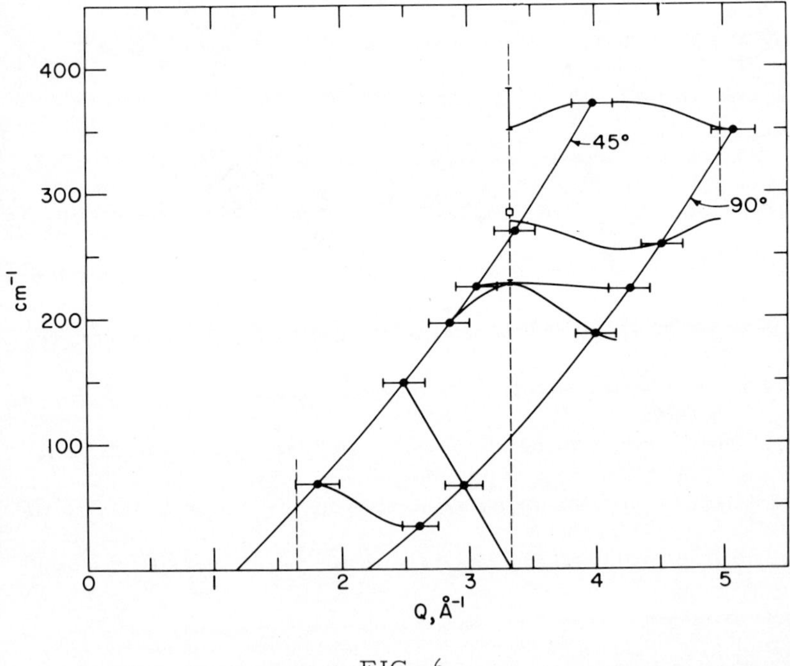

FIG. 6

The frequency versus momentum transfer plot obtained from the neutron spectra of molten zinc chloride at 400°C.

in on Figure 6 as shown. In Figure 7 these dispersion curves are reduced to the first zone.

The lowest frequency mode is assigned to a doubly degenerate transverse mode derived from the transverse acoustic mode in the solid and is likened to a type of swirling motion. The high frequency modes are all assigned to optical modes, the infrared active vibration being a transverse optical mode and the highest mode being its corresponding longitudinal counterpart. The Raman active vibration is here considered triply degenerate at $Q = 0$ splitting into a longitudinal and transverse component at other values of Q. This then accounts for all the nine vibrations expected on the basis of a "primitive," triatomic $ZnCl_2$ entity. It is furthermore consistent with a reasonably symmetrical "coordination" about the zinc atom in which there are two or more distinct three-fold axes (e.g., tetrahedral, octahedral, etc.).

If the separation between the high frequency longitudinal optical mode and the infrared active mode is due solely to the retardation effect (i.e., they would be degenerate if it wasn't for this effect), then they should be related by the Lyddane, Sachs, and Teller relationship[17]

$$\omega_{LO}^2 = (\epsilon_s/n^2)\,\omega_{TO}^2$$

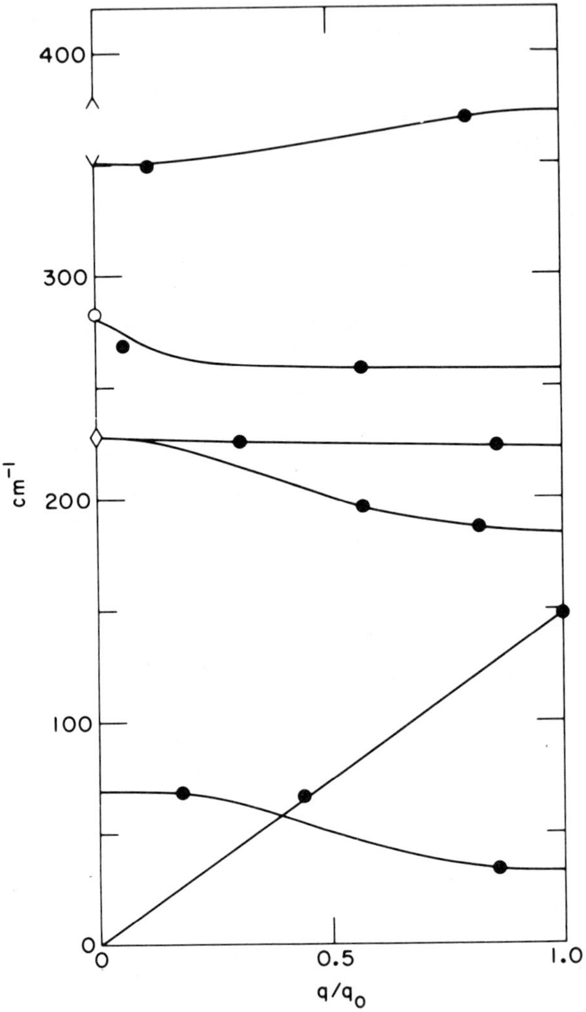

FIG. 7

The average dispersion relations for molten zinc chloride at 400°C.

where n is the refractive index of zinc chloride in the optical region and ϵ_s is the long wavelength dielectric constant. Extrapolation of the infrared refractive index data obtained[10] for $ZnCl_2$ gives n = 1.54 and hence we may now calculate the long wavelength dielectric constant of $ZnCl_2$ to be 3.59.

Recent Raman data[8, 9, 11] have been interpreted as showing the presence of a pair of bands at ≈250↔270 and 305↔310 cm^{-1}. This is to be compared with the early work of Salstrom and Harris[18] who found only one band at 285 cm^{-1}. Analysis of this very weak region of the Raman spectrum is difficult and we prefer to consider just the one band present in this region at 285 cm^{-1} coincident with the infrared active transverse optical mode.

The frequencies for the scattering maxima of the fine structure observed in the low energy region of the scattering spectrum are tabulated in Table 1. Although at present the origin of this fine structure is unclear, we believe it is real and is associated with the transverse acoustic mode. The intensity pattern of the fine structure is typical of rotational fine structure, yet clearly the "line" spacing is much too large to be explained on the basis of any entity involving zinc and chlorine atoms. It is of interest, however, that some of the observed frequencies appear to correlate with bands reported in the Raman effect.

In Figure 8 the ingoing neutron spectrum, scattered elastically by molten zinc chloride at 400°C, is compared with that scattered elastically by vanadium. The absence of any broadening of this ingoing spectrum suggests that either the diffusive process in molten zinc chloride is slow ($>10^{-12}$ sec) or that there is a relatively long time delay ($>10^{-12}$ sec) before the scattering nucleus undergoes diffusion.

TABLE 1

Frequencies of the Fine Structure "Lines" Observed in the Low Energy Region of the Scattering Spectra

Frequency in cm^{-1} at a scattering angle of		Raman data from		
90°	45°	Ref. 8	Ref. 9	Ref. 11
45				
52	54			
59	61			
66	67			
76	74	80		75
89	91	90	95	
104	—		110	
	118			
	129			
	145			
	160			

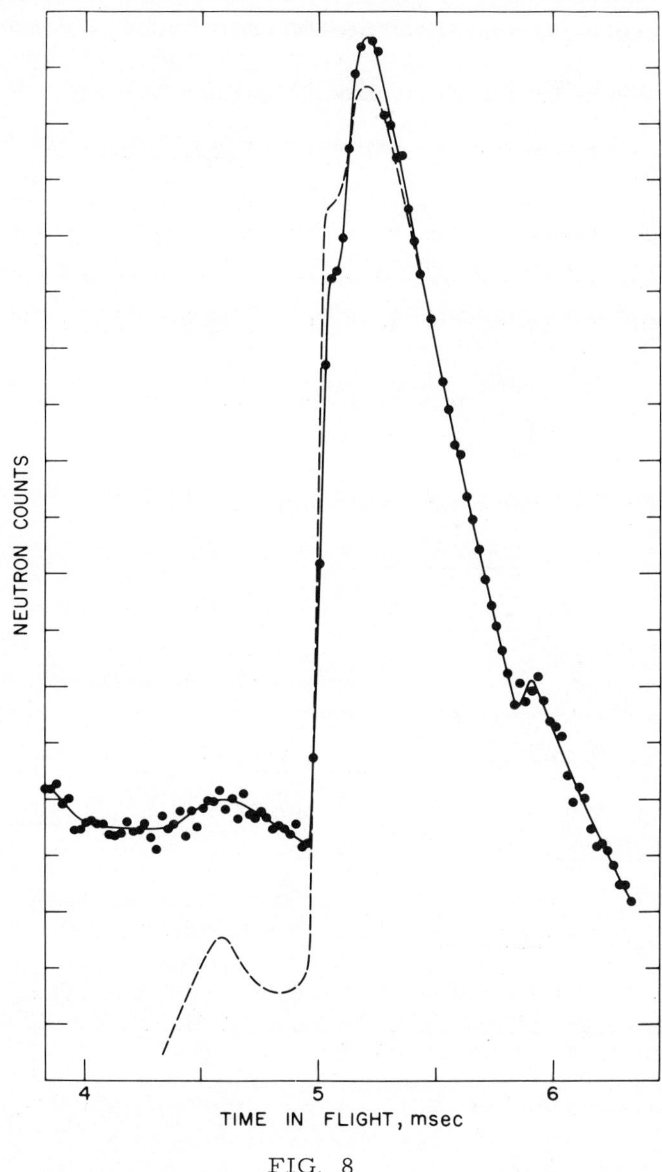

FIG. 8

Comparison of the incident neutron spectrum elastically scattered at 90° from molten zinc chloride at $400^{o}C$ with that scattered elastically at 90° from vanadium.

REFERENCES

1. W. Bues, Z. anorg. u. allgem. Chem., 279, 104 (1955).

2. G. J. Janz and D. W. James, J. Chem. Phys., 35, 739 (1961).

3. J. K. Wilmshurst and S. Senderoff, J. Chem. Phys., 35, 1078 (1961).

4. J. S. Fordyce and R. L. Baum, J. Chem. Phys., 44, 1159 (1966).

5. A. Bandy, J. P. Devlin, R. Burger, and B. McCoy, Rev. Sci. Instr., 35, 1206 (1964).

6. J. K. Wilmshurst, J. Chem. Phys., 39, 2545 (1963).

7. W. Bues, Z. physik. Chem., 10, 1 (1957).

8. R. B. Ellis, J. Electrochem. Soc., 113, 485 (1966).

9. D. E. Irish and T. F. Young, J. Chem. Phys., 43, 1765 (1965).

10. J. K. Wilmshurst, J. Chem. Phys., 39, 1779 (1963).

11. J. R. Moyer, J. C. Evans, and G. Y-S. Lo, J. Electrochem. Soc., 113, 158 (1966).

12. C. M. Eisenhauer, I. Pelah, D. J. Hughes, and H. Palevsky, Phys. Rev., 109, 1046 (1958).

13. P. F. Zweifel and J. M. Carpenter, Proc. Symp. on Inelastic Scattering of Neutrons in Solids and Liquids, p. 199, IAEA, Vienna, 1961.

14. W. M. Lomer and G. G. Low, Thermal Neutron Scattering, p. 16, P. A. Egelstaff, Ed., Academic Press, Inc., London, 1965.

15. D. J. Hughes and R. B. Schwartz, Neutron Cross Sections, BNL 325, U. S. Government Printing Office, Washington, D.C., 1958.

16. H. A. Levy, P. A. Agron, M. A. Bredig, and M. D. Danford, Ann. N. Y. Acad. Sci., 79, 762 (1960).

17. R. H. Lyddane, R. G. Sachs, and E. Teller, Phys. Rev., 59, 673 (1941).

18. E. J. Salstrom and L. Harris, J. Chem. Phys., 3, 241 (1935).

TRANSPORT PROCESSES IN LOW-MELTING MOLTEN SALT SYSTEMS

C. A. Angell and C. T. Moynihan

Department of Chemistry
Purdue University
Lafayette, Indiana 47907

INTRODUCTION

It is not to be expected, for liquids with wide temperature ranges of stability, that a single concept of the molecular mechanism by which transport of mass occurs will suffice to interpret measurements performed over the entire range. It is clear for instance that near the critical temperature all molecular motions will contribute to mass transport and the concept of "activation" will be without value, i.e., the entire system will be "activated." On the other hand, the drastic fall-off in particle mobility encountered at low temperatures, in the same range in which sound velocities and thermal conductivities increase, demonstrate that here the motions leading to mass transport are of an uncommon type. Thus approaches based on calculation of the probability of uncommon energy states or volume fluctuations would be reasonable for the low temperature regime.

It is unfortunate that the melting points of most common molten salts are such that transport studies have generally been performed in a "middle" or overlap region where neither the approaches suitable for the high temperature "gas-like" range nor those designed for the more "solid-like" low temperature region can be expected to apply. Accordingly, it is not too surprising that progress in understanding transport properties of fused salts has been slow. For the most part this paper restricts itself to a discussion of the low temperature region of transport behavior in fused salt systems, the authors feeling justified in this choice by the fact that most molten salt media of interest to analytical and preparative chemists are chosen on the basis of their low-lying liquidus regions. These melts are generally multi-component liquids of about eutectic composition, the transport properties of which usually exhibit departures from the Arrhenius behavior used to characterize the middle temperature region.

The low temperature region of behavior, it must be emphasized immediately, is not bounded by any liquidus line on a phase diagram. Rather, it is to be thought of as extending (metastably) down to a low temperature limit imposed by the glass transition temperature T_g. (There is nothing hypothetical in this statement--most of the common nitrate and, less simply, chloride eutectics may be obtained in the glassy condition by squirting the liquids onto a copper sheet cooled in liquid nitrogen.) According to Gibbs and Dimarzio[1] the true thermodynamic low temperature limit, T_o, lies a little below the experimental glass transition temperature, T_g, but for kinetic reasons cannot be reached in a finite time scale experiment. Because it approximately defines the temperature region of obviously non-Arrhenius transport behavior, the low temperature region has been conveniently, but quite arbitrarily, defined as the region falling between T_o and $2T_o$.[2] Few of the data referred to in this paper will fall far outside this region. To show its relevance to melts of common experience, some phase diagrams are presented in Figure 1 with limits T_o and $2T_o$ demarcated as well as possible from available data.

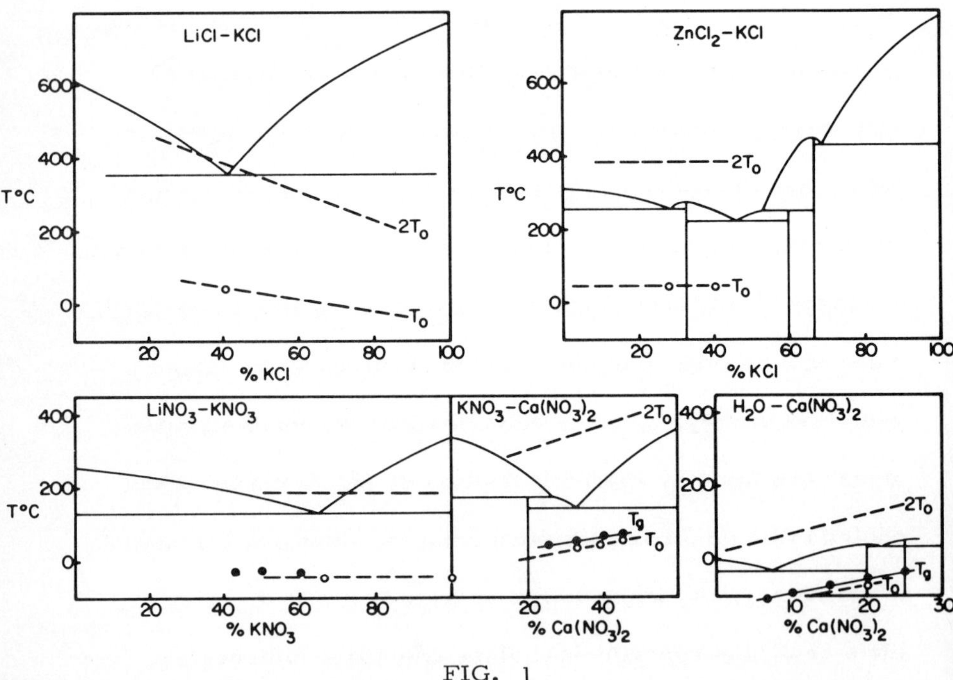

FIG. 1

Low temperature regions, T_o-$2T_o$, in relation to equilibrium lines for common binary systems with low liquidus temperatures.

LOW TEMPERATURE LIQUID STATE BOUNDARY. DISTINCTION BETWEEN LIQUID AND SOLID AMORPHOUS PHASES

If we assert the existence of a low temperature limit on the liquid state which lies above $0^{o}K$, then it is necessary to consider the distinction between the liquid state and the state into which it passes at lower temperatures.

The properties of the "ideal glass" produced by transition of the equilibrium liquid at T_o cannot be ascertained (the concept may even be placing undue weight on the reality of T_o), however, they would evidently not be greatly different from those of the glass produced at the normal experimental glass transition. This transition is characterized by a more or less sudden decrease in the intensive thermodynamic properties heat capacity, expansion coefficient, and compressibility from liquid-like values to values characteristic of the crystalline phase of the substance. In the temperature region where these changes occur, the viscosity increases rapidly, but not discontinuously, to values in the vicinity of 10^{15} poise, at which point the material is to all ordinary tests a solid.

The quasi-thermodynamic attributes of the glass transition are not generally appreciated outside glass and polymer science circles. It must be emphasized that changes in thermodynamic properties do not occur merely because the viscosity becomes high. Rather the changes must occur in order to avert the occurrence of a thermodynamic catastrophe (the imminence of which was first pointed out by Kauzmann),[3] viz., the production of an amorphous phase of lower entropy than the crystalline phase of the same substance at the same temperature. The fact that the occurrence of the changes is coupled closely with the development of a high viscosity permits us to learn something important about the basis of viscous flow, and, by association, of transport processes in general.

The thermodynamic necessity for the glass transition, or its thermodynamic equivalent, is illustrated here for the cases of $ZnCl_2$ and the simple molecular fluid, 2-methylpentane, for which more complete thermodynamic data are available. In Figure 2 the heat capacities of the liquid and crystalline phases of $ZnCl_2$[4] are shown as a function of temperature using a logarithmic temperature scale. Thus the entropy of the phase is represented directly by the area under the appropriate curve plus any entropy increments introduced during first order phase changes. In the case of

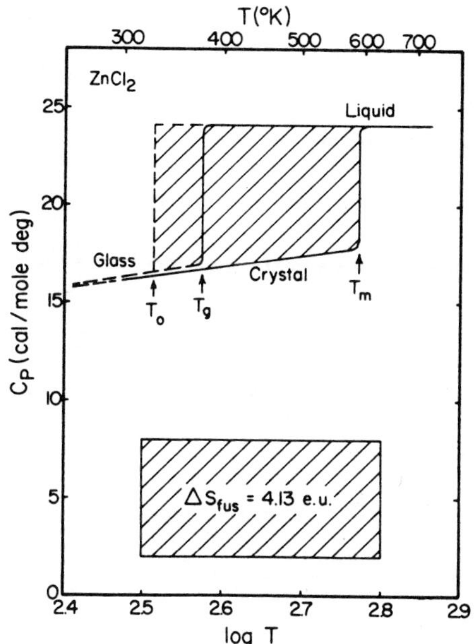

FIG. 2

Liquid and crystal heat capacity-temperature curves for $ZnCl_2$, showing estimation of T_o from thermodynamic data.

$ZnCl_2$, as far as is known, the only such increment is the entropy of fusion, ΔS_{fus}, and this is shown by the separate shaded area at the bottom of the figure. The Kauzmann paradox can be described as follows.

Since the liquid has a considerably higher heat capacity than the solid phase, the liquid loses entropy at a greater rate than does the solid when both are cooledover the same temperature interval. Quantitatively, the difference between the entropy losses of the liquid and crystalline phases during cooling over a particular temperature interval is given by the area in Figure 2 between the liquid and crystal heat capacity curves, and the vertical lines defining the temperature interval in question. For example, between the melting point, 591°K (318°C), and T_g, 376°K (103°C), the (now supercooled) liquid loses about 3 cal/mole-deg more entropy than the crystal, as can be seen from the fact that the area between the liquid and crystal curves in this temperature interval is about 3/4 the area representing the entropy of fusion. The heat capacity of the equilibrium[5] liquid phase cannot be determined below about 100°C in feasible experiments because a non-equilibrium (rate dependent) phenomenon, the glass transition, occurs and the heat capacity changes value. Note however that the liquid heat capacity is independent of temperature over the entire range in which it can be measured. (For some liquids this may be many hundreds of degrees). At this point the problem becomes clear: if the equilibrium liquid were to continue to lose entropy at the rate observed above $100^{\circ}C$, then by the time a temperature of $336^{\circ}K$ (marked T_o) was reached an excess entropy increment equal to the entropy of fusion would have been lost by the liquid (shaded area in Figure 2). Any further cooling would then result in an amorphous $ZnCl_2$ phase with a total entropy less than that of the crystalline phase at the same temperature--which, barring some extraordinary, loose crystal structure, is inconceivable. In most cases continued cooling at the liquid entropy loss rates would even yield negative entropies before $0^{\circ}K$ is reached.

The paradox, therefore, is that it appears that only the timely intervention of a non-equilibrium phenomenon spares us the trauma of an observation which would strike at the foundations of thermodynamics. We believe in the firmness of these foundations, however, and since we recognize the possibility in principle of performing a cooling experiment at an infinitely slow rate, it is clear that the equilibrium heat capacity of the liquid $ZnCl_2$ must change value at some temperature not less than $336^{\circ}K$.

It cannot of course be said that the decrease of C_p at equilibrium would be more or less sharp than that of the non-equilibrium transition observed at T_g. On the other hand it should be noted from Figure 2 (and this seems to be a general finding) that the difference between liquid and glass heat capacities is increasing rather than decreasing at the lowest temperatures where equilibrium behavior can be observed, so a rather sudden decrease at equilibrium seems inevitable. The retention of the term "transition" for the equilibrium phenomenon thus seems appropriate. For the transition to be thermodynamically second order, as required by the lattice model of Gibbs and Dimarzio[1] in which the configurational entropy vanishes at T_o, the change in C_p at the transition temperature must be discontinuous. However, except from theory, the order of the transition cannot be ascertained, since behavior at T_o can never be observed.

In the case of 2-methylpentane, two different assessments of the temperature before which C_p must change can be made since in this case not only the entropy of fusion but also the residual entropy at 0^oK has been determined.[6] In the estimation of T_o for $ZnCl_2$, for which the residual entropy is unknown, we assumed that the entropy of fusion was all configurational. However if part of the fusion entropy is vibrational (as might be expected from the decreased instantaneous "lattice" rigidity)[7] then our procedure of finding T_o by equating area between liquid and crystal heat capacity curves to the area representing the entropy of fusion would place T_o too low. The residual entropy determined calorimetrically, on the other hand, is a direct measure of the configurational entropy frozen in at T_g. Therefore placing an area equal to the residual entropy between crystal and extended liquid heat capacity curves starting at T_g should give a more accurate estimate of T_o.

In Figure 3 these alternatives are shown. It turns out that in this case T_o estimated from the entropy of fusion is the same as T_o estimated from the residual entropy within one or two degrees and equal to about 58^oK.

Other examples are possible, but it should be clear from the two discussed that the equilibrium phase tends, with falling temperature, to a configurational ground state from which even in principle no further entropy can be lost by configurational changes.

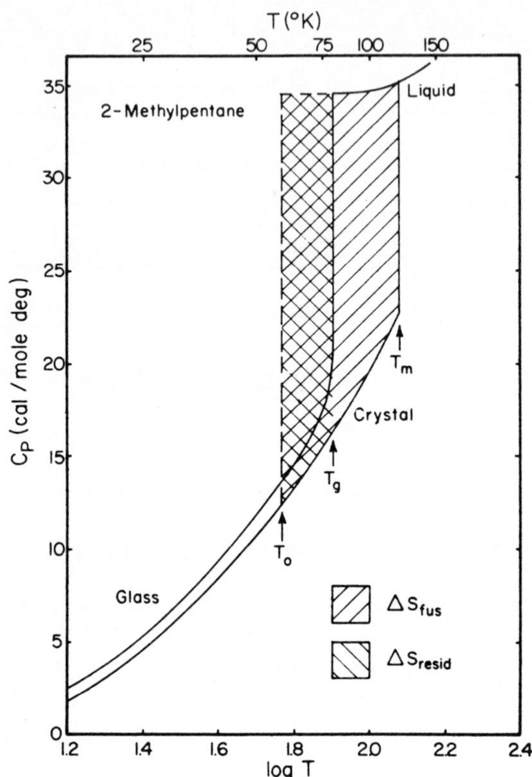

FIG. 3

Liquid, crystal and glass heat capacity curves for 2-methyl pentane showing estimate of T_o from entropy of fusion and residual entropy data.

It is the arrival at this state which, in principle, terminates the liquid regime for an amorphous phase. This will usually be associated with the falling of temperature below a characteristic value, but can equally well be conceived as being reached by rise in pressure above a characteristic value at constant temperature or by change of composition beyond a characteristic value at constant temperature and pressure (for multicomponent systems.)

TRANSPORT PROPERTIES AND THE COULOMB POTENTIAL. THE QUESTION OF CORRESPONDING TEMPERATURES

In the previous section we showed how thermodynamic data on molten salts and other liquids point to the existence, in principle, of a ground state for amorphous packing. In this section we consider the temperature characteristic of this ground state in the context of the general problem of establishing a meaningful corresponding temperature scale for use in comparison of fused salt properties, in particular, transport properties.

An ion in a molten salt is acted upon by a number of classifiable electrical forces, of which the long range attractive and repulsive Coulomb force is the only one which cannot be considered as solely a nearest neighbor or next-nearest neighbor interaction. The short range forces to be considered are polarization forces, ion-dipole and ion-multipole forces, Van der Waals forces, and, most important, short range repulsive forces. It has been stated as a conclusion by Rice[8] and others[9] that the transport properties of fused salts, like other liquids, are determined by the short range forces, and not by the Coulomb potential. This is intended to mean that when, for example, KCl is melted a liquid with transport properties much like those of other simple liquids is obtained. What must be emphasized here is that the melting points are assumed to provide a suitable basis for a corresponding states comparison of properties. It is essential then to keep in mind that it is the Coulomb potential which determines the high melting point for KCl, and thus the Coulomb potential which sets the corresponding temperature. This becomes of importance when dealing with mixed systems, in which the crystallization temperatures are of no value as reference points, or with salts containing asymmetric ions in which packing considerations can lead to large fluctuations in the relation between potential energies of ordered and disordered particle arrangements and hence to rather irregular melting point sequences.

The problem may be avoided to an extent by reducing physical properties by means of an ionic radius parameter[10] or by an (anion + cation) radius sum parameter as Reiss et al. have proposed in their corresponding states theory for highly ionic liquids.[11] While reduction by size parameters has yielded consistent results with asymmetric as well as spherical ions,[12] difficulties are to be expected as covalent bond character becomes appreciable. In general it seems that a corresponding state internal to the amorphous phase must be preferable to those imposed from without. The critical temperature, of course, provides such a reference state, but, even ignoring the great number of salts of interest which decompose before even boiling, the relevance of critical or boiling constants to data obtained near and frequently below common single component melting temperatures is doubtful. On the other hand, the temperature of the configurational ground state, T_o, or, in absence of data, the experimental glass transition temperature, T_g, provides an internally based reference state, the value of which can now be determined experimentally for a great many systems of common interest and can be estimated for others. Lying of necessity below all pure component melting points, it is a reference state primarily for use with data acquired at low rather than high (corresponding) temperatures. As the temperature below which it is impossible for the substance to exhibit any liquid-like characteristics other than that of long range positional disorder, it seems as well qualified to serve as a reference temperature as does the critical temperature. In the case of ionic liquids we can expect its value to be strongly dependent on the Coulombic energy of the substance.

The question of the Coulomb potential in the description of transport properties of fused salts is therefore easy to answer. Since mass transporting motions must become impossible below the ground state temperature, T_o, as discussed further below, this temperature must enter into any equation describing the transport process and its temperature dependence. In this way, through T_o, the Coulomb potential clearly does largely determine the transport properties, the influence becoming rapidly stronger the more closely T_o is approached.

It is worth emphasizing the parallel between T_o for liquids and absolute zero for solids. At 0^oK the vibrational particle motions characteristic of equilibrium solids are frozen out (except for zero point motions), while at T_o the mass-transporting motions, i.e., the particle rearrangements characteristic of equilibrium liquids, are frozen out. Below T_o only the vibrational motions of the amorphous phase remain, subsequently to be frozen out at 0^oK.

THE NATURE OF THE TRANSPORT PROCESS IN THE LOW TEMPERATURE REGION

Although the thermodynamic changes accompanying the glass transition are, as we saw earlier, those necessary to hold off the entropy crisis, the glass transition itself is a time-dependent phenomenon, such that the temperature of the transition is lower the more slowly the experiment used to determine it is performed. The rate of relaxation of the system towards the equilibrium distribution of states would thus appear to depend rather directly on the number of states remaining into which it can relax. On this basis in the immediate vicinity of the ground state the relaxation rate would become infinitely slow.

From ultrasonic studies[9] one finds that bulk relaxation processes of this type, and the shear relaxation process measured in a viscosity experiment, have identical temperature coefficients and closely similar relaxation times. The molecular rearrangement processes involved in each relaxation are evidently the same. Hence the finding that the glass transition occurs at a constant shear viscosity, viz. 10^{13} poise at "normal" cooling rates, carries with it the implication that the fluidity of a liquid is critically dependent on the number of configurational states available to it. This is essentially the idea expressed by Gibbs and Dimarzio[1] in relating their statistical thermodynamic result for polymer system, $S_c \to 0$ for $T \to T_o$, to the observed glass transition. It finds quantitative statement in the theory of Adam and Gibbs[13] for relaxation rates at low temperatures in which the configurational entropy S_c appears in the expression for the probability w(T) of the mass-transporting cooperative rearrangement:

$$w(T) = A \exp\left(\frac{-N \Delta\mu\, s_c^*}{R\, T\, S_c}\right) \qquad (1)$$

Here N is Avogadro's number, s_c^* is the minimum configurational entropy which a region of the liquid must possess in order to undergo cooperative rearrangement ($s_c^* = k \ln 2$), and $\Delta\mu$ is the free energy barrier per mole of particles opposing the cooperative rearrangement.

We shall not go into the derivation of this expression in this paper. Note however that the development of the equation by approximating

$$S_c = \Delta C_p \ln T/T_o$$

where ΔC_p is the change in heat capacity at the glass transition, leads to an expression which, near T_o, is equivalent to the well-known Vogel-Tammann-Fulcher (VTF) form for the temperature dependence of relaxation processes in glass-forming liquids:

$$w(T) = A \exp\left(\frac{-k}{T-T_o}\right) \qquad (2)$$

k is now a conglomerate of presumably constant terms:

$$k = \frac{N\,\Delta\mu\; s_c^*}{R\,\Delta C_p} \qquad (3)$$

A and k (but not T_o) take different values depending on the transport process being considered.

The form of temperature dependence shown in equation 2 is also derived from the free volume model of Cohen and Turnbull[14] and the theory of Bueche.[15] Of the three, that of Adam and Gibbs seems best supported by the experimental data, but it seems clear that none are adequate to explain in full detail all the observations on liquids. In particular, the conclusion that the departures from Arrhenius behavior of the viscosity of supercooled liquids can be attributed directly to the temperature dependence of S_c in equation (1) has been tested recently using thermal data for o-terphenyl,[16] 1, 3, 5-tri-α-naphthylbenzene,[17] and for B_2O_3.[18] The agreement with experimental data was not complete in the first case, but fairly good for the latter two.

It has been common in the application of equation (2) to fused salts to include, following Cohen and Turnbull, a

$T^{1/2}$ temperature dependence in the frequency factor.

Thus

$$D_i = A_{D_i} T^{1/2} \exp\left(\frac{-k_{D_i}}{T-T_o}\right) \tag{4}$$

where D_i is the diffusion coefficient of species i,

$$\phi = A_\phi T^{-1/2} \exp\left(\frac{-k_\phi}{T-T_o}\right) \tag{5}$$

where ϕ is the fluidity, and

$$\Lambda = A_\Lambda T^{-1/2} \exp\left(\frac{-k_\Lambda}{T-T_o}\right) \tag{6}$$

where Λ is the equivalent conductance. The negative pre-exponential temperature exponents in equations (5) and (6) arise from the Nernst-Einstein and Stokes-Einstein relations between diffusion coefficients and conductance and fluidity.

The manner in which the parameters in these equations vary with transport process, composition changes, and pressure, and the insight into the details of the transport mechanism which these variations provide will be discussed in the next section.

There is a further consequence of the "dearth of states" hypothesis which should be discussed, as it is important to one's way of thinking about the nature of the transport process.

Since the notion of a configurational state (as opposed to a "hole", for instance) requires joint consideration of a rather large number of particles, changes of configuration, i.e., transitions between states, would seem to require the cooperative movements of a comparatively large number of particles. Evidence that motion at low corresponding temperatures is indeed cooperative in character is provided by the ultrasonic studies of Litovitz and co-workers.[19,20] These show that where departures from Arrhenius behavior occur (i.e., in the low temperature region) it is at the same time necessary to invoke a spectrum of relaxation times in order to account for the data. In fact Macedo[21] shows that initially the departure from Arrhenius behavior is solely due to the development of spectrum of relaxation times. The most probable relaxation time remains that predicted from the single relaxation time Arrhenius plot, but because the spectrum is symmetrical for log τ, the average relaxation time, which determines the measured fluidity, becomes non-Arrhenius. At lower temperatures the most probable value also becomes non-Arrhenius.

That transport should occur by such group cooperative rearrangements is intuitively more plausible than the hole-to-hole jumping motions which have been given so much attention in the past. In some respects plausibility is gained at the expense of simplicity, but also some conceptual difficulties encountered with hole mechanisms are resolved. One such is the problem of how the complex ions, much discussed by molten salt chemists, could possibly exhibit any independent mobility if their motion depended on a hole mechanism. This matter is taken up in the general context of the complex ion as a kinetic entity discussed in a later section.

APPLICATION OF THE VOGEL-TAMMANN-FULCHER EQUATION TO MOLTEN SALTS

In Figure 4 are shown Arrhenius plots for the equivalent conductances of some nitrate melts with low liquidus temperatures; the non-linearity of these plots clearly demonstrates the inadequacy of the Arrhenius equation,

$$\Lambda = B_\Lambda \exp\left(\frac{-E_\Lambda}{RT}\right) \qquad (7)$$

to account for transport properties in these temperature regimes. (B_Λ and E_Λ in equation (7) are empirical constants, E_Λ being the so-called activation energy or Arrhenius coefficient.)

In Figure 5 the same data are plotted according to equation 6, where T_o for each melt has been chosen to give the best linear plot of $\Lambda T^{1/2}$ vs $1/(T-T_o)$. It is seen that an

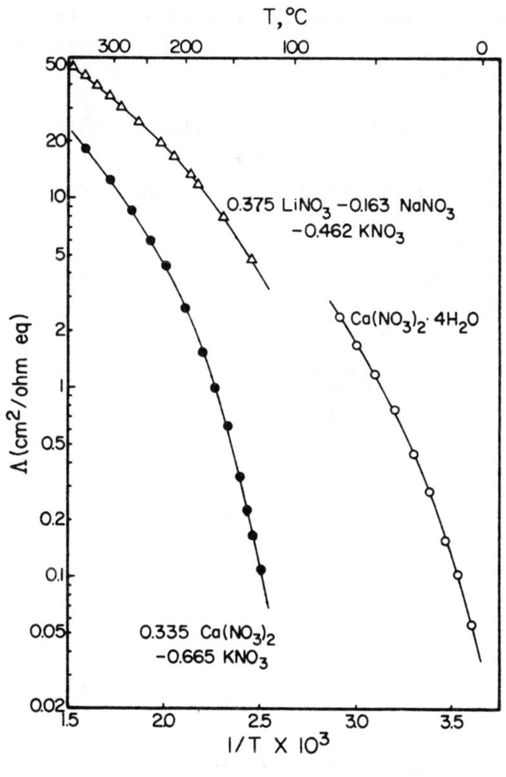

FIG. 4

Arrhenius plots for equivalent conductance of 0.335 $Ca(NO_3)_2$-0.665 KNO_3,[22] $Ca(NO_3)_2 \cdot 4H_2O$,[23] and $LiNO_3$-$NaNO_3$-KNO_3 ternary eutectic.[24]

adequate representation of the data does result from the use of this equation. The improvement is not surprising, however, since equations (4) - (6) contain each three adjustable parameters, while equation (7) contains only two. In fact there are other three-parameter equations[25] which will

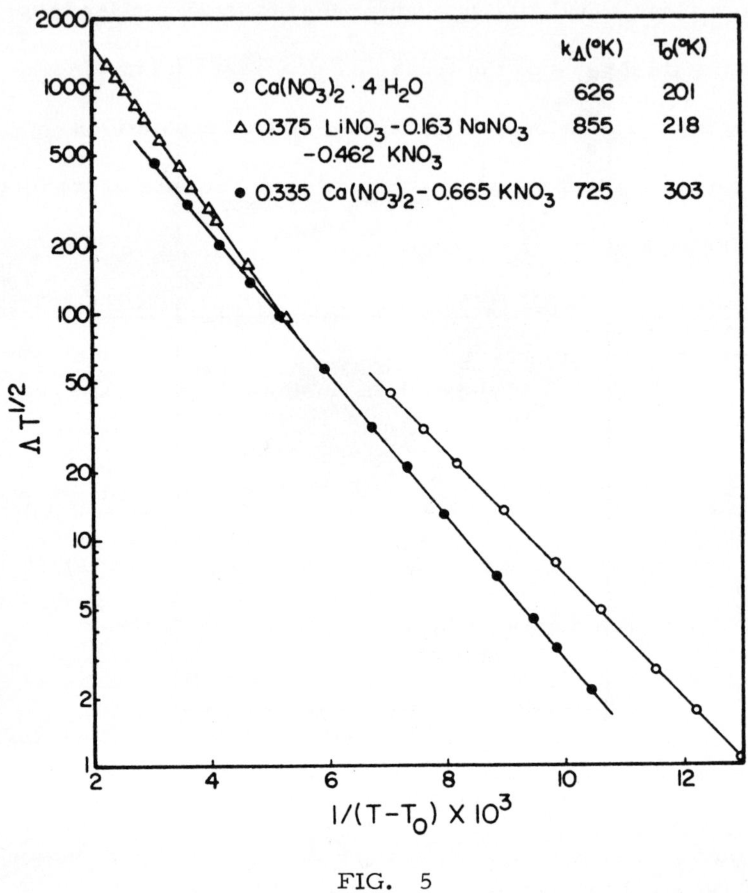

FIG. 5

Best fit Vogel-Tammann-Fulcher plots for data in Fig. 4.

describe transport properties equally well over a few orders of magnitude variation. Thus from a more or less pragmatic point of view it is necessary to demonstrate the superiority of equations (4) - (6) in terms of a) the physical reasonableness of the model implied in the theoretical derivations of the equations and the insight which these theories offer into the intimate details of the transport processes themselves, and b) the correlations, predictions, and extrapolations that can be accomplished by casting transport data in the form of these equations.

EVIDENCE FOR THE PHYSICAL INTERPRETATION OF THE T_o PARAMETER

An immediate question is whether for a liquid T_o(cal) determined from thermodynamic data according to the method outlined in the previous section agrees with T_o(transport) determined via equations (4) - (6). For several systems the agreement appears to be fairly good, if T_o(transport) is calculated from data in the first three orders of magnitude after the onset of non-Arrhenius behavior. For $ZnCl_2$ T_o(cal) of 336°K may be compared with T_o(transport) of 320-330°K estimated from conductance data by Angell.[26] For 2-methylpentane T_o(cal) of 58°K likewise agrees well with T_o(transport) of 59°K determined from the viscosity of n-hexane.[27] Finally, for $NaNO_3$ T_o(cal) is 229°K, close to the value of T_o(transport) of 226-230°K estimated from conductance data for the $Ca(NO_3)_2 \cdot 4H_2O$-$NaNO_3$ system.[28]

On the other hand, recent fluidity data for a number of systems[17,18,29,30] show that at low fluidities below 10^{-4} poise^{-1} the curvature in the Arrhenius plots becomes somewhat less pronounced than that predicted from equations (4) - (6) for the first three orders of magnitude of non-Arrhenius data. This leads to an apparent decrease in the value of T_o(transport) needed to fit the data in these low fluidity regions, and the agreement between T_o(transport) and T_o(cal) becomes less satisfactory than that noted for the examples above. The apparent decrease in T_o at low fluidities in turn implies that equations (4) - (6) will not give an exact fit to experimental data when the data extend over many orders of magnitude, although an approximate fit with an "average" T_o is possible. At this point it appears that it is simply asking too much to require that these simple three-parameter equations describe within experimental accuracy data that may range over as many as 15 orders of magnitude.

Another test, however, of the reasonableness of equations (4) - (6) in terms of their theoretical basis discussed in the preceding sections is in fact possible in that the T_o parameter is considered to be a thermodynamic rather than a kinetic quantity. Consequently one predicts that for a given liquid the T_o parameters derived from the temperature dependence of different transport properties should be identical, if the different transport properties are compared over identical temperature ranges. At the moment there is a great paucity of data of this sort, due partly to the fact that a precise determination of T_o for a liquid demands that the transport properties be determined over a range covering at least 1 1/2 orders of magnitude and with a high degree of precision. A summary of the available data is given in Table 1 and is consistent within experimental uncertainty with the requirement that different transport properties be fitted with the same T_o. A further significant fact is that T_o for all these liquids lies slightly below the experimental glass transition temperature, T_g, as is required by the theory, although the transport data from which the T_o values were obtained are in most cases several orders of magnitude from their values at the glass transition.

For multicomponent systems in which the composition can be varied continuously, T_o will in general vary with composition. For the general case in which k (Equation (2))

TABLE 1

Comparison of T_o Values Derived from Different Transport Properties

Liquid	Property	$T_o(^oK)$	$T_g(^oK)$
isobutyl bromide	fluidity	86.5[14,31]	95[33]
	dielectric relaxation time	86.5[14,32]	
n-propanol	fluidity	73.5[34]	98[33]
	dielectric relaxation time	73.5[34]	
glycerol	fluidity	132[34]	186[33]
	dielectric relaxation time	132[34]	
$Ca(NO_3)_2 \cdot 4H_2O$	conductance	201[23]	220[36]
	fluidity	205[23]	
	chronopotentiometric diffusion coefficient of Cd^{+2}	203[35]	
	chronopotentiometric diffusion coefficient of Tl^{+}	203[35]	
$Na_2S_2O_3 \cdot 5H_2O$	conductance	204[23]	231[37]
	fluidity	203[23]	
$0.40\ Ca(NO_3)_2$-$0.60KNO_3$	conductance	320[2,22]	338[38]
	fluidity	318-320[30]	

changes little with composition, the composition dependence of T_o should in turn closely parallel that of T_g, since the experimental glass transition is presumably a kinetically imposed reflection of the reversible second order thermodynamic transition which would occur at T_o for infinitely slow cooling rates. To date sufficient data are available to compare the composition dependence of T_o (determined from transport properties) with the composition dependence of T_g (determined from DTA experiments) for three systems: $Ca(NO_3)_2$-KNO_3,[2,22,38] $Ca(NO_3)_2$-H_2O,[28,36] and $Ca(NO_3)_2\cdot 4H_2O$-$Cd(NO_3)_2\cdot 4H_2O$.[39] The results for these systems (cf. Figure 1) indeed show that changes in T_o with composition are paralleled within experimental uncertainty by changes in T_g, a further confirmation of the essential correctness of the theoretical interpretation of T_o and T_g in Equations (4) - (6).

DEPENDENCE OF T_o AND T_g ON COMPOSITION

The T_o (or T_g) value for a substance is the quantity which is of cardinal importance in determining the temperature range in which the liquid will exhibit more or less normal fluidities. For instance, to account for the fact that at room temperature water is a free flowing liquid and Pyrex a rigid solid it is quite sufficient merely to note that for the former T_o lies some 170° below and for the latter some 320° above room temperature. Qualitatively, there appear to be a number of factors which determine the value of T_o for a substance.

One obviously important factor is simply molecular size and complexity (compare gasoline and polypropylene). For molten salts, however, the constituent particles are quite small, so that molecular complexity has only a secondary influence on T_o. Rather it appears that for systems of small molecular or ionic aggregates (which we will discuss here) the most important factor in setting the magnitude of T_o (or T_g) is the overall cohesive energy of the liquid (cf. Figure 1 and Table 1). For liquids in which the cohesive energy arises from London and dipole-dipole forces T_o generally falls below 100°K, although it may range somewhat above this value for strongly hydrogen-bonded systems such as water, polyols, or hydroxylic acids. The strong coulomb forces in molten salts and highly concentrated aqueous solutions generally give rise to T_o's in the range 200-500°K. Finally, T_o's for network liquids such as silica glasses, B_2O_3, etc. range between 400° and 700°K, a reflection of the very strong coulomb and covalent bonding forces at work in these systems.

The effect of cohesive energy changes on T_g (and hence on T_o) is shown clearly in Figure 6 for several binary molten salt systems. The coulomb energy of an electrolyte is lower the lower the average charge:radius ratio of the anions or cations of the melt. Hence one finds that melts with divalent anions have higher T_g's than melts with monovalent anions and that an increase in the concentration of the salt with the cation of lower charge: radius ratio brings about a corresponding decrease in T_g. The similarity in the slopes of the T_g vs. mole percent plots for the K-Ca/NO_3 and Tl-Cd/NO_3 systems is particularly striking, since the relative cation charge:radius ratios are quite similar for these two melts.

On the basis of cation potentials, the Coulomb energies of these latter two systems should be similar at equal mole fractions, and the marked difference in T_g's is therefore a matter of interest. An important factor may well be the greater masses of the component particles of the Tl-Cd/NO_3 system, an effect which would be in accord with the correlation that has been noted[36, 40, 41] between T_o or T_g and the

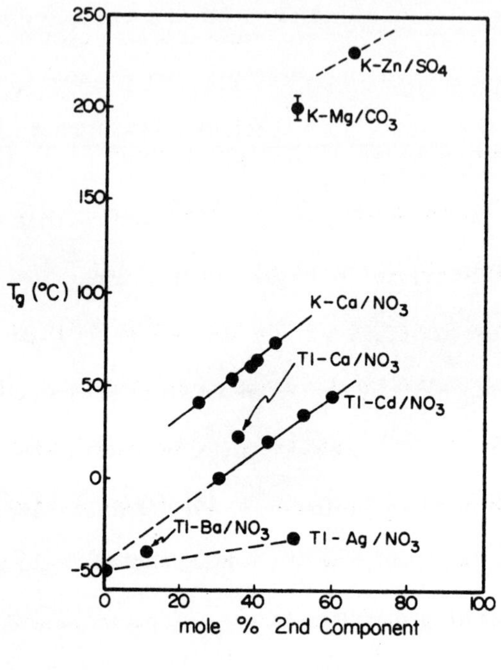

FIG. 6

Composition dependence of the glass transition temperatures in binary molten salt systems.

characteristic Debye temperature, θ_D, for a substance. This correlation has been rationalized[41] by postulating that cooperative rearrangements are induced among the particles of the amorphous phase by interaction of energetic (Brillouin zone boundary) phonons. In any event, the Debye temperature, θ_D, which is directly proportional to the characteristic Debye frequency, ν_D, should show an $m^{-1/2}$ dependence on the effective masses of the component particles of the amorphous phase. Hence one predicts the observed decreases in T_o and T_g with increasing ionic masses for fused salts.

DEPENDENCE OF THE k TERM ON COMPOSITION AND TRANSPORT PROPERTY

The k terms in equations (4) - (6) are formally analogous to the activation energies in the Arrhenius equation and in fact contain a term (cf. equation (3)), $\Delta\mu$, which corresponds quite directly to a transport-impeding free energy barrier. On the face of it, one would expect $\Delta\mu$ and hence k to change quite markedly from system to system. In fact, however, if one compares substances which are not markedly dissimilar, one finds often that k values are very much the same from one system to the next.

Adam and Gibbs[13] suggest that changes in $\Delta\mu$ from system to system may be offset by parallel changes in ΔC_p (cf. equation (3)). They note, for instance, that the temperature dependence of the viscosities of a variety of polymers and organic liquids can be described at low temperatures fairly well by a "universal" value of k 480°K.[42] Similarly the temperature dependence of the conductance of a large variety of nitrate and chloride fused salts and aqueous solutions with liquidus temperatures below $2T_o$ are consistent with a "universal" k_Λ of about 690°K.[2] A plot of this sort for the data in Figure 4 is shown in Figure 7. Conductance and viscosity temperature dependence studies of the systems $Ca(NO_3)_2 \cdot 4H_2O$-$Cd(NO_3)_2 \cdot 4H_2O$[39] and $Ca(NO_3)_2$-H_2O($H_2O:Ca(NO_3)_2$ = 4 to 7)[28] have yielded results consistent with constant $k_\Lambda = 590^\circ K$ and $k_\phi = 690^\circ K$, using T_o values which show the same composition dependences as the

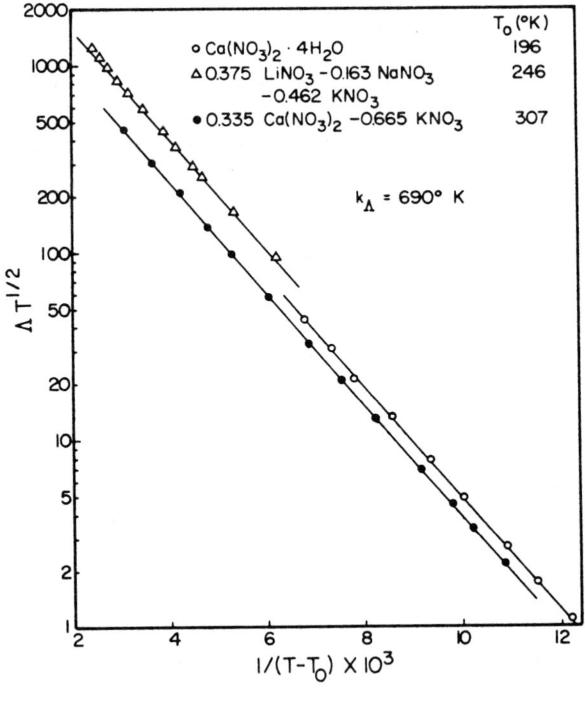

FIG. 7

Vogel-Tammann-Fulcher plots for data in Fig. 4 with "Universal Slope" of $k_\Lambda = 690$.

measured T_g's. Angell[44] has shown that the viscosity temperature dependence of several concentrated aqueous strong acid solutions are consistent with a "universal" $k_\phi = 418^\circ K$.

It appears fairly certain that, rigorously, the "universal values" of k for similar systems are in fact an artifact of the curve fitting procedure. If the available transport data do not extend over too great a range (say 2-3 orders of magnitude), there is generally a small range of T_o values which will give a reasonably good fit to the data. The corresponding range in k values is much greater, since the k parameter derived from the curve fitting procedure is extremely sensitive to the choice of T_o. For example, the conductance results for $Ca(NO_3)_2 \cdot 4H_2O$ shown in Figure 4 can be fitted with a standard deviation of 1% or less with T_o values anywhere in the range 196-206°K; the corresponding range in k_Λ is 690-560°K.

More precise data over larger temperature ranges will no doubt in the future lead to more precise k values for transport properties of these low liquidus melts and reveal variations in k from system to system. Indeed, currently available data are sufficient to disclose variations in k with composition for the $ZnCl_2$-KCl and a number of silicate systems, since in these cases k is apparently changing quite markedly with composition.[41] From an empirical standpoint, however, the use of "universal" values of k has obvious advantages in extrapolating, interpolating, or correlating fused salt transport data in low temperature regimes. It appears that there are a number of types of similar systems whose transport properties can be reduced to a family of lines like those shown in Figure 7, when T_o is used as a scaling or corresponding states parameter.

The implication of the Adam-Gibbs theory that transport at low temperatures occurs by means of cooperative rearrangements of large groups of particles suggests that the transport impeding free energy barrier, $\Delta\mu$, and hence the k terms in equations (4) - (6) should be identical for different transport properties of the same system. This has been found to be the case for the organic liquids listed in Table 1. For fused salts and concentrated electrolyte solutions, however, the presumed equality of k terms does not hold true in all cases. In Figure 8 the temperature dependence of four transport properties in $Ca(NO_3)_2 \cdot 4H_2O$ are shown in the form of an Arrhenius plot. It is significant

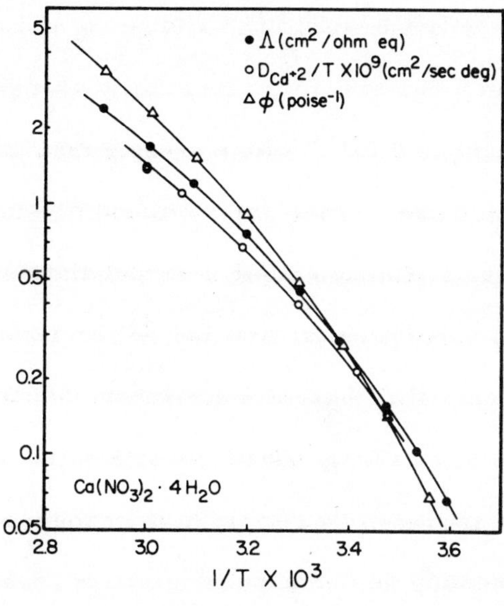

FIG. 8

Arrhenius plots for conductance,[23] fluidity,[23] and chronopotentiometric diffusion coefficient[35] of Cd^{+2} and Tl^{+} in $Ca(NO_3)_2 \cdot 4H_2O$.

that the Λ, D_{Tl^+}, and the $D_{Cd^{+2}}$ curves are virtually parallel showing that the k_Λ and k_D terms for this system are nearly identical.[45] The fluidity plot clearly has a larger slope; for this system $k_\phi/k_\Lambda = k_\phi/k_D = 1.17$, if the k values are compared at equal T_o's. This value of k_ϕ/k_Λ is identical with that for the $Ca(NO_3)_2 \cdot 4H_2O$-$Cd(NO_3)_2 \cdot 4H_2O$[39] system and quite close to those found for the $Na_2S_2O_3 \cdot 5H_2O$ (k_ϕ/k_Λ =

1.18)[23] and recently for the 0.40 $Ca(NO_3)_2$-0.60 KNO_3 melt (k_ϕ/k_Λ = 1.2).[25]

The finding of k_ϕ/k_Λ ratios greater than unity is a reflection of the general result for molten salts that the ratio of the Arrhenius coefficients $E_\phi:E_\Lambda$ at higher temperatures is likewise invariably greater than one.[46,47] In fact, for nitrates and other melts with complex anions the $E_\phi:E_\Lambda$ ratios are quite low (< 1.5), but for simple halide systems $E_\phi:E_\Lambda$ ratios are more commonly in the range 2-7. The answer as to whether for fused salts the $k_\phi:k_\Lambda$ ratio can be generally expected to be fairly close to unity or not must await measurements on some halide melts with low liquidus temperatures. The large values of the $E_\phi:E_\Lambda$ ratios for the halide melts at temperatures above $2T_0$ provide no clue, since they could arise simply from the onset of Arrhenius behavior at a lower temperature for fluidity than for conductance. There is in fact good evidence in the case of $ZnCl_2$ system that this is indeed the case, even though $ZnCl_2$ is thought structurally to partake of many of the characteristics of network liquids such as fused silicates and borates. The fluidity of molten $ZnCl_2$[48] shows marked departures from Arrhenius behavior only at temperatures below 325°C, while the conductance appears to be non-Arrhenius up to temperatures some 300° higher than this.[49,50] At high temperatures, where both Λ and ϕ show an Arrhenius temperature dependence, the $E_\phi:E_\Lambda$ ratio appears to be about 2.0. At 400°C, however, $E_\phi:E_\Lambda$ has dropped to about 1.35, and at 350°C, to about 1.15.

A reason frequently advanced for $E_\phi:E_\Lambda$ ratios greater than unity in molten salts is that one of the ions of the ionic liquid is more mobile than the other. Electrical conduction is presumed to occur chiefly by migration of the more mobile ion. In viscous flow, however, the ions are constrained to move along together, and the case of such flow is determined primarily by the resistance afforded to the motion of the less mobile ions. For transport in network liquids such as silicates and borates, this explanation may be essentially correct, even if the transport is thought to occur via a cooperative rearrangement mechanism, since the anions may be firmly anchored by the complex network to regions outside a region with sufficient energy to allow cation transport. Indeed $k_\phi >> k_\Lambda$ for several such systems.[41] For simple fused salts in which the attractions are primarily coulombic, however, there is no good reason to believe in the existence of anion domains of relative immobility or rigidity which are more extensive than similar cation domains. The apparently general result[51] for fused salts that the Arrhenius activation energies for self-diffusion of anions and cations in the same melt are very nearly equal and that mobility differences arise mainly from differences in pre-exponential terms seems to support this latter point of view, that the free energy barrier to anion migration should be no greater than that for cation migration. It is significant here to note that the rule of equality of anion and cation self-diffusion activation energies applies even in the cases $ZnCl_2$ and $ZnBr_2$.[52-54] Consequently, reasonable qualitative interpretation of the cooperative rearrangement transport mechanism at low temperatures seems to demand at the moment $k_\phi : k_\Lambda$ ratios which do not depart significantly from unity. Further, the k_D for diffusion coefficients of all ions in the same melt should prove the same. This is shown by the data in Figure 8 and also by the work of Francini and Martini,[55] who found the activation energies for the diffusion coefficients of Tl^+, Pb^{+2}, Cd^{+2}, Zn^{+2}, Ni^{+2} were all very nearly equal (about 8.5 kcal/mole) in a low-melting ternary nitrate mixture (0.30 $LiNO_3$-0.17 $NaNO_3$-0.53KNO_3). (Although the latter measurements were carried out in a non-Arrhenius region of the melt, the data showed sufficient scatter and covered so small a temperature range that the diffusion coefficients could be best represented by an average Arrhenius activation energy).

DEPENDENCE OF THE PRE-EXPONENTIAL TERM ON COMPOSITION

Of the parameters in equations (4) - (6), the pre-exponential terms A_{D_i}, A_ϕ, and A_Λ are the most difficult to assess. The primary reason for this is that A_{D_i}, A_ϕ, and A_Λ are extremely sensitive to the choice of T_o. For instance, if T_o for the $Ca(NO_3)_2 \cdot 4H_2O$ system is allowed to vary from 196-206°K, A_Λ undergoes a corresponding variation of nearly 100%.[23]

The free volume derivation[14] of equation 4 suggests that

$$A_{D_i} \propto \frac{r_i}{m_i^{1/2}}$$

where r_i and m_i are the radius and mass of particle i. In fact some diffusion data have been presented for cations in molten nitrates[55, 56] which tend to bear out the $A_{D_i} \propto r_i$ prediction, but the $m^{-1/2}$ dependence seems not to be supported by experiment. Rather conductance and fluidity measurements on the $Ca(NO_3)_2 \cdot 4H_2O$-$Cd(NO_3)_2 \cdot 4H_2O$ system,[39] in which the hydrated Ca^{+2} and Cd^{+2} cations have been shown to have equal sizes but have masses in a ratio of about 0.61, show that the pre-exponential constants A_Λ and A_ϕ increase by about 30% in going from pure $Ca(NO_3)_2 \cdot 4H_2O$ to pure $Cd(NO_3)_2 \cdot 4H_2O$--an increase in the direction opposite from that predicted on the basis of the relative masses.

Clearly the question of the factors determining the pre-exponential constant in the VTF equation is at the moment an open one. It should be noted, however, that for purposes of correlating transport data in systems at temperatures corresponding to low $T:T_o$ ratios the composition dependence of the transport properties is in general dominated by the $1/(T-T_o)$ term and hence by the composition dependence of T_o. Indeed for two systems, $Ca(NO_3)_2$-KNO_3[22] and $Ca(NO_3)_2$-H_2O,[57, 58] the composition dependence of transport data could be correlated over fairly large concentration spans by assuming a constant value of A_ϕ or A_Λ. On the other hand, at high temperatures the pre-exponential terms may become dominant in determining the composition dependence of transport properties. For instance, a maximum at intermediate compositions in diffusion coefficients in the Li_2CO_3-Na_2CO_3 system occurs in the face of a monotonous, linear decrease in Arrhenius activation energies in going from pure Na_2CO_3 to pure Li_2CO_3.[59,60] The maximum is hence assignable purely to the influence of the pre-exponential constants. An analogous case of the influence of the pre-exponential constants on the composition dependence of conductance in the $LiNO_3$-$AgNO_3$ system has been detailed by Angell.[58,61]

THE COMPLEX ION PROBLEM

The questions of relevance to this paper are (a) are there groups of ions in molten salts of such kinetic stability that they can be ascribed an independent mobility and must accordingly be considered as "species" in any overall discussion of the transport properties of fused salts, and (b) if so, what are their transport properties?

To answer (a) we need to discuss certain general aspects of the complex ion problem, a matter which has been a source of some controversy among molten salt chemists for a number of years. While semantic problems have been evident in this controversy, the major difficulty has been non-trivial and lies in the problem of recognizing a group of anions clustered around a particular cation as being somehow distinct from all other groups in a liquid phase which is essentially close-packed with respect to these anions, e.g., the $CdCl_4^{-2}$ group in dilute solution in the LiCl-KCl eutectic. The terms "complex ion" and "species" clearly imply such a distinction and although the thermodynamic definitions of complexes contain no connotations concerning time, it has been popular to suppose that a complex ion will be distinguished from other groups in, e.g. pure chloride media, by maintaining its identity over "long" time intervals, hence behave as a kinetic entity, exhibit a unique mobility, etc. Complexes of high lability are, of course, recognized in inorganic chemistry, so the question of what constitutes a "long" time interval arises. This is generally answered by reference to the characteristic diffusion times of order 10^{-11} sec, a complex ion in principle being required to maintain its identity longer by a respectable number of orders of magnitude than the rearrangement times of its immediate environment. The question has remained academic for complex ions in molten salts, as no lifetimes or exchange times have been determined.

Despite this recognition of the need for consideration

of the kinetic lifetime in the use or otherwise of the term "complex ion", much weight has often been given to spectroscopic evidence for groups of ions of well-defined symmetry. Such groups can sometimes be defined quite unambigously from the ligand field spectra for particular transition metal ions. However recognizing the experiment is performed on a time scale orders of magnitude faster than the diffusion time, investigators have tended to avoid the term "complex ion" in describing the observed local structure adopting instead such terms as "configuration", "center" and "group".[62,63]

Generally, more conviction seems to be associated with observations of vibrational spectra in molten salts, the results of Raman spectra studies in molten salt mixtures being particularly liable to interpretation in terms of "complex ions." For this reason it seems important to emphasize here that the Raman and I.R. bands attributed to complex ions are strictly speaking only the long wavelength limits of the various longitudinal and transverse branches of the optical phonon spectrum characteristic of the vibration of the entire molten salt quasi-lattice. That the quasi-lattice can sustain such long wavelength modes despite the absence of long range order is clearly established by the observations of Wilmshurst,[64] supported by recent, more detailed studies of mixed nitrate lattice vibrations by Angell and Wong.[65] As such, the observed number and frequencies of the bands are dependent on the force constants, masses, and geometrical arrangement of <u>all</u> the ions in the quasi-lattice, and it is only in the limit where the force constants within a particular group are very high with respect to those between the group and its surroundings that one can expect the number of bands observed in a given spectral region to correspond with that anticipated on the basis of group theory predictions for the isolated molecule ion. Thus it can be understood that a tendency in $ZnCl_2$ + KCl melts to have $ZnCl_4$ tetrahedra share a common chlorine to give Zn_2Cl_7 groups in the composition range 60-66% KCl[66] need not be reflected in the generation of the additional Raman bands which the (3n-6) rule for the isolated molecule-ion would require. Experimentally one observes in this case[67] only a shift of the most intense, (and already broad), band to <u>lower</u> energies with increasing % KCl in this composition range, consistent with the far I. R. observations on the composition dependence of lattice band energies in mixed nitrate glasses.[65]

Vibrational spectra therefore not only fail to give information on kinetic stability, but may also give misleading information on local co-ordination geometry.

The wealth of spectral information available for fused salts therefore does not provide any clear basis for the use of the terms "complex ion" or "species" with kinetic connotations, giving only the most indirect indications of whether such quasi-lattice structural elements as are observed will behave as kinetic entities in an appropriate experiment. The interesting position then is that in a given system the presence of a complex ion, whose transport properties it may be important to understand in detail, can apparently only be established in the first place by some form of transport experiment. (N.M.R. spectral techniques, in which the interaction time is long compared to diffusion times, offer a possible but unexploited exception to this statement.) What types of transport experiment, then, would provide unambiguous evidence for the presence of long-lived complex ions?

We discuss two types, the most obvious of which is an electrolysis experiment in which the complex is observed to move, e.g., towards the anode, when the probe cation itself (e.g. Co^{+2} in $CoCl_4^{-2}$ when color is followed or $^{115}Cd^{+2}$ when radioactivity is followed) should otherwise move towards the cathode. Such an experiment is beset by such problems as bulk flow of salt and reference frame choices, but by suitable choice of internal markers (e.g., anions such as CrO_4^{-2} in a color-based experiment, $^{36}ClO_3^-$ in a radioactivity-based experiment) such problems can presumably be eliminated.

For such an experiment to be feasible, however, a definite mobility for the complex is required. If one believes that ions in molten salts move by jumping into ion-sized holes, then it is difficult to see how an ion such as $CdCl_4^{-2}$ could have an appreciable mobility in a melt such as LiCl-KCl eutectic, since the probability of volume fluctuations (or "holes" or "critical voids") several times the size of those required for chloride ion motion must be extremely small. Such species, one would have to conclude, could lead to conductance decreases, or even minima, by immobilizing otherwise charge-transporting ions, but could not itself move towards the anode in an observable manner.

On the other hand, observable motion of stable complexes seems entirely feasible in the light of the cooperative rearrangement transport mechanism discussed earlier. Improbable density fluctuations are not required, since a complex ion, like any other particle, is considered to move as a result of its involvement in the cooperative rearrangement of a larger group of particles. In the absence of applied fields such rearrangements lead to random diffusive displacements. The presence of a field imposes, in the usual fashion, a directional bias, predisposing a tightly bound group of ions with a net negative charge to post-rearrangement positions nearer the anode. Thus a large group can have an appreciable mobility. At the same time it must be recognized that motion at all times be such that the "complex" only moves in concert with other particles (whose net motion during the rearrangement may be in opposite directions). The concept of independent motion for a complex ion thus remains clouded. The observed movement of a metal ion toward the anode would perhaps be better described as the motion of a "trapped cation."

Attempts to observe such movements have usually failed. Laity and Duke[68] failed to observe motion of Pb^{+2} towards the anode; Kwak[56] found the Cd^{+2} mobilities in $CdCl_2$-KCl solutions to be similar to those of divalent cations in mixtures where no complexes are expected. On the other hand Alberti et al.[69] reported briefly observations on the motion of various cations in LiCl-KCl on a soaked glass fiber strip in which it appeared that Co^{+2} and Zn^{+2}, distinct from Pb^{+2}, Cd^{+2}, and others, moved towards the anode, suggesting that similar experiments, performed with suitable internal marker ions, may prove of great interest.

It might be expected that such "trapping" of cations will be more effective at lower temperatures, and one can anticipate that observations similar to those of Alberti et al. will be possible in various low-melting mixtures, hydrate melts such as $ZnCl_2$-$Mg(H_2O)_6Cl_2$ being particularly amenable to study. In such melts the reverse motion of even the larger cations such as Cd^{+2} should be observable.[70]

Clear evidence for the presence in a given system of stable complex ions could also be obtained from a diffusion experiment in which two radioisotopes, e.g., $^{60}Co^{+2}$ and $^{36}Cl^-$ in LiCl-KCl introduced into a capillary tube (or to one side of porous plug) as active $CoCl_2$, would diffuse into an inactive bath of the same composition (or into an inactive salt across the plug). If a stable complex is present the relative concentrations of isotopes must remain unchanged in the capillary (or on either side of the plug). Such an experiment, intended to give only qualitative information, should be relatively simple to perform.

Information on complexes of intermediate lifetime, i.e., temporarily trapped cations, may be more difficult to acquire, though there seem good reasons for expecting sound absorption-frequency studies to yield positive information in this area. This would again be a form of transport experiment in which relative motion of particles in response to cyclic stress is detected. Transport experiments in which the frequency of an electric field is matched to the complex lifetime also seem feasible in certain cases.

To consider briefly the transport properties of complex ions, and systems containing them, we distinguish two cases; (a) complex ions in dilute solution, where the thermodynamic factors influencing transport processes, principally T_o, are determined by the solvent, (b) complex ions present as a major component, e.g., $AlCl_4^-$ in molten $KAlCl_4$, when the nature of the complexes themselves determine the value of T_o.

In the first case, little can be said until more experimental studies of the behavior of individual ionic species in non-crystallizing melts have been performed. As far as can be told from the work described in the previous section, mobility differences will show up principally in the pre-exponential term of the transport equation. The factors determining the magnitude of this term are however quite unclear at the moment. Chronopotentiometric studies of diffusion coefficients in solutions where the presence of strongly trapped cations ("stable complexes") has been demonstrated in a prior experiment would be valuable in this respect.

The second case is the more interesting. It is also the more studied, though most investigations of such systems have not given any information on kinetic stability of the postulated groupings.

In line with the dominance of the T_o parameter over the transport properties, we expect the mobilities of complex ions as well as other species at a given temperature will be largely determined by the behavior of T_o with composition. This in turn will depend on whether or not the trapping of a cation is accompanied by changes in bond character. If the complex formation is essentially an electrostatic phenomenon involving decreases in nearest neighbor repulsions and more efficient packing geometry, then the net cohesion of the liquid should increase, and T_o should exhibit a maximum. Consequently complex formation will cause a net decrease in mobility at a given temperature for all species. If, on the other hand, the complexes are formed with an increase in overlap and greater covalent character in the metal-ligand interaction, then the effective coulombic cohesive energy of the liquid will decrease, and the behavior of T_o will be complicated and will depend on the opposing effects of asymmetry of the complex formed (increase in T_o at constant cohesion) and loss of coulombic energy (decrease of T_o at constant asymmetry). In the formation of highly covalent complexes with quasi-spherical symmetry a decrease in T_o is anticipated, hence at constant temperature an increase in fluidity and diffusivity (though presumably not of conductivity) would be expected. These considerations apply to systems with weakly trapped as well as strongly trapped cations (short and long-lived complexes) and their experimental elucidation in terms of the parameters of equations (4)-(6) may be considered an urgent experimental problem.

BEHAVIOR AT HIGHER TEMPERATURES

It is appropriate to complete this paper by commenting briefly on the characteristics of transport processes at temperatures outside the low temperature region, e.g., at temperatures in the range 3-4T_o, the temperature region of common single component salt melting points.

Most directly interesting from the point of view of this paper is the fact that the temperature dependence again becomes Arrhenius in form, although very precise measurements[71] can still detect a temperature dependence of the Arrhenius coefficient. Moynihan and Cantor[72] have suggested in discussion of the Arrhenius temperature dependence of the fluidity of BeF_2 that the rate of increase of configurational entropy decreases with increasing temperature, so that the term S_c in equation (1) approaches a constant value. This would lead to Arrhenius expression with a many-termed Arrhenius coefficient. It is interesting in this connection that alkali halides melt with only small increases in heat capacity,[73] whereas the heat capacity of a $Ca(NO_3)_2$ + KNO_3 glass increases by no less than 55% at the glass transition.[74]

Despite these considerations, it is not expected that the assumptions underlying the Adam-Gibbs theory should be valid at temperatures much outside the low temperature region. Generally it is by no means clear why unassociated liquids should obey an Arrhenius type law at higher temperatures. Indeed statistical mechanical formulations for transport properties at medium packing densities provide no justification for this form. While it can be pointed out that in cases of low Arrhenius coefficient the data are equally well fitted by a linear temperature dependence, the Arrhenius form does gain support from the very general finding that Arrhenius coefficients for diffusion coefficients of the ions of each salt are equal within experimental error, even though the diffusion coefficients themselves at a given temperature may be very different. This impressive pattern of behavior would be lacking in a comparison of linear temperature coefficients.

Again it is also notable at higher temperatures that the Arrhenius coefficients for conductance decrease relative to those for diffusion and viscosity, a fact which leads to deviations from the Nerst-Einstein equation (for major components of a melt) which increase with increasing temperature. This has been attributed by various authors[75, 76] to the increasing importance of cation-anion interactions, and accords with observations by Grantham and Yosim[77] that certain salts actually show conductance maxima at very high temperatures. It seems probable that, whereas the glass transition and low temperature flow properties are controlled by excess entropy rather than volume, volume *per se* becomes increasingly important at higher temperatures, and the concept of mean free path should play a role in theoretical developments.

It is to be hoped that as more detailed studies of the very high and very low temperature regions are extended downward and upward respectively into the "normal" region the answers to some of these questions may become clearer.

ACKNOWLEDGMENTS

This work was performed with the assistance of a grant from the Office of Saline Water, U. S. Department of the Interior.

REFERENCES

1. J. H. Gibbs and E. A. Dimarzio, *J. Chem. Phys.*, *28*, 373 (1958).

2. C. A. Angell, *J. Phys. Chem.*, *70*, 2793 (1966).

3. W. Kauzmann, *Chem. Rev.*, *43*, 219 (1948).

4. D. Cubiciotti and H. Eding, *J. Chem. Phys.*, *40*, 978 (1964).

5. It must be stressed that the liquid phase, although metastable with respect to the crystal, still has clearly defined equilibrium properties. The melting point after all is only that point in temperature at which the *independent* liquid and crystal free energy functions

happen to intersect. Below the glass transition, on the other hand, the liquid is no longer in equilibrium, as the relaxation times are too long to permit internal equilibrium to be established.

6. D. R. Douslin and H. M. Huffman, J. Am. Chem. Soc., 68, 1704 (1946).

7. T. A. Litovitz in Non-Crystalline Solids, V. D. Frechette, ed., J. Wiley and Sons, Inc., New York, N.Y., 1960, p. 252.

8. S. A. Rice, Trans. Faraday Soc., 58, 499 (1962).

9. C. A. Angell and J. W. Tomlinson, Trans. Faraday Soc., 61, 2312 (1965).

10. C. A. Angell, J. Phys. Chem., 69, 399 (1965).

11. H. Reiss, S. W. Mayer, and J. L. Katz, J. Chem. Phys., 35, 820 (1961).

12. B. B. Owens, J. Chem. Phys., 44, 3918 (1966).

13. G. Adam and J. H. Gibbs, J. Chem. Phys., 43, 139 (1965).

14. M. H. Cohen and D. Turnbull, J. Chem. Phys., 31, 1164 (1959).

15. F. Bueche, J. Chem. Phys., 36, 2940 (1962).

16. R. J. Greet and D. Turnbull, J. Chem. Phys., 47, 2185 (1967).

17. J. H. Magill, J. Chem. Phys., 47, 2802 (1967).

18. P. B. Macedo and A. Napolitano, J. Chem. Phys., to be published.

19. T. A. Litovitz and G. McDuffie, J. Chem. Phys., 39, 729 (1963).

20. P. Macedo and T. A. Litovitz, Phys. Chem. Glasses, 6, 69 (1965).

21. J. Tauke, T. A. Litovitz, and P. B. Macedo, J. Am. Ceram. Soc., to be published.

22. C. A. Angell, J. Phys. Chem., 68, 1917 (1964).

23. C. T. Moynihan, J. Phys. Chem., 70, 3399 (1966).

24. L. A. King, C. L. Bissell, and F. R. Duke, J. Electrochem. Soc., 111, 720 (1964).

25. E. Rhodes, W. E. Smith, and A. R. Ubbelohde, Proc. Roy. Soc., A285, 263 (1965).

26. C. A. Angell, J. Am. Ceram. Soc., 51, 125 (1968).

27. O. G. Lewis, J. Chem. Phys., 43, 2693 (1965).

28. C. T. Moynihan and C. R. Smalley, unpublished results.

29. A. J. Barlow, J. Lamb, and A. J. Matheson, Proc. Roy. Soc., A292, 322 (1966).

30. P. B. Macedo, paper presented at 156th National ACS Meeting, Atlantic City, N. J., Sept. 8-13, 1968.

31. D. J. Denney, J. Chem. Phys., 30, 159 (1959).

32. D. J. Denney, J. Chem. Phys., 27, 259 (1957).

33. M. R. Carpenter, D. B. Davies, and A. J. Matheson, J. Chem. Phys., 46, 2451 (1967).

34. D. W. Davidson and R. H. Cole, J. Chem. Phys., 19, 1484 (1951).

35. C. T. Moynihan and C. A. Angell, unpublished results.

36. C. A. Angell, E. J. Sare, and R. D. Bressel, J. Phys. Chem., 71, 2759 (1967).

37. M. Samsoen, Ann. Phys., 9, 35 (1928).

38. C. A. Angell and D. Helphrey, unpublished results.

39. C. T. Moynihan and C. R. Smalley, paper presented at 156th National ACS Meeting, Atlantic City, N.J., Sept. 8-13, 1968.

40. R. H. Cole and D. W. Davidson, J. Chem. Phys., 20, 1389 (1952).

41. C. A. Angell, J. Am. Ceram. Soc., 51, 117 (1968).

42. The Adam-Gibbs "universal value" of k_ϕ is in fact an alternative form of the older WLF equation[43] for polymer and organic liquid viscosities.

43. M. L. Williams, R. F. Landel, and J. D. Ferry, J. Am. Chem. Soc., 77, 3701 (1955).

44. C. A. Angell, J. Electrochem. Soc., 114, 1033 (1967).

45. In deference to the Nernst-Einstein equation, $D_{Cd^{+2}}/T$ rather than $D_{Cd^{+2}}$ has been plotted. Over this temperature range, however, the influence of the $1/T$ factor for the diffusion coefficient has very little effect on the parallelism of the Λ and $D_{Cd^{+2}}/T$ plots.

46. J. P. Frame, E. Rhodes, and A. R. Ubbelohde, Trans. Faraday Soc., 55, 2039 (1959).

47. G. J. Janz, Molten Salts Handbook, Academic Press, New York, N. Y., 1967, pp. 54-58, 287-289.

48. G. J. Gruber and T. A. Litovitz, J. Chem. Phys., 40, 13 (1964).

49. F. R. Duke and R. A. Fleming, J. Electrochem. Soc., 104, 251 (1957).

50. J. O'M. Bockris, E. H. Crook, H. Bloom, and N. E. Richards, Proc. Roy. Soc., A255, 558 (1960).

51. B. R. Sundheim, ed., Fused Salts, McGraw-Hill Co., Inc., New York, N. Y., 1964, pp. 224-227.

52. C. -A. Sjoblom and A. Lunden, Z. Naturforsch., 18a, 942 (1963).

53. C. -A. Sjoblom, Z. Naturforsch., 18a, 1247 (1963).

54. C. -A. Sjoblom and A. Behn, Z. Naturforsch., 23a, 495 (1968).

55. M. Francini and S. Martini, EUR Report 1908e, (1964).

56. J. C. Th. Kwak, Ph.D. Thesis, University of Amsterdam, 1967.

57. C. A. Angell, J. Phys. Chem., 70, 3988 (1966).

58. C. A. Angell, J. Chem. Phys., 46, 4673 (1967).

59. P. L. Spedding and R. Mills, J. Electrochem. Soc., 112, 594 (1965).

60. P. L. Spedding and R. Mills, J. Electrochem. Soc., 113, 599 (1966).

61. H. C. Cowen and H. J. Axon, Trans. Faraday Soc., 52, 242 (1956).

62. J. Brynestad, C. R. Boston, and G. P. Smith, J. Chem. Phys., 47, 3179 (1967).

63. C. A. Angell and D. M. Gruen, J. Am. Chem. Soc., 88, 5192 (1966).

64. J. K. Wilmshurst and S. Senderoff, J. Chem. Phys., 35, 1078 (1961).

65. C. A. Angell and J. Wong, unpublished results.

66. C. A. Angell and D. M. Gruen, J. Phys. Chem., 70, 1601 (1966).

67. J. R. Moyer, J. C. Evans, and G. Y-S. Lo, J. Electrochem. Soc., 113, 158 (1966).

68. R. W. Laity and F. R. Duke, J. Electrochem. Soc., 105, 97 (1958).

69. G. Alberti, G. Grassini, and R. Trucco, J. Electroanal. Chem., 3, 283 (1962).

70. A. Berlin, F. Menes, S. Forcheri, and C. Monfrini, J. Phys. Chem., 6, 2505 (1963).

71. E. R. Van Artsdalen and I. S. Yaffe, J. Phys. Chem., 59, 118 (1955).

72. C. T. Moynihan and S. Cantor, J. Chem. Phys., 48, 115 (1968).

73. G. J. Janz, Molten Salts Handbook, Academic Press, New York, N. Y., 1967, p. 201.

74. J. de Neufville and D. Turnbull, private communication.

75. W. Fisher, K. Heinziger, W. Herzogg, and A. Klemm, A. Naturforsch., 17a, 799 (1962).

76. R. W. Laity, Disc. Faraday Soc., 32, 172 (1962).

77. L. F. Grantham and S. J. Yosim, J. Chem. Phys., 45, 1192 (1966).

TRANSPORT PROPERTIES: CONDUCTIVITY, VISCOSITY, AND ULTRASONIC RELAXATION

P. B. Macedo and R. A. Weiler

Vitreous State Laboratory
The Catholic University of America
Washington, D. C. 20017

INTRODUCTION

In this paper we will consider the inter-relation of the various transport parameters in liquids. The structure of the liquid will be considered arrested for most of the time. When a fluctuation in local energy large enough to permit "melting" of a local region occurs, the atoms or ions will be able to rearrange themselves. In so doing, they will relieve any local stresses that may be present. They will also permit the transport of matter by random walk. The kinetics of this local melting can best be studied by examination of various transport properties. Special emphasis will be given to viscoelastic studies since they give detailed information on the melting of this structure.

The discussion is limited to a comparison between the molten oxides and molten salts in which systems only visco-elastic and conductivity data are available. In addition to considering the inter-relation of these properties, we shall also investigate their temperature dependence. It is shown that the well-known free volume and excess entropy theories not only do not fit the data but have the wrong physical significance.

VISCO-ELASTIC CONSIDERATIONS

The fundamental question in visco-elastic behavior is: what and where is the Glass Transition? In Figure 1 from Ref. (1), the increase in length is plotted versus temperature for a GeO_2 glass sample. In this experiment the temperature was raised at $2^oC/min$.

The material's behavior can be described in terms of three regions: a) below the critical point (C.P.); b) between

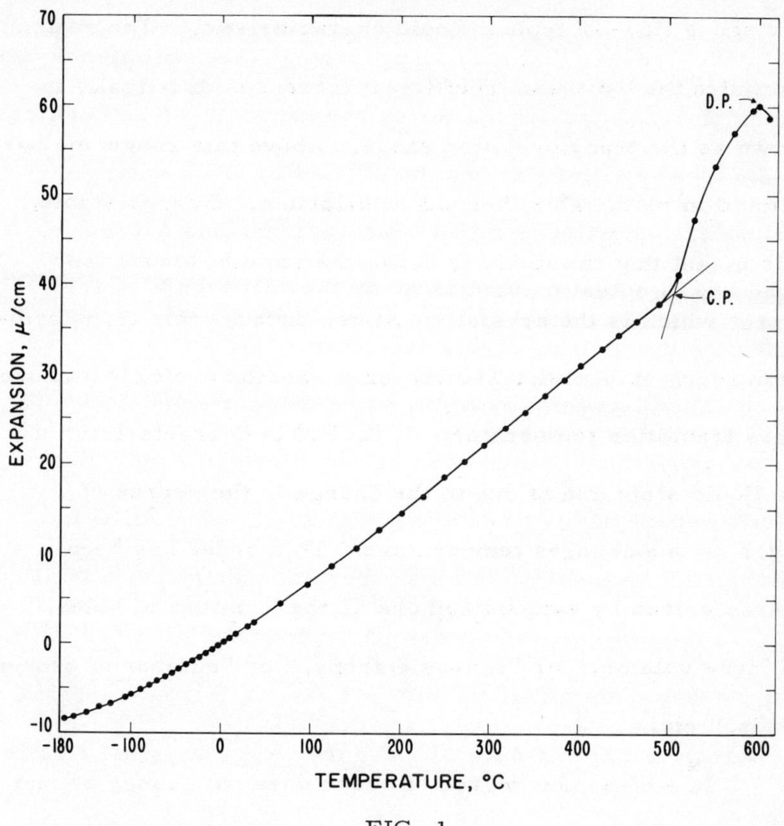

FIG. 1

Thermal expansion of vitreous GeO_2 showing the critical point and the deformation point. Figure taken from N & M, Ref. 3.

the critical point and the deformation point (D. P.); c) above the deformation point. In region (a) the sample is considered a glass and has solid-like properties such as a modulus of rigidity. The deformation point is associated with the onset of viscous flow--a typical liquid characteristic. The region (b) where the expansion coefficient increases drastically is known as the transformation range. Above this range one has a liquid in metastable thermal equilibrium. By metastable, it is meant that the liquid is not in the state of lowest free energy which is the crystalline state, because this transformation requires nuclei. The larger expansion coefficient above glass transition temperature T_g (C.P.) is characteristic of the liquid state and is due to the change in the degree of order as one changes temperature. This order has been characterized by various authors as the "number of holes," or "free volume," or "excess entropy," or "number of broken bonds," etc.

In a glass the degree of order does not change as one raises or lowers the temperature; thus all the second order thermodynamic parameters (thermal expansion, $\alpha_P = (1/V)(\partial V/\partial T)_P$; specific heat $C_P = (\partial H/\partial T)_P$; compressibility $\kappa_T = -(1/V)(\partial V/\partial P)_T$) have a value lower than the corresponding parameters in the liquid. Glasses are thermodynamically unstable so that if one observes a glass for long periods of time, one expects to see the degree of order change as it tries to approach equilibrium. This process is known as relaxation and measurements of its time dependence are commonly called annealing. In order to mathematically describe the annealing process, one has to study the time dependence of the approach to equilibrium as a function of both temperature and one or more order parameters. Let us define as linear an experiment where the time dependence of the normalized stress or strain relaxation is independent of the initial stress or strain. One of the major problems is to design an annealing experiment which is linear. Ultrasonics provides probably the best solution to both of these problems. Sound waves can be generated and detected with only infinitesimal stresses and strains, and over a wide range of frequencies. Thus, by choosing ultrasonics as an experimental technique one has converted the problem from measuring the time dependence of the compressibility to measuring the frequency dependence of the modulus. The choice of modulus is one of mathematical convenience. The bulk modulus, K, is the reciprocal of the compressibility, κ. The consequence of converting the time dependence has to be examined in detail. Any time dependent property can be transformed into a frequency dependent property by the Fourier Theorem. However, this frequency dependent property will have a real and imaginary part. The real part of the modulus is what one would normally consider a modulus, while the imaginary part is related to viscosity. Thus both the shear modulus G and the bulk modulus K are complex:

$$K^* = K' + iK'' \tag{1}$$

$$G^* = G' + iG'' \tag{2}$$

Both G and the longitudinal modulus, M = K + 4/3G,[2,3] can be calculated from the appropriate data according to the equation:

$$M = \rho V_m^2 \frac{\left[1 - \left(\frac{\alpha_m V_m}{\omega}\right)^2 + 2i \frac{\alpha_m V_m}{\omega}\right]}{\left[1 + \left(\frac{\alpha_m V_m}{\omega}\right)^2\right]^2} \tag{3}$$

where M is the modulus associated with the appropriate absorption α_m and velocity V_m at the angular frequency ω, and ρ is the density.

Figure 2(a) shows a plot of the longitudinal modulus, M, versus frequency. At low frequencies where the structure of the liquid has plenty of time to rearrange itself between successive crests of the sound wave, M is frequency independent and represents the reciprocal of the liquid compressibility, i.e., the modulus of compression of the liquid. As the sound period becomes shorter and shorter

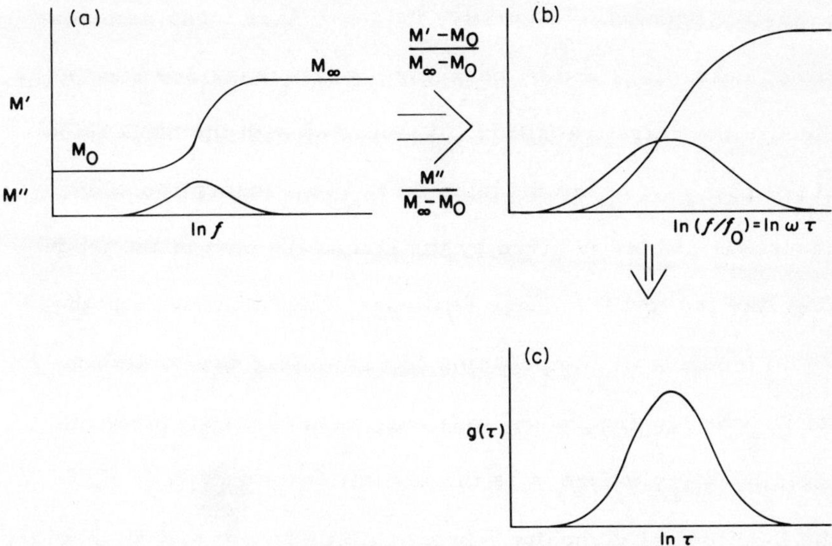

FIG. 2

a) Plot of real and imaginary part of modulus versus log frequency. b) Reduced plot. c) Spectrum of relaxation times.

the structure cannot keep up with the cycling pressure, and the liquid becomes progressively less and less compressible. Finally, at high frequencies the liquid exhibits glass-like moduli and

$$M_\infty = K_\infty + 4/3\ G_\infty \tag{4}$$

where K_∞ is the unrelaxed modulus of compression and G_∞ is the shear modulus. Notice liquids do not exhibit shear rigidity, thus the high frequency behavior is definitely a solid-like property. Actually, it is the glass modulus that one is observing, and by changing the frequency one can go through the glass transition. Associated with the dispersion of the real part of the modulus there is an absorption of the sound wave which is given by the imaginary part of the modulus. Just like the low frequency real part of the modulus was the liquid modulus of compression (reciprocal of the compressibility) the low frequency imaginary part of the modulus is given by $\omega\eta_L$, where ω is the angular frequency and η_L is the longitudinal viscosity. In a liquid there are two elementary and independent stress-strain processes, one associated with shear and the other with compression. The shear component is characterized by the well-known shear viscosity, η_s while the compressive component is characterized by an analogous volume viscosity, η_v. The longitudinal viscosity is represented by

$$\eta_L = \eta_V + 4/3\,\eta_S \quad . \tag{5}$$

By measuring the propagation of both shear and longitudinal sound waves, one can subtract G from M and obtain the relaxation curves for pure compression. By calculating the reduced or normalized moduli Figure 2 (b)

$$N_k' = \frac{K' - K_o}{K_\infty - K_o} \; ; \quad N_k'' = \frac{K''}{K_\infty - K_o} \quad \text{for K} \tag{6}$$

at each frequency one can compare

$$N_G' = \frac{G'}{G_\infty} \; ; \; N_G'' = \frac{G''}{G_\infty} \tag{7}$$

for G the experimental curves with theoretical results for a single relaxation time, τ,

$$N' + iN'' = \frac{(\omega\tau)^2}{1 + (\omega\tau)^2} + i \frac{\omega\tau}{1 + (\omega\tau)^2} \tag{8}$$

In general, the data does not fit, and one has to assume a distribution of relaxation times $g(\tau)$. Thus, one has integrals of the form

$$N' + iN'' = \int_0^\infty \left[\frac{(\omega\tau)^2}{1 + (\omega\tau)^2} + i \frac{\omega\tau}{1 + (\omega\tau)^2}\right] g(\tau)d\tau \tag{9}$$

where various $g(\tau)$'s may be used depending upon the liquid being investigated. For molten salts and oxides the form of $g(\tau)$ used most commonly is calculated from an assumed Gaussian distribution of activation energies. Because the relaxation times are exponential functions of the activation energies, $g(\tau)$ is a Gaussian on the logarithm of the relaxation times

$$g(\tau)d\tau = (b/\sqrt{\pi})\exp\left[-b^2 \ln^2(\tau/\tau')\right] \; d\ln\tau \tag{10}$$

where τ' is the most probable τ, and b is the width.

Schematically $g(\tau)$ is given in Figure 2(c). From these equations one obtains a very useful expression

$$\eta_S = G_\infty \bar{\tau} = G_\infty \int_0^\infty \tau \, g(\tau) d\tau \tag{11}$$

where $\bar{\tau}$ is the average shear relaxation time.

OXIDE GLASSES

Initial Departures from Arrhenius Region

The apparent activation energy[4] can be calculated from the "raw" viscosity data by making a least squares fit of the equation

$$\ln \eta = A + E_{app}/RT \tag{12}$$

Since, in general, the viscosity does not obey this equation, the temperature intervals should be limited. For example, in Figure 3 the apparent activation energy versus temperature for B_2O_3[5] is not a constant.

Because of the finite temperature intervals used, the onset of non-Arrhenius behavior is exaggerated. For example, B_2O_3 obeys Eq. (12) down to 800°C in Figure 4, even though the apparent activation energy increases below 800°C in Figure 3. A much more significant point is the behavior at 650°C. Here in both plots the behavior is definitely non-Arrhenius. Tauke et al.[6] from shear relaxation

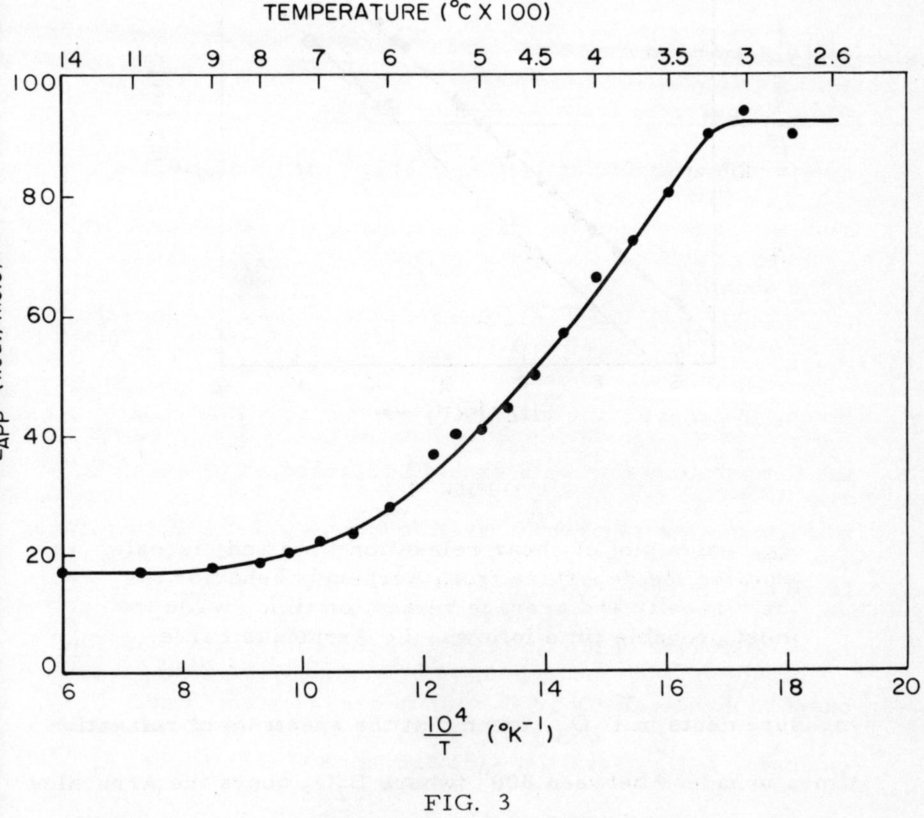

FIG. 3

Plot of apparent activation energies versus temperature for B_2O_3 glass from Ref. 10.

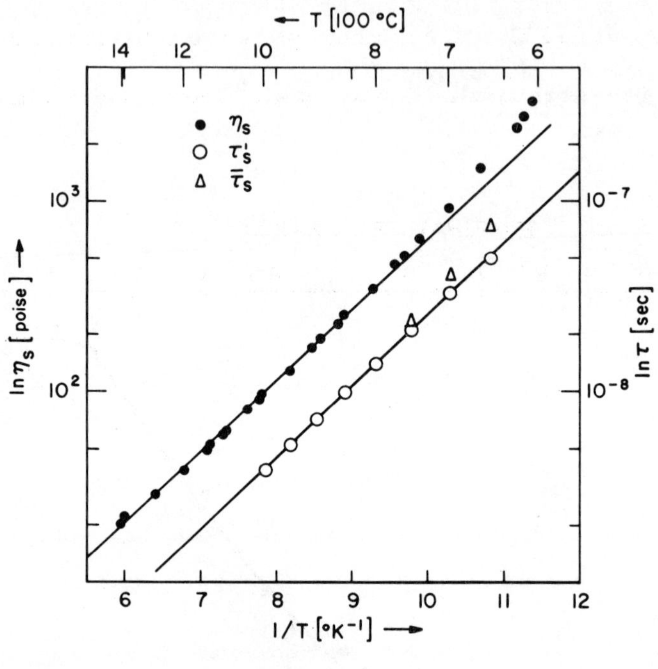

FIG. 4

Arrhenius plot of shear relaxation time and viscosity showing the departure from Arrhenius behavior for the viscosity and average relaxation time while the most probable time follows the Arrhenius curve.

measurements in B_2O_3, found that the spectrum of relaxation times broadens between 800° (where B_2O_3 obeys the Arrhenius equation) and 650^oC. Plots of the distribution of relaxation times are shown in Figure 5. By using Eyring's rate equation

$$\tau = A\exp[E/RT] \tag{13}$$

where A is a constant, E the "activation free energy," T is

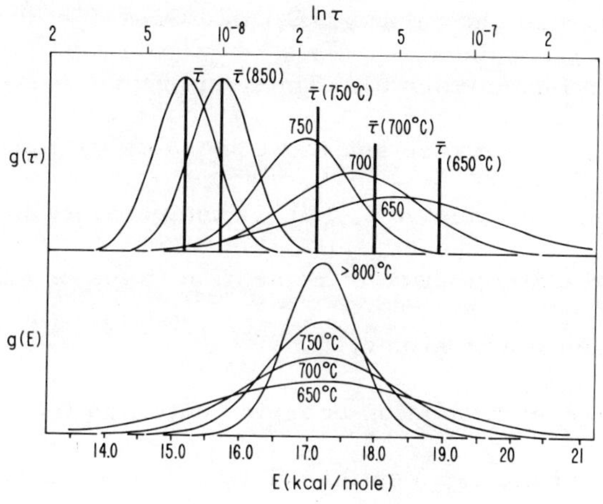

FIG. 5

Plot of the relaxation time spectrum and the activation energy distribution.

the temperature in ^{o}K, and R the gas constant, and by assuming that A is the same constant for all the relaxation spectra one can calculate the distribution of activation free energies. The fractional number of molecules which relax with relaxation times τ_i between $[\ln\tau_i - 1/2\Delta(\ln\tau)]$ and $[\ln\tau_i + 1/2\Delta(\ln\tau)]$ is

$$g_{\tau}(\tau_i)d\ln\tau = g_E(E_i)dE \tag{14}$$

where $g_E(E_i)dE$ is the fractional number of molecules with activation free energy between $E_i - 1/2dE$ and $E_i + 1/2dE$. The activation energy E_i is given by

$$E_i = RT(\ln\tau_i - A) \tag{15}$$

From this one can show that $g_E(E_i)$ is a Gaussian in E, rather than in lnE (Figure 8). Thus, unlike the relaxation time spectra, the average and most probable values coincide. Also, the average value of E, ($\bar{E}$), is temperature independent even though the distribution of relaxation times broadens to both higher and lower values.

The larger activation energies outweigh the lower ones because of the exponential relation between activation energies and relaxation times. The apparent activation energy further emphasizes non-Arrhenius behavior[7] because

$$E_{app} = E - T(\partial E/\partial T) \qquad (16)$$

between 700 and 800°C, E starts to become temperature dependent, i.e., large $\partial^2 E/\partial T^2$. Thus, for a small increase in E with decreasing temperature there is a large increase in E_{app}. (Note: $\partial E/\partial T$ is negative.)

Therefore, the cause of the non-Arrhenius region may not be a higher average activation energy but rather the appearance of a distribution of energies some of which are larger than $\bar{E}$. It should be emphasized that in previous work on GeO_2[1] where the spectrum was measured by direct time dependence technique the same behavior was observed, i.e., the onset of a non-Arrhenius region in which smaller as well as larger activation energies appeared. The present theories[8,9,10] for a non-Arrhenius behavior generally depend upon some increase in cooperative behavior, or a loss in certain diffusing degrees of freedom as the temperature is lowered. These theories lead to an increase in apparent activation energies, but it is not clear how any of these can lead to the presence of smaller activation energies.

As an example of this disagreement let us consider the Adam and Gibbs[10] viscosity theory. They assume that each flow unit has to overcome a potential barrier having an activation energy E in order to perform a diffusion "jump." They further assume that in order that the liquid go from one equilibrium configuration to the next, more than one flow unit may have to simultaneously jump. The number of simultaneous jumps required depends on the number of equilibrium configurations available, which in turn is a function of the excess entropy of the liquid. Adams and Gibbs suggested that the activation energy would have form

$$E = \mu c / S_{ex} \tag{16a}$$

where μ is the activation energy for one flow unit, S_{ex} is the excess entropy that a liquid has above its solid counterpart and c is a constant.

This theory, although not necessarily Eq. (16a), predicts that at high temperatures, when the excess entropy is large enough to permit individual jumps, the relaxation time will have an Arrhenius temperature dependence. Under this condition one would expect a single relaxation time since all the flow units would have the same potential barrier. As the temperature is lowered the number of configurations available would decrease and the same micro regions would require two or more simultaneous jumps. This would predict the appearance of higher activation energies. If one believed the model literally one would expect the appearance of activation energies 2μ, 3μ, etc. Thus, the Adam and Gibbs' model predicts an Arrhenius high temperature behavior with an appearance of higher activation energies, simultaneously with the onset of non-Arrhenius behavior. It cannot account for the appearance of lower activation energies so clearly demonstrated in the experiments.

Conductivity in Oxide Glasses

Einstein's relations do not hold for the temperature dependence of conductivity, σ, and viscosity, η, in oxide melts. In fact the product $\sigma\eta$ varies in many oxides by six orders of magnitude or more, indicating that the conductivity and viscous flow have entirely separate mechanisms in the oxides.[11] This can best be understood if one considers the molecular nature of these materials. Using the standard naive, but yet very useful model, one can divide the cations into two groups: First, the network formers which are covalently bonded to the oxygen atoms i.e., Si, Al, B, Ge, and second, the network modifiers which are ionic bonded to the oxygen atoms, i.e., Na, K, Li, Ca, Mg, etc. One measures the slowest moving mechanism in the shear viscosity, that is the deformation of the covalently bonded network. Conductivity measurements reflect the motion of the fastest moving charged particle, that is, the network modifiers. As a result of this dual mechanism conductivity measurements can be made below the glass transition permitting one to measure the conductivity as a function of annealing.

Further analysis is possible if one considers a model which was developed[12] to analyze annealing experiments on a borosilicate crown glass.[13] In that model various alternative distributions were considered to represent the spectrum of relaxation times. It was noted that even though a given distribution of relaxation times has a unique frequency representation of the modulus, the inverse is not true for data with some experimental uncertainty. It was found that a Gaussian distribution could be fairly well represented by two appropriately chosen delta functions. The two-relaxation time model was chosen in that case for mathematical convenience in determining boundary values and for giving physical significance to the resulting parameters.

In this model, $g(\tau)$ is assumed to be two delta functions one at τ_1 and the other at τ_2. In order to obtain a G' which is anti-symmetric and a G'' which is symmetric in the logarithm of the frequency it is necessary to have equal weighting factoring for each τ_i. If τ' (from the Gaussian) is set equal to the geometric average of the two times (i.e., $\tau' = \sqrt{\tau_1 \tau_2}$), then $\tau_1 = \tau'/\theta$ (where $\theta^2 = \tau_2/\tau_1$), and $\tau_2 = \tau'\theta$. Inserting this relation in equation (9) one has

$$N' = \frac{(\omega\tau'\theta)^2}{2[1+(\omega\tau'\theta)^2]} + \frac{(\omega\tau'/\theta)^2}{2[1+(\omega\tau'/\theta)^2]} \tag{17}$$

and

$$N'' = \frac{\omega\tau'\theta}{2[1+(\omega\tau'\theta)^2]} + \frac{\omega\tau'/\theta}{2[1+(\omega\tau'/\theta)]} \tag{18}$$

In the limit of the width parameter $\theta = 1$ this model gives the usual single relaxation equations. For $\theta<3$ this model predicts the frequency dependencies of both G' and G'' similar to those of the Gaussian model. Above $\theta = 3$ there is a significant splitting of the relaxation region.

In the borosilicate annealing experiments,[12] the model provided a convenient framework in which to expand the failing fictive temperature concept of Toole.[14] Two parameters were thus available to characterize the sample history. For example, it was found that the ionic conductivity below the glass transition temperature is completely controlled by the parameter associated with the fast volume relaxation time, while index measurements are controlled by both parameters (fast and slow volume relaxation times). In the case of ionic conductivity the short ionic jump time was directly connected to the existence of regions of short volume relaxation time. The index of refraction or density measurements while not as positively linked to the structure of glass showed that the only consistent conclusion was the direct relationship between the spectrum of relaxation times and a spatial distribution of environments. These results at the same time contradict models based on non-exponential decay for each flow unit.

Relaxation Spectrum in Oxide Mixtures

Macedo, Simmons, and Haller[15] measured the spectrum of relaxation times in an alkali borosilicate of a composition which is known to undergo a liquid-liquid immiscibility at 749°C. The spectrum was a very narrow peak at high temperatures which broadens as one approaches the immiscibility temperature. Since the only other molten oxide, B_2O_3, in which the spectrum of shear relaxation was measured, has a temperature dependent spectrum, a comparison is somewhat ambiguous. But within the present equipment limitations the soda borosilicate glass has a much broader spectrum of shear relaxation times than B_2O_3.[6]

A correlation between the spectrum of relaxation times and fluctuation theory was developed in which the activation energy was considered to be a function of the environment within a radius r_o about the molecule (the sphere of influence). For immiscible systems where the differences in environment are due to composition fluctuations, numerical results could be obtained using the supercritical fluctuation theory of Ornstein and Zernike.[15] These calculations not only indicated that the temperature dependence of the spectrum was consistent with the theory, but also an estimate for r_o was obtained (44Å). This fluctuation model was qualitatively applied to GeO_2 glass, where the spectrum of volume relaxation is temperature dependent within the glass transformation region.[1] In germania the differences in environment are not due to composition fluctuations but rather association fluctuations. From this theory it was predicted that these association fluctuations would have a radius of about 40Å in the thermally arrested glass. Electron micrographs of Zarzycki and Mezard[16] on GeO_2 support this point.

MOLTEN SALTS

$ZnCl_2$

Gruber and Litovitz[17] have studied the ultrasonic relaxation in $ZnCl_2$. They found both volume and shear to exhibit a single relaxation time. In fact, it was because $ZnCl_2$ showed an Arrhenius dependence of the viscosity, that Litovitz and McDuffie[18] proposed that non-Arrhenius behavior was associated with a distribution of relaxation times. Since the work on the oxides has answered the above question, one wonders why the spectrum of relaxation times in $ZnCl_2$ is comparable with that of the oxides and very much narrower than the organic associated liquids like glycerine. Gruber and Litovitz, and Angell[19] ascribe this to the partially covalent nature of the $ZnCl_2$ bonds. In this way one can picture $ZnCl_2$ as a network former equivalent to B_2O_3 or GeO_2. Like these materials if one lowers the temperature sufficiently it becomes non-Arrhenius and exhibits a distribution of relaxation times.[20]

KNO_3-$Ca(NO_3)_2$ Viscosity

A study[21] of 60% KNO_3- 40% $Ca(NO_3)_2$ by the authors was undertaken because no covalent bonding is expected between ions in these salts. According to Angell[19] the ratio T_g/T_o, where T_o is taken from the Fulcher[22] equation

$$\ln \eta = A + B/(T - T_o), \tag{19}$$

when considered as a function of various chemical systems, is a minimum for the nitrates (1.02). This should indicate a glass with smaller excess entropy than either the organic or the oxide glasses. Our study encompasses three parts: first, examining the validity of the Fulcher equation for viscosity (Angell[19] had tested it for conductivity over a limited temperature range); second, measuring the spectrum of relaxation times; and third, comparing the conductivity with the visco-elastic parameters.

If the viscosity data are limited to the same temperature range as that of Angell's[19] conductivity it follows a Fulcher equation, see Figure 6. However, if instead of stopping 76°C above T_o, one goes to within 36°C of T_o, the Fulcher equation overestimates the data by a factor of 500. The whole temperature range can be fitted to the Fulcher

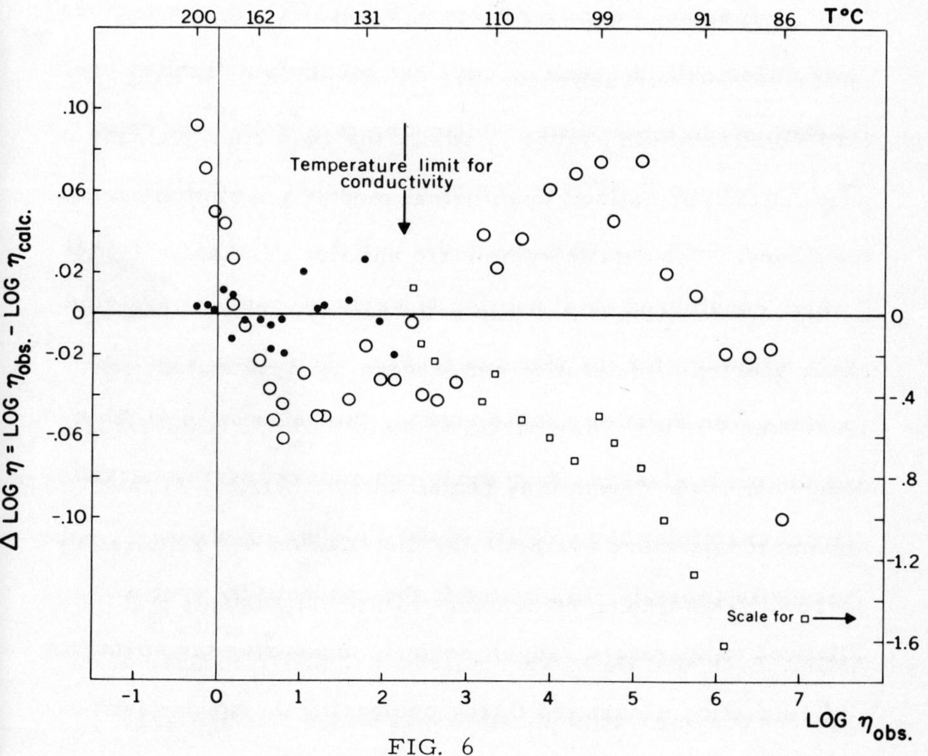

FIG. 6

Deviation of the Fulcher Equation for the KNO_3-$Ca(NO_3)_2$ mixture. □ Fit 1; ○ Fit 2.

equation resulting in a lower T_o (19°C below the old) and a 54% larger B. Even so this new fit has the characteristic S shape indicating that the discrepancy of the Fulcher equation is well outside the experimental uncertainty. The failure of the Fulcher equation of course raises the question as to how much excess entropy do these glasses have.

KNO_3-$Ca(NO_3)_2$ Spectra of Relaxation Times

Weiler and Macedo[23] measured the shear and longitudinal moduli versus frequency. A sample of their data (117°C) is shown in Figure 7, where the real and imaginary parts of the normalized longitudinal moduli are plotted versus frequency. The data is symmetric and fits a Gaussian distribution of relaxation times (the solid curve). The volume and shear spectra of relaxation times were determined between 126°C and 100°C. The spectra were found to be temperature dependent, even though they remained symmetric. At the highest temperature available the distribution was remarkably narrow. This finding contradicts earlier statements that $ZnCl_2$, B_2O_3 and GeO_2 were narrow primarily because of covalent bonding. It is possible that because of the large asymmetry associated with the nitrate ion directional bonding may cause the observed narrow spectrum.

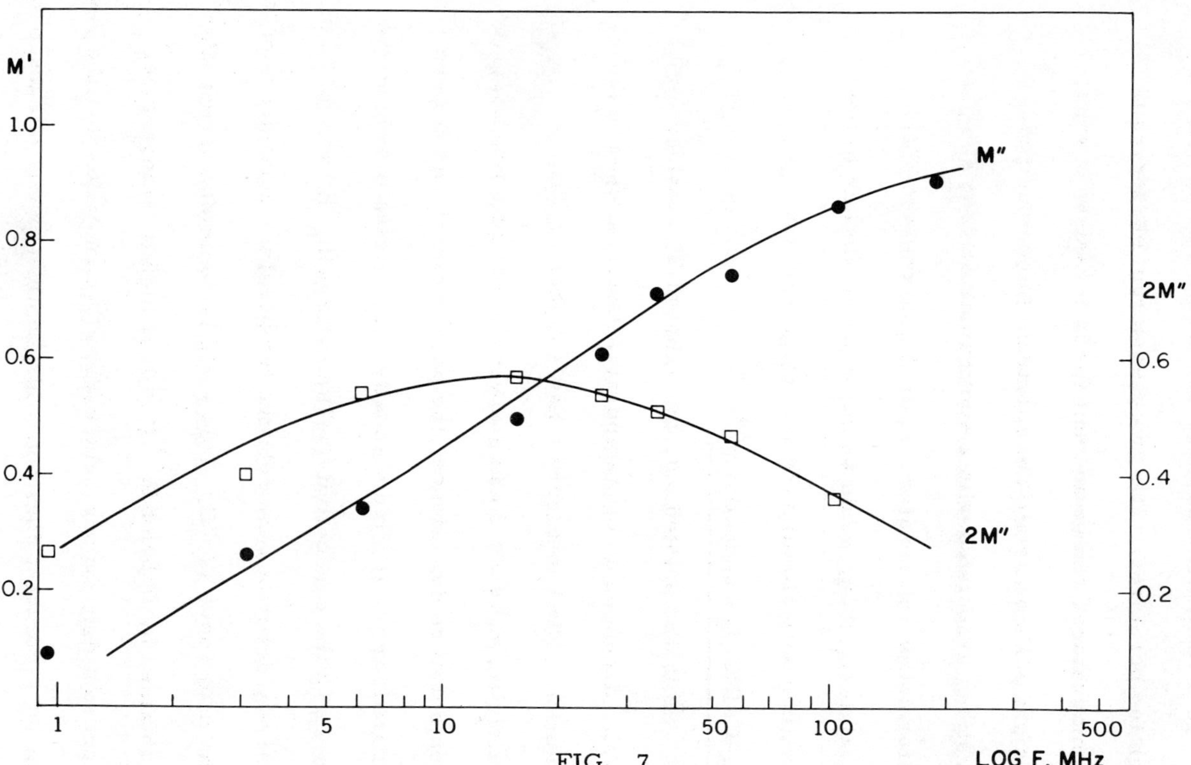

FIG. 7

Real (M') and imaginery (M'') part of the reduced modulus for KNO_3-$Ca(NO_3)_2$ mixture.

It is difficult to account for this very narrow spectrum at high temperatures. The broad spectrum obeserved at low temperatures can be account for in a variety of ways. One might perhaps relate the spectrum to composition fluctuations associated with a subliquidus or even subglass transformation temperature liquid/liquid immiscibility. This mixture, then, would be remarkably similar to the sodium borosilicate mixture described earlier.

<u>KNO_3-$Ca(NO_3)_2$ Conductivity</u>[24]

As it was pointed out, the value of T_o obtained for viscosity depended on the temperature interval used in the calculation. Fit 1 in Figure 6 had a higher T_o than the conductivity, while Fit 2 had a lower T_o. In order to compare the temperature dependence, the parameters A and B were calculated for the limited viscosity data forcing it to have the same T_o as the conductivity. The ratio of $B_\eta / B_\sigma = 1.30 \pm .02$ effectively describes the relative temperature dependence of all the data above 125°C. This ratio is somewhat larger than that obtained by Ubbelohde[25] (1.20) at higher temperature. The present data support the earlier[19] conclusion that the relation $\sigma\eta$ = constant does not hold. Instead of this microscopic expression let us consider Eyring's rate theory Eq. (18)

$$\sigma = \frac{F\,|Z|\,na^2}{ckT\tau_\sigma} \tag{20}$$

in which F is Faraday constant, Z is the average charge per ion, n is the number density, c is a geometrical constant usually set at 6, a is the root mean square diffusion length, τ_σ is the relaxation or jump time associated with the conductivity process. This relates the conductivity to a structural relaxation time τ_σ and a jump distance, a, while the shear viscosity is proportional to both the average shear relaxation time $\bar{\tau}_s$ and the shear modulus of rigidity, G_∞. One should identify the temperature dependence of the conductivity directly to that of structural time. Unfortunately, at this point one is faced with a large number of viscoelastic relaxation times as possibilities. The two most promising being τ_v' and τ_s' (see Section 2 for definition of most probable time τ'). It is proposed to choose the proper τ by matching the temperature dependences. From preliminary data $E_{\tau'_s}/E_\sigma = 1.2$, $E_{\tau'_v}/E_\sigma = 0.8$, and $E_{\tau'_L}/E_\sigma = 1$. The longitudinal relaxation in this mixture represents an equally weighted addition of the shear and volume relaxations. This indicates that both kinds of elementary motions contribute about equally to conductivity as one might have expected. Substituting τ_L' in Eq. 20 one can solve for

$$a^2 = \frac{ckT\,\tau_L'}{n\sigma F|Z|} \qquad (21)$$

The geometric constant c was set at 6, and all ions were considered disassociated and contributing to the conductance in the calculation of n and Z. The jump distance was found to be 2.6Å. This large distance is compatible to a weighted average of the K^+, Ca^{++} and NO_3^- ion diameters.

Let us now consider various types of transport model.

A) The simulated models tend to predict a jiggling type of ionic motion. The vibration and diffusional motions are not distinguished. In each ionic "collision" there is a small permanent displacement of the ion's equilibrium position. Since this model would have not only a very small "a" but also a poorly defined activation barrier, i.e. a broad spectrum of relaxation times, it must be ruled out.

B) There are various molecular descriptions of the jump model. However, our data does not permit us to distinguish between them. We will describe our best guess[24] of the jump model, recognizing that most jump models will fit the above data. Let us define the period in which the ion performs no diffusional motion as the residence time λ_R. In this period the ion's motion is arrested by the solid like structure about it. An energy fluctuation may cause the structure to "melt". During the molten time the ion may undergo many diffusion motions. This time will be denoted as λ_m. If $\lambda_R >> \lambda_m$ then we have a jump model. The mean square diffusion distance, a, should be comparable with ionic diameters. If it was much smaller the new positions would be unstable compared with the old positions, because the lattice about this microregion is "frozen." A much larger "a" would require a very large microregion to melt and an exceedingly large activation energy. The microregion concept could also lead to a well definied activation energy since they may be large enough to average out environmental differences (see relaxation spectrum in oxides mixture). Two pertinent questions which cannot be answered at present are: (a) how long is λ_m and how many collisions occur during this period? (b) how large are these activated regions in the molten salts?

SUMMARY

Relaxation times may be obtained directly from visco-elastic measurements. These can be used to provide a critical test of the applicability of viscosity theories. Data on oxide melts contradict the predictions of the well-known viscosity models. Additionally, in these substances there appears to be no direct relationship between electrical conductivity and viscosity.

The topological model permits one to correlate the ionic conductivity below the glass transition temperature with the thermal history of the glass. In so doing it predicts a geometric distribution of relaxation mechanisms.

The flow mechanism in ionic and covalent molten salts was found to be remarkably similar to that of oxides. From viscoelastic and conductivity data one can calculate the average diffusion jump distance, a. The narrow spectra of relaxation times in KNO_3-$Ca(NO_3)_2$ mixture can be described in terms of a well-defined diffusion jump which is not expected in a totally ionized melt.

ACKNOWLEDGMENT

This research was sponsored by A.F.O.S.R. under Grant #68-1376.

REFERENCES

1. A. Napolitano and P. B. Macedo, J. Res. NBS 72A, 425-33 (1968).

2. K. F. Herzfeld and T. A. Litovitz, Absorption and Dispersion of Ultrasonic Waves, Academic Press, New York and London (1959).

3. T. A. Litovitz and C. M. Davis, Physical Acoustics, Vol. II, Ch. 5, Academic Press, New York and London (1965).

4. S. Glasstone, H. Eyring, and K. Laidler, Theory of Rate Processes, McGraw-Hill, New York (1941).

5. P. B. Macedo and A. Napolitano, "The Inadequacy of Viscosity Theories for B_2O_3," J. Chem. Phys., (to be published).

6. J. Tauke, T. A. Litovitz, and P. B. Macedo, J. Am. Ceram. Soc., 51, 158-63 (1968).

7. P. B. Macedo, W. Capps, and T. A. Litovitz, J. Chem. Phys., 44, 3357-64 (1966).

8. P. B. Macedo and T. A. Litovitz, J. Chem. Phys., 42, 245-56 (1965).

9. T. S. Ree, T. Ree, and A. Eyring, Proc. Natl. Acad. Sci., U.S., 48, 501-16 (1962).

10. G. Adam and J. H. Gibbs, J. Chem. Phys., 43, 139-46 (1965).

11. J. D. MacKenzie, Modern Aspects of the Vitreous State, Vol. 1, Butterworths, Washington, D. C. (1960).

12. P. B. Macedo and A. Napolitano, J. Res. NBS 71A (Phys. and Chem.), 3, 231038 (1967).

13. S. Spinner and A. Napolitano, J. Res. NBS 70A (Phys. and Chem.), 2, 147-152 (1966).

14. A. Q. Toole, J. Amer. Ceram. Soc., 29, 240 (1948).

15. P. B. Macedo, J. H. Simmons, and W. Haller, "Spectrum of Relaxation Times and Fluctuation Theory--Ultrasonic Studies on Alkali--Borosilicate Melt," J. Phys. and Chem. of Glasses (to be published).

16. J. Zarzycki and R. Mezard, J. Phys. and Chem. of Glasses, 3, 163 (1962).

17. G. Gruber and T. A. Litovitz, J. Chem. Phys., 40, 13026 (1964).

18. T. A. Litovitz and G. McDuffie, J. Chem. Phys., 39, 729 (1963).

19. C. A. Angell, J. Phys. Chem., 65, 1917 (1964). (For further reference see also chapter in this volume written by C. T. Moynihan and C. A. Angell.)

20. M. Goldstein, in J. D. MacKenzie, "Modern Aspects of the Vitreous State," Vol. 3, Butterworths, Washington, D. C. (1964).

21. R. A. Weiler, S. Blaser, and P. B. Macedo, "Viscosity of KNO_3-$Ca(NO_3)_2$ Mixture," J. Am. Ceram. Soc. (to be published).

22. G. S. Fulcher, J. Am. Ceram. Soc., 8, 339-55 (1925).

23. R. A. Weiler and P. B. Macedo, "Structural Relaxation and Ionic Mobility in Molten Salt," J. Chem. Phys. (to be published).

24. P. B. Macedo, R. A. Weiler, and T. A. Litovitz, "Correlation Between Structural Relaxation and Ionic Mobility in Molten Salt," J. Chem. Phys. (to be published).

25. E. Rhodes, W. E. Smith, and A. R. Ubbelohde, Proc. Royal Soc. London, 285A, 263-74 (1965).

ELECTRONIC CONDUCTION IN FUSED SALTS[1]

L. F. Grantham and S. J. Yosim

Atomics International Division of
North American Rockwell Corporation
Canoga Park, California 91304

INTRODUCTION

In general, molten salts conduct electricity by the transport of ions. However, it has been shown that electronic conduction can be an important mechanism, particularly in the case of molten metal-metal halide solutions. Above the consolute temperature where the metal and salt are miscible in all proportions, the conductivity of the solution must approach that of the pure metal as the metallic content of the solution is increased. It has been experimentally shown that above the consolute temperature the conductivities of these solutions increase monotonically as a function of composition and there is a gradual transition from ionic to electronic conduction. The systems in which electrical conductivities have been measured across the entire composition range from pure salt to pure metal are K-KBr[2], Bi-BiI_3[3,4] and Bi-$BiBr_3$[5].

In addition to metallic conduction, at least one other mechanism of electronic conduction must occur, especially in salt-rich regions where the magnitude of conductivity is too low to be metallic, but too large to be ionic. One such mechanism of electronic conduction is "electron hopping" or electron exchange, in which the localized electron is transferred between species of different valences. This mechanism, originally proposed by Rice[6] for alkali-alkali halide systems, was developed further by Raleigh[7] to explain the almost exponential rise in the specific electrical conductivity κ with metal composition in the salt-rich region of the Bi-BiI_3 system[4]. Grantham[5] also applied this model to the Bi-$BiCl_3$ and Bi-$BiBr_3$ systems.

As in the Bi-BiX_3 system where Bi^{+}-Bi^{+++} are the exchanging species, electron exchange might be expected to be an important mechanism of conduction in other molten salt systems which contain two valence states of the same element. Examples of such systems are the neodymium-neodymium halide systems investigated by Bredig et al[8-10]. Since a positive deviation from additivity was observed for conductance in these systems (maxima occurred in the Nd-$NdBr_3$ and in the Nd-NdI_3 systems), they suggested that electron exchange between the Nd^{++} and Nd^{+++} species may occur. However, the total specific conductivities were not unusually large in these systems[11]; in no case did the conductivities exceed $2(ohm\text{-}cm)^{-1}$. Therefore a search was made for a system in which the conductivities at intermediate concentrations would be both higher than the pure salts and higher than that expected for ionic solutions. As part of this search, the electrical conductivities of the $CuCl$-$CuCl_2$ system were measured over the entire composition range up to 800°C.

EXPERIMENTAL

Materials

Cuprous chloride was prepared by dissolving reagent grade material in hot concentrated HCl which contained copper filings to assure that no cupric chloride was present. The solution was poured into cold water and the insoluble Cu-CuCl mixture was filtered from the solution. The dried Cu-CuCl mixture was distilled at 1100° in vacuo in a quartz "λ" tube; the distillate was subsequently filtered through a quartz frit. The purified CuCl was a clear colorless solid. Anhydrous reagent-grade cupric chloride was dehydrated by passing dry HCl through the powdered material at 110°C for several hours. Subsequently Cl_2 gas was passed through the powder at 100°C to assure complete oxidation of the copper to the cupric state.

Apparatus

One complicating aspect of this system is the excessive decomposition pressure of Cl_2 over the molten $CuCl_2$. For example, at the melting point of $CuCl_2$ (625°C) the decomposition pressure of Cl_2 is about 25 atm[12], while at 800°C it may be as high as 175 atm. These high pressures required a special pressurized enclosure, which was described previously[13], to prevent the high internal Cl_2 pressure at high temperatures from rupturing the quartz cells. In the initial phase of this study, the external argon pressure in the vessel exceeded the internal chlorine decomposition pressure by only 10-20 atmospheres; however this required numerous pressure adjustments at the higher temperatures for the $CuCl_2$-rich region. Later it was found that a properly designed conductivity cell[13] could withstand 150 atmospheres excess external pressure. Therefore 150 atmospheres argon was applied externally and only a single pressure adjustment was required to prevent cell rupture.

In order to correct the composition of the melt due to thermal decomposition of $CuCl_2$ (see below), a determination of the total internal volume of the cell was measured by determining the difference in weight of the cells when empty and when filled with water to a predetermined mark on the fill tube. After the conductivity cells were filled with weighed amounts of $CuCl_2$ in an argon atmosphere drybox, the desired amounts of CuCl were added; the cells were evacuated to 10^{-5} mm of Hg and sealed off at the predetermined mark on the fill tube. The error in the internal volume of the cell associated with sealing the quartz fill tube was less than 0.5 cc (i.e. < 2%).

The conductivity cell[13], Jones Bridge[14], and associated equipment[4] have been described previously. The quartz cells were calibrated with .01N HCl and with mercury, as in previous metal-metal salt conductivity studies[4].

RESULTS

It is desirable to present the data in terms of the most likely species present. Unfortunately it is not known whether cuprous chloride consists mainly of CuCl, Cu_2Cl_2 or Cu_3Cl_3. The gas phase in equilibrium with the melt is known to consist mainly of the trimer[15]; however, the presence of this species in the gas phase is not sufficient evidence that it exists in the melt. In addition, one would not expect the trimer to possess as high a specific conductivity as is observed (κ = 4 ohm^{-1} cm^{-1} at the melting point). In view of the stability of the dimer of other subhalides, such as Hg_2Cl_2[16,17], formed by reaction of a metal with its divalent salt, it is tempting to consider Cu_2Cl_2 as the component for cuprous chloride as suggested by Fontana et al[12]. However, AgCl which is in the same group as cuprous chloride is not considered to be dimerized; on that basis one might expect cuprous chloride to exist as monomeric ions. In view of the uncertainty of the species in cuprous chloride, the results will be discussed mainly in terms of CuCl and $CuCl_2$; however the effect on the results, if cuprous chloride should exist as Cu_2Cl_2, will also be given.

The specific conductivity results of the molten CuCl-$CuCl_2$ system are tabulated as a function of temperature in Table I. These data are uncorrected for the decomposition of $CuCl_2$ at elevated temperatures. The conductivity data for the more dilute $CuCl_2$ compositions vs. temperature are shown in Fig. 1 while those for the more concentrated $CuCl_2$ compositions are shown in Fig. 2. (Two figures were used for purposes of clarity.) The conductivities increase monotonically with increasing temperature in the temperature range studied (up to 800°C). In general the specific conductivities vary linearly with temperature.

The specific conductivities are plotted as a function of composition in Fig. 3 for several isotherms, 400°, 500°, 600°, 700° and 800°C. The shaded portion of the curves were obtained by extrapolation of the data shown in Figures 1 and 2. A maximum in the conductivity is observed at each temperature. As the temperature increases, the maximum becomes more pronounced and occurs at higher $CuCl_2$ compositions. For example the maximum at 400°C occurs at 25 mole % $CuCl_2$ while that at 800°C occurs at 31%. If one

TABLE 1

Specific Conductivity $(\Omega \text{cm})^{-1}$ of Molten CuCl-$CuCl_2$ System as a Function of Temperature (°C)*

CuCl		5.0%		9.9%		13.6%	
t	κ	t	κ	t	κ	t	κ
424	3.42	408	4.15	440	4.86	422	4.97
459	3.54	432	4.19	460	5.07	436	5.15
474	3.59	441	4.26	490	5.42	491	5.86
495	3.64	453	4.33	522	5.75	508	6.06
512	3.67	489	4.52	556	6.09	541	6.71
554	3.73	504	4.70	594	6.42	566	6.95
591	3.80	532	4.79	624	6.70	591	7.17
607	3.83	551	4.87	645	6.90	620	7.59
620	3.85	613	5.26	654	7.04	649	7.97
672	3.89	633	5.41				
692	3.89	656	5.58				
728	3.92	684	5.75				
769	3.93	700	5.85				
799	3.93	726	5.90				

* Compositions are expressed as mole % $CuCl_2$ in CuCl; data are not corrected for thermal decomposition of $CuCl_2$.

TABLE 1 (continued)

19.3%		31.2%		39.1%		49.5%	
t	κ	t	κ	t	κ	t	κ
385	4.76	405	5.10	523	6.91	556	6.4
395	4.99	438	5.86	545	7.36	600	7.4
442	5.75	455	6.10	568	7.86	649	9.0
469	6.25	467	6.43	591	8.46	671	9.6
478	6.41	488	6.79	618	9.10	704	10.5
492	6.69	501	7.00	642	9.42	738	11.2
501	6.75	519	7.44	659	9.93	764	11.7
505	6.81	543	8.05	689	10.5	792	12.1
528	7.31	564	8.57	704	10.7	805	12.2
532	7.39	584	8.53	723	11.3		
558	7.89	599	9.30	748	11.8		
592	8.36	617	9.71	764	12.1		
621	8.76	640	10.0	776	12.3		
639	9.10	656	10.3	788	12.5		
646	9.18	688	10.9	802	12.7		
		698	11.2				
		720	11.6				
		746	12.0				
		772	12.4				
		802	12.7				

TABLE 1 (continued)

60.0%		79.5%		79.9%		$CuCl_2$	
t	κ	t	κ	t	κ	t	κ
571	5.3	593	3.9	598	4.3	613	3.2
584	5.7	600	4.0	629	5.1	624	3.6
603	6.0	613	4.4	654	5.9	640	4.2
618	6.5	630	4.9	674	6.5	659	4.7
636	7.0	645	5.3	701	7.3	676	5.1
650	7.4	659	5.7	727	8.1	690	5.5
662	7.7	672	5.8	749	8.7	708	6.1
683	8.1	685	6.4	762	8.9	733	6.7
694	8.4	697	6.6			756	7.4
						764	7.6
						779	8.0

plots these results on the basis of Cu_2Cl_2 instead of CuCl, the general features of the curves remain the same except that the maxima occur at higher compositions of $CuCl_2$, increasing from 40 mole % at 400°C to about 50 mole % at 800°C.

Due to the extensive thermal decomposition of $CuCl_2$ at the higher temperatures,

$$2\,CuCl_2(l) \rightleftharpoons 2CuCl(l) + Cl_2(g) \qquad (1)$$

it was necessary to correct the melt compositions for the amount of $CuCl_2$ decomposed and the amount of additional CuCl formed by the decomposition. To carry out this correction it was assumed that the CuCl-$CuCl_2$ system was ideal,

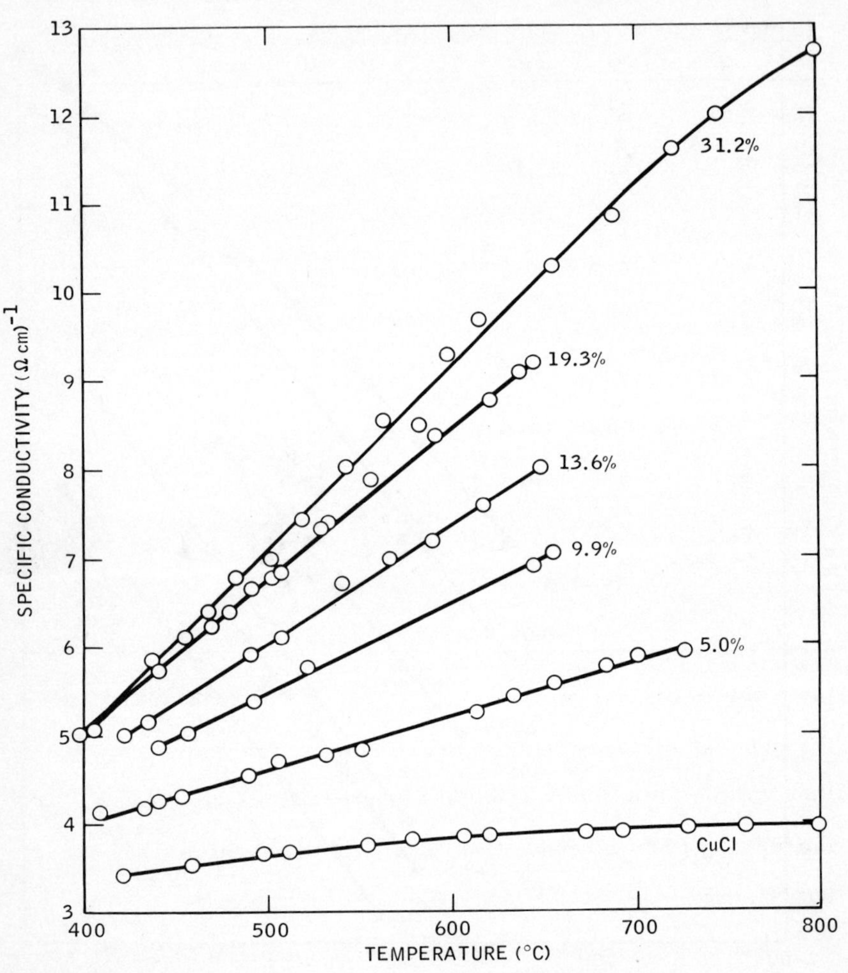

FIG. 1

Specific conductivities of molten CuCl-$CuCl_2$ solutions as a function of temperature. The concentrations are given in mole % $CuCl_2$ in CuCl.

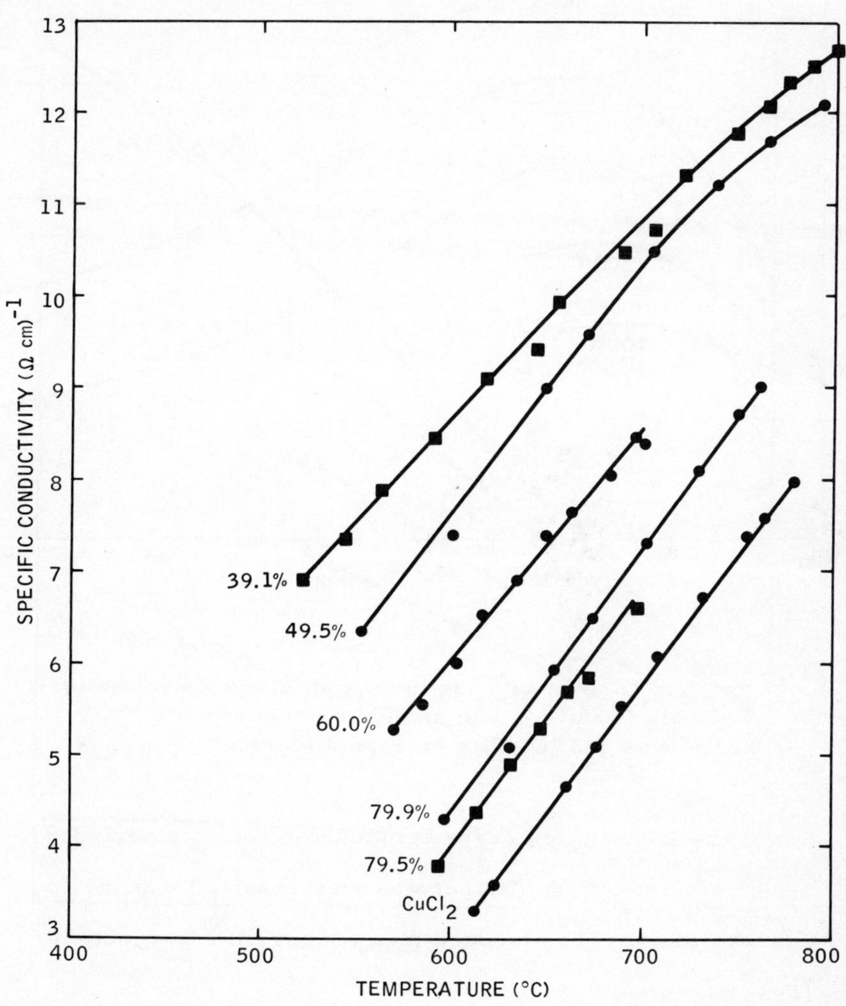

FIG. 2

Specific conductivities of molten $CuCl-CuCl_2$ solutions as a function of temperature. The concentrations are given in mole % $CuCl_2$ in CuCl.

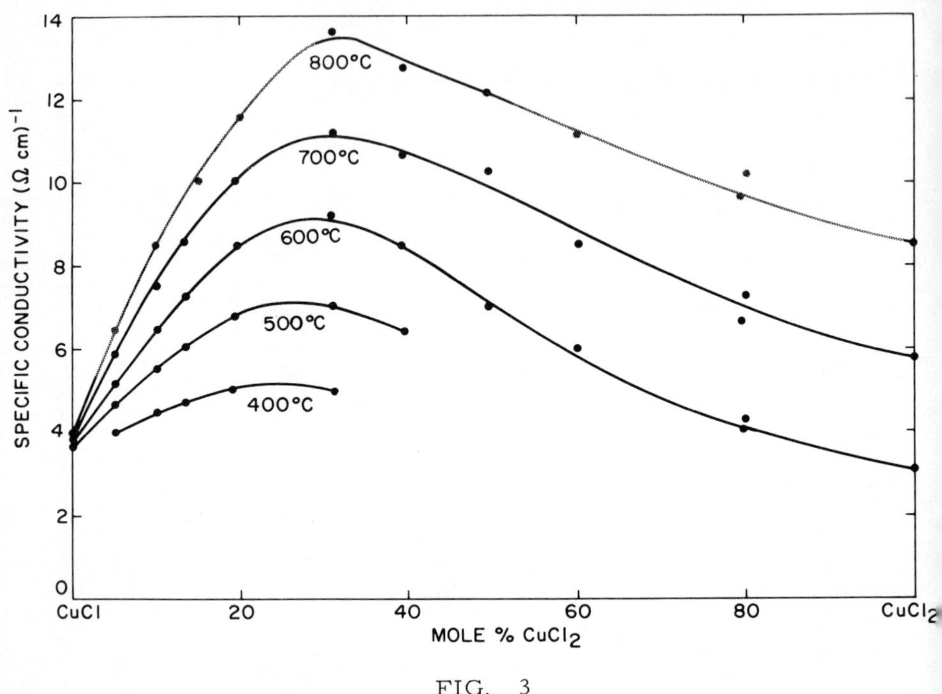

FIG. 3

Specific conductivity isotherms as a function of mole % $CuCl_2$ in CuCl. The shaded portion of the 800° C curve was plotted with extrapolated data.

and thus the activities of the components could be replaced by the mole fractions (X). It was also assumed that the Cl_2 gas was ideal. The standard free energy ΔF^o of reaction (1) is given by

$$\Delta F^o = -NkT \ln(X^2_{CuCl} P_{Cl_2} / X^2_{CuCl_2}) \quad (2)$$

where k and T have their usual significance, N is Avogadro's number, and P is the pressure. By substituting the Cl_2 pressure data for solid $CuCl_2$ tabulated by Fontana et al.[12]

(4.57, 0.309, 0.0209 atm at 833°, 769° and 714°K respectively), and the solubilities of solid $CuCl_2$ in molten CuCl (X_{CuCl_2} = 0.64, 0.38 and 0.22 at these temperatures), values of -0.300, +0.147, and +0.953 were obtained for ΔF^o at these temperatures. The values of ΔF^o for the desired higher temperatures were obtained by a linear extrapolation which yielded the following equation for ΔF^o as a function of temperature (t) in °C: $\Delta F^o = -0.0109t+5.70$. Then from the use of the following equation

$$\Delta F^o = -NkT \ln \left\{(b+2x)^2(0.083xT/V)/(a-2x)^2\right\} \quad (3)$$

where $\underline{a}$ and $\underline{b}$ equal the initial moles of $CuCl_2$ and CuCl respectively, $\underline{x}$ is the number of moles of Cl_2 above the melt, $\underline{V}$ is the volume of gas in liters, and 0.083 is the gas constant, one can obtain values of $\underline{x}$ and thus the true composition of $CuCl_2$, $(a-2x)/(a+b)$, and that of CuCl, $(b+2x)/(a+b)$. Values of $\underline{V}$ were obtained by subtracting the volume of the melt from the total volume. This required a knowledge of the density of the melt. While the densities of CuCl are known, those for $CuCl_2$ are not. Therefore the density of molten $CuCl_2$ was estimated by increasing the molar volume of solid $CuCl_2$ at 298°K by 10%. The density of molten $CuCl_2$ was thus estimated to be 2.9 g/cm^3. It was further assumed that the volumes were additive. The data required to correct the melt composition for thermal decomposition are given in Table II.

TABLE II

Data Required to Correct the Composition of Molten CuCl-$CuCl_2$ Solutions for Thermal Decomposition of $CuCl_2$

Mole % $CuCl_2$*	Moles of $CuCl_2$*	Moles of CuCl*	Volume of Cl_2 (liters)
5.0	.015	.290	.0191
9.9	.030	.272	.0190
13.6	.038	.242	.0194
19.3	.039	.162	.0216
31.2	.088	.192	.0219
39.1	.102	.158	.0223
49.5	.128	.131	.0204
60.0	.127	.085	.0237
79.5	.154	.040	.0238
80.0	.170	.043	.0259
$CuCl_2$	.220	.000	.0198

* Original uncorrected composition and moles present.

The electrical conductivity curves at constant temperature plotted as a function of corrected composition are given in Fig. 4. A similar plot where cuprous chloride is assumed to exist as the dimer instead of the monomer is shown in Fig. 5. The maximum concentration correction due to $CuCl_2$ decomposition can be seen by examining the extreme points of the 600°, 700° and 800° isotherms; these

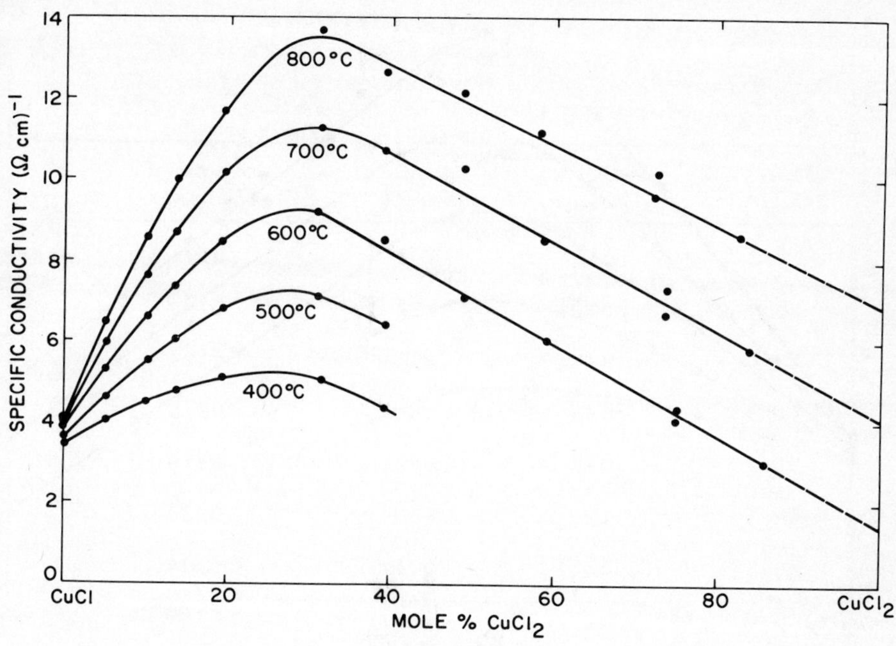

FIG. 4

Specific conductivity isotherms as a function of mole % $CuCl_2$ in CuCl; compositions are corrected for thermal decomposition of the $CuCl_2$.

points represent compositions which were originally 100% $CuCl_2$. The curves are extrapolated to pure $CuCl_2$ (dotted portion of the curves in Figures 4 and 5). The conductivity of pure $CuCl_2$ was taken as 1.4, 4.0 and 6.9 $(\Omega\,cm)^{-1}$ at 600, 700 and 800°C respectively. These values were used to obtain the excess conductivity (see below).

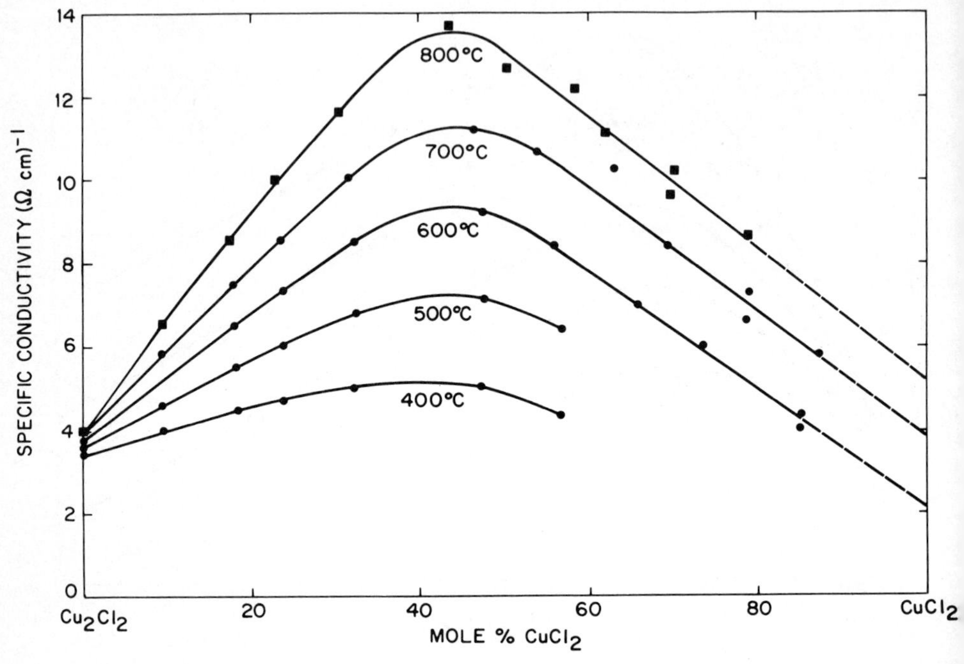

FIG. 5

Specific conductivity isotherms as a function of mole % $CuCl_2$ in Cu_2Cl_2 (dimer); compositions are corrected for thermal decomposition of $CuCl_2$.

At 800°C the corrected composition for the "pure" $CuCl_2$ is about 80 mole % $CuCl_2$. However the correction was quite small for compositions less than 60% $CuCl_2$. Thus, the compositions of maximum conductivities remain essentially unchanged.

ELECTRONIC CONDUCTION

The activation energy for conduction, obtained from the data in Figures 4 and 5, varies considerably with both composition and temperature. For example at 750°C the activation energy varies from 1 kcal for pure CuCl to 12 kcal for pure $CuCl_2$. At 40 mole % $CuCl_2$ the activation energy is 1 kcal at 450°C and 4 kcal at 650°C.

DISCUSSION OF RESULTS

The fact that the current efficiency of electrolysis of molten CuCl near the melting point is 100 ± 3%[18] indicates that pure molten CuCl is a typical ionic conductor. Since no data are available to indicate the mechanism of conductivity in $CuCl_2$ it is assumed that ionic conduction predominates. However, since the conductivity is high for a divalent cationic salt, a portion of the current transport may be by an electronic mechanism.

The high values of specific conductivity of CuCl-$CuCl_2$ solutions, especially at elevated temperatures (e.g. up to 13.6 $ohm^{-1} cm^{-1}$ at 800°C), indicate that electronic conduction is playing a significant role in the transport of current in these solutions; no other fused salt system has been reported to have such high conductivity and still be considered as a purely ionic conductor. The fact that the conductivities of the mixtures are considerably greater than those of the pure components suggests that transport by electron exchange is a likely mechanism for electronic conduction. Thus, the conductivity behavior of this system resembles that of the Nd-$NdBr_3$ and Nd-NdI_3 systems in which electron exchange between Nd^{+2} and Nd^{+3} was proposed to explain the conductivity maxima[9, 10]. However, as stated above, the magnitudes of the maximum conductivities of those systems were those expected of ionic systems; therefore there was some question as to whether a maximum in the conductivity was true evidence of electron exchange[11]. In the case of the CuCl-$CuCl_2$ system, the conductivities are so high that there is little doubt that the conductivity maxima of the mixtures are due, at least in part, to electron exchange.

The conductivity behavior of the salt-rich region of several metal salt solutions is compared in Fig. 6. The solid and dotted lines are the conductivity isotherms of the same system but 200°C apart, i.e. K-KI at 700 and 900°C, Bi-BiI_3 at 400 and 600°C, CuCl-$CuCl_2$ at 500 and 700°C and the Cd-$CdBr_2$ system at 600 and 800°C. Since it was postulated that electron exchange also occurs in the Bi-BiX_3 systems, the difference between the conductivity vs composition curve for Bi-BiI_3 and CuCl-$CuCl_2$ should be discussed. In the former system the specific conductivity increases almost exponentially with metal composition, whereas in the latter system the conductivity reaches a

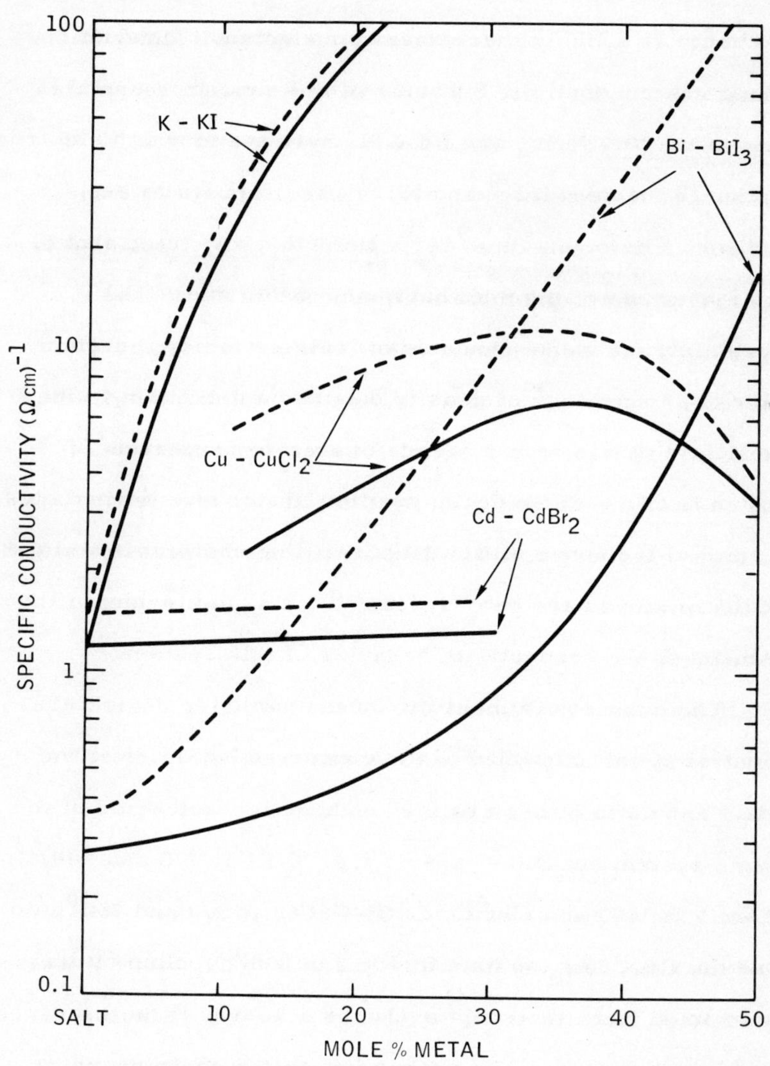

FIG. 6

Specific conductivities of four molten metal-halide systems as a function of mole % metal at temperatures near the melting point of the halide (solid line) and 200°C higher (broken line).

maximum at some intermediate composition. Thus, it appears that if the model proposed by Raleigh[7] to explain the conductivity behavior of the bismuth systems is to be applied to the copper systems, the treatment must be modified. However, it is recognized that any treatment of this system may be somewhat questionable due to the uncertainty as to the predominant species in pure cuprous chloride, to the lack of density data of molten $CuCl_2$, and to the lack of data on other high temperature properties of molten $CuCl_2$ such as decomposition pressure. Nevertheless in spite of the above difficulties it will be shown that a slight modification in Raleigh's treatment can explain many of the features of the conductivity behavior of this system.

In Raleigh's treatment the concentration n of carriers (electrons) corresponded to the concentration of lower valent salt. Thus n is defined by

$$n = NX_{CuCl}/v \tag{4}$$

where v is the molar volume. However, one must take into consideration that the exchanging electron supplied by the Cu^+ donor must have an acceptor Cu^{+2} ion nearby if the exchange is to be completed. The probability that such an acceptor is available is $X_{Cu^{+2}}$. Thus a factor $X_{Cu^{+2}}$ should be included in equation (4).

$$n = NX_{CuCl}X_{CuCl_2}/v \tag{5}$$

(Actually a corresponding factor should have been included in Raleigh's treatment, especially in the more concentrated Bi solutions.) The carrier mobility μ is as defined in Reference 7

$$\mu = \frac{e}{6kT} \nu R^2 \tag{6}$$

where e is the electronic charge, ν is the jump frequency of the electron and R is the distance between the exchange sites.

The excess conductivity κ_e is defined by

$$\kappa_e = \kappa - X_{CuCl}\kappa_{CuCl} - X_{CuCl_2}\kappa_{CuCl_2} \tag{7}$$

and is that portion of the conductivity that can be attributed to electron exchange. The magnitude of κ_e is shown as a function of composition at 100°C intervals in Figure 7. κ_e is also defined[7] as

$$\kappa_e = ne\mu \tag{8}$$

Since the jump frequency cannot be determined independently it can be calculated substituting equation 5 and 6 into equation 8 and rearranging. Thus,

$$\nu = 6KTv\kappa_e/e^2NR^2X_{CuCl}X_{CuCl_2} \tag{9}$$

where $R = (N/v)^{-1/3}$.

A test of the electron exchange model is to plot ln ν vs R at constant temperature and ln ν vs 1/T at constant R.[7]

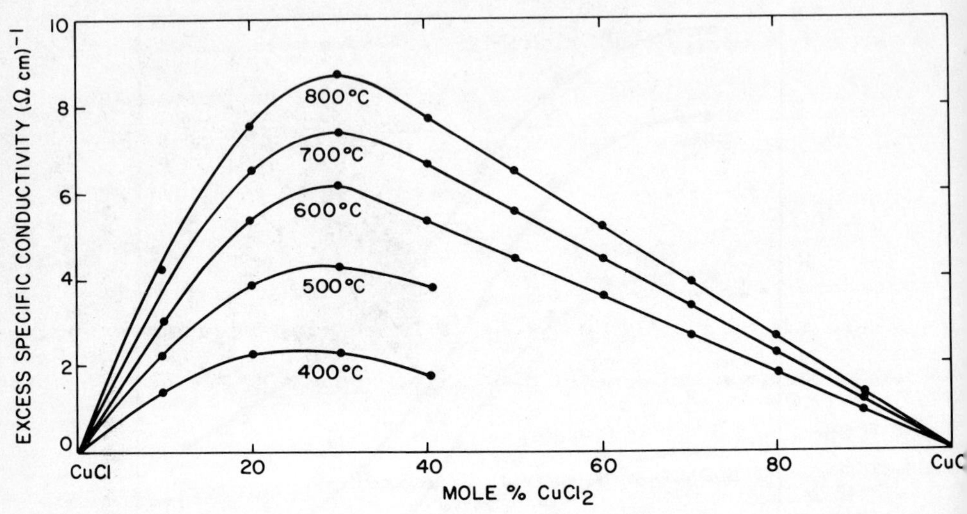

FIG. 7

Excess specific conductivity isotherms as a function of mole % $CuCl_2$ in CuCl; compositions are corrected for thermal decomposition of $CuCl_2$.

Linear plots are evidence that this model describes the conductivity behavior. Plots of ln ν vs R and plots of ln ν vs 1/t are shown in Figures 8 and 9 respectively.

In Figure 8 the jump frequency is plotted as a function of R at 10 mole % intervals from 10 mole % $CuCl_2$ to 90 mole % $CuCl_2$ in CuCl. The average distance between exchanging species has not been corrected for density changes in the melt as the temperature is increased, since a rather crude approximation was made to obtain the density of molten pure $CuCl_2$. In spite of these assumptions and the assumptions

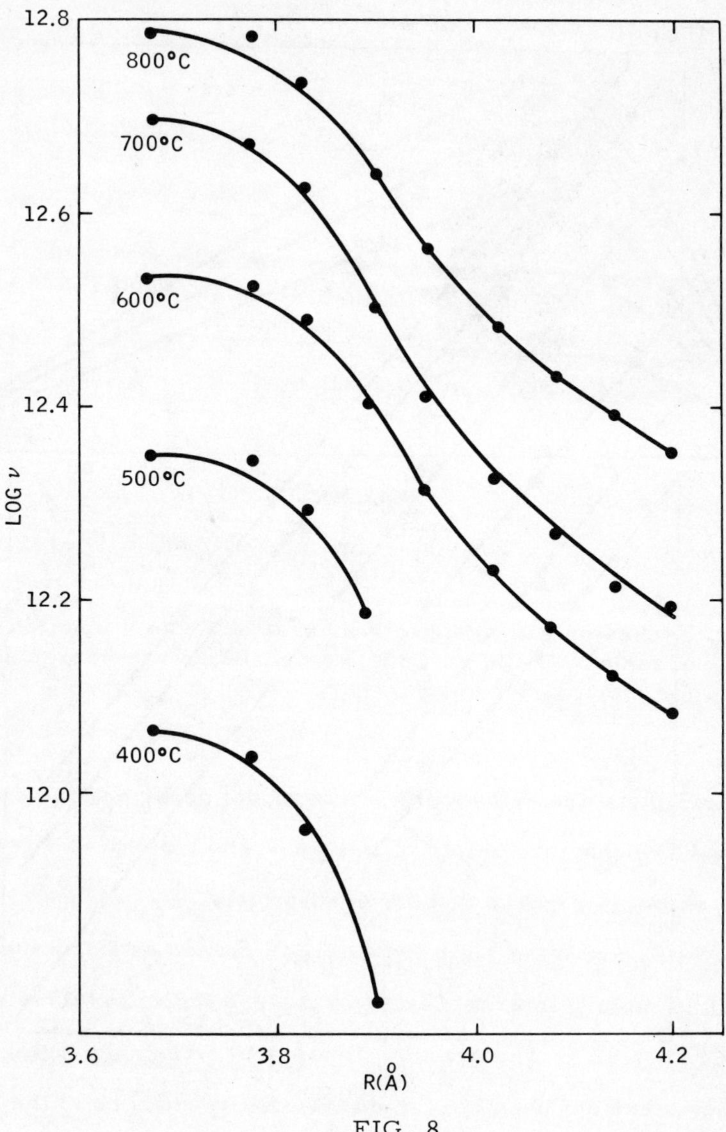

FIG. 8

Logarithm of the jump frequency (ν) as a function of the average distance (R) between Cu^{+}-Cu^{++}

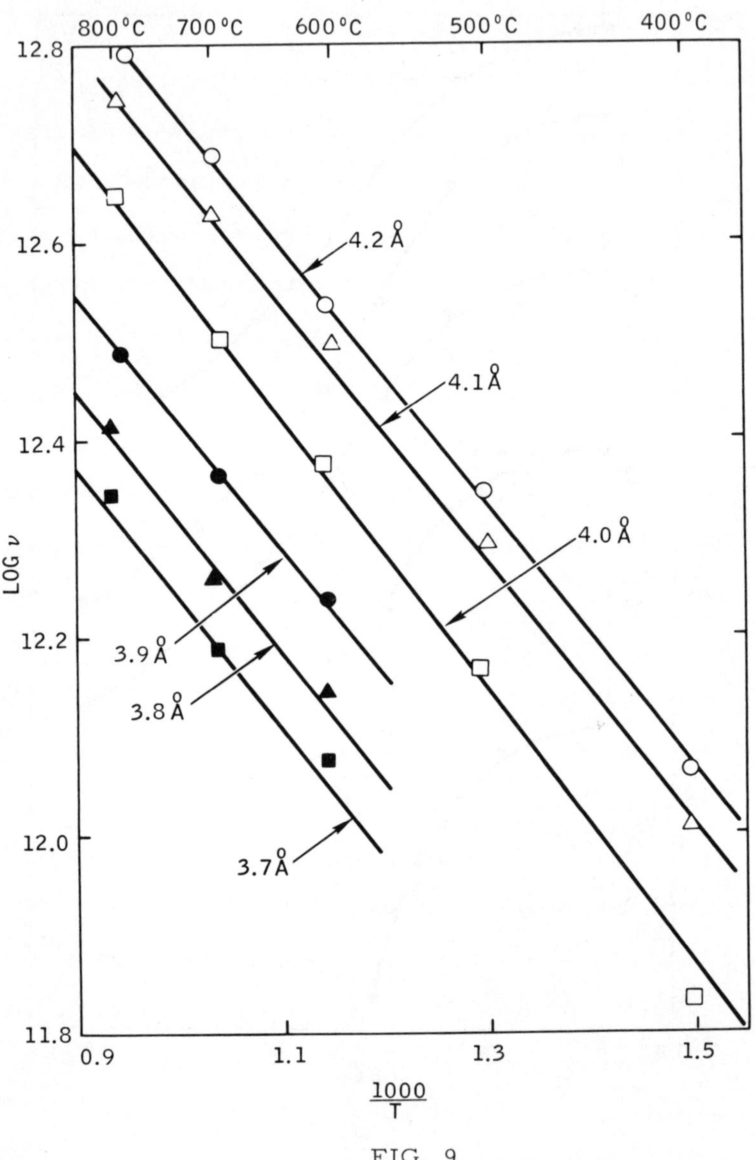

FIG. 9

Logarithm of the jump frequency (ν) as a function of reciprocal temperature.

associated with correcting the melt composition, the jump frequency is nearly an exponential function of the average distance between the exchanging species except for low values of R (CuCl rich solutions). Thus it appears that the jump model is applicable to the CuCl-$CuCl_2$ system. The slope of these isotherms (Fig. 8) in the "linear" region corresponds to an average value of 3 $\mathring{A}^{-1}$ for the Gamow penetration factor.[7] This value for a single electron process corresponds to a barrier height for electron transition of over 9 electron volts compared to 2.4 eV per electron for the Bi-BiI_3 system and to less than 4 eV per electron in the Bi-$BiCl_3$ system. The magnitude of this barrier height may be the critical factor in limiting the electronic conductivity in the CuCl-$CuCl_2$ system.

In Fig. 9 the logarithm of the jump frequency ν is plotted as a function of reciprocal temperature at six different values of average exchange distance (R). The curves are essentially linear with a constant slope, even for R values corresponding to the nonlinear portion of Fig. 8. Thus it appears as if the exchange model is applicable for the CuCl-$CuCl_2$ system. The slope corresponds to an enthalpy of symmetrization of 6.3 kcal/mole, which is to be compared with about 20 kcal/mole for the Bi-BiX_3 systems; this suggests that the symmetrization energy is much less in the Cu System than in the Bi systems. This might be expected since the symmetrization of a monovalent-divalent cation should require less energy than the symmetrization of a monovalent-trivalent cation.

No unequivocal explanation can be offered for the large barrier height found in the $CuCl$-$CuCl_2$ system compared to that found in the Bi-$BiCl_3$ system. Perhaps this is related to the difference in the size of the copper and bismuth cations. The anion polarizability (which controls the bridging action of the anion in electron transfer) is less in the field of the smaller copper cations than in the field of the large bismuth cations. Thus one might expect a larger barrier to electron exchange in the copper system.

The fact that the conductivity maxima occur at less than 50 mole % $CuCl_2$ dissolved in CuCl can be attributed to two possible factors. Since the molar volume of CuCl is less than that of $CuCl_2$, the average exchange distance is less in CuCl-rich solutions. In addition, it is possible that a fraction of the cuprous chloride may be dimerized. Formation of the dimer subhalide Cu_2Cl_2 would be expected to prevent the ionic atmosphere around Cu^{+2} and Cu_2^{+2} from symmetrizing, a necessary condition stipulated by the Franck-Condon principle. Thus, any cuprous chloride that is dimerized would not be available for electron exchange. Both these effects would result in a shift of the maximum to the CuCl-rich region.

ELECTRONIC CONDUCTION

In view of the apparent success of the electron exchange model in explaining the general behavior of the $CuCl-CuCl_2$ system as well as the bismuth systems, it would be instructive to examine the results of other systems in light of this model.

The electrical conductivities of molten $Hg_2Cl_2-HgCl_2$ and $Hg_2I_2-HgI_2$ systems have recently been investigated by Grantham[19]. A plot of κ vs composition of Hg_2X_2 showed negative deviations from additivity at the $HgCl_2$-rich region and positive deviations at the Hg_2Cl_2-rich region, although the conductivities increased monotonically with Hg_2Cl_2 content; no maxima were observed. In addition the conductivities were relatively low, never exceeding 1 $(ohm\text{-}cm)^{-1}$. One possible explanation of the positive deviation from additivity is that some electron exchange may be taking place in the Hg_2Cl_2-rich region. The fact that the mercury iodide system exhibited a substantially larger positive deviation than the chloride is in accord with this model since one would expect the iodide ion to act more effectively as an electron bridge between exchanging sites, as was observed in the bismuth-bismuth halide systems[5]. However, it is more probable that conduction is mostly ionic in this system, especially in view of the relatively low conductivity. The fact that little or no electron exchange was observed in this system is probably due to the stability of the Hg_2^{+2} dimer ion. It is known that little or no Hg^+ ions exist in the melt.[20]

The same considerations apply to the cadmium-cadmium halide systems (see Fig. 6), the conductivities of which were measured by Grantham[21]. In these systems the conductivities did not vary appreciably with metal composition. Grantham attributed the absence of electronic conduction to the lack of discrete Cd^+ ions in the melt, due to the dimerization of Cd^+ to form Cd_2^{+2}. The decrease in the electrical conductivity of Bi-$BiCl_3$ and Bi-$BiBr_3$ with metal composition near the melting points of the salts has also been explained on the basis of subhalide polymer formation which impedes electron exchange[4,5].

Fadeev and Fedorov[22] measured the electrical conductivities of In-$InCl_3$ melts from 32 atom % indium to 67.7% indium and from 340-380°C. Unfortunately, no data are tabulated; however, a plot of the conductivity isotherm in the InCl-$InCl_2$ region shows negative deviation from additivity, which indicates little or no electron exchange in this region. This lack of exchange can be attributed to the fact that $InCl_2$ probably exists as the complex[23] $In^+(InCl_4)^-$.

Extensive studies of the electrical conductivity behavior of the rare earth-rare earth halides have been carried out by Bredig and co-workers[16,17]. The systems can be classified in two general categories. In the first set, which includes the Nd systems, the conductivities are not unusually large but exhibit positive deviation from additivity and often have conductivity maxima. The results have been explained in terms of electron exchange. In the second set, which includes La and Ce, a rapid increase in conductivity is observed with increasing metal composition. The accelerated increase in conductivity with metal composition was attributed by Bredig and co-workers[8, 24] to increasing orbital overlap and gradual establishment of a conduction band for electrons. However, as Corbett[17] points out, the electron exchange model could also explain the high conductivity of the lanthanum-lanthanum halide systems as due to the active electron on La^{+2} (presumably $5d^1$) which is more responsive to anionic fluctuations. That is, symmetrization can occur more readily. On the other hand, the small extent of electron exchange in the Nd case is attributed to the highly shielded character of the exchangeable 4f electron in Nd^{+2}.

The conductivities of alkaline earth metals dissolved in their halides have been measured both by Dworkin et al.[8, 25] and by Emons and Richter[26, 27]. The conductivities increase with increasing temperature and with increasing metal composition. The results are discussed in terms of electron exchange. If electron exchange is to take place, monovalent subhalide ions (e.g. Ca^+) would be expected to be involved. Dissolution of the alkaline earth metals as the monovalent ion is consistent with the vapor pressure results of Van Westenburg and Peterson[28], although Dworkin et al.[25] consider the formation of such an ion to be less likely than the dissolution of alkaline earth metal as Ca^{+2} and mobile F-centers in equilibrium with Ca_2^{+2}.

The electrical conductivities of molten alkali-alkali halide systems have been extensively investigated by Bredig and co-workers[16,17]. The theoretical aspects of these systems have also been extensively investigated.[16,17] The most noteworthy of these are the electron-exchange model by Rice[6] and a model by Pitzer[29] in which F-center electrons exist in cavities within the irregular lattice of positive ions. If the assumption is made that electron exchange occurs between F-centers and anion vacancies in the pseudo lattice (concentration calculated from the volume of fusion), Pitzer's model offers an alternate mechanism of conduction to that proposed by Rice (electron exchange between dissolved metal and the cation). As will be shown in a subsequent paper[30], more linear plots of log ν vs R at constant T and log ν vs 1/T at constant R are obtained when Pitzer's model of metal dissolution is used.

The vast difference in conductivity dependence on temperature and compositions for four metal-salt solutions is illustrated in Fig. 6. In some systems such as the alkali-halide solutions there is a large concentration dependence but little temperature effect; the opposite is found in the $Cd\text{-}CdX_2$ systems. In the $Bi\text{-}BiI_3$ systems, the conductivity increases monotonically with temperature and concentration while a conductivity maximum is found at all temperatures as the concentration is varied in the Cu-$CuCl_2$ systems. In spite of these differences it appears as if the electron exchange model of semiconduction may be applicable to all.

ACKNOWLEDGMENTS

The authors wish to thank Dr. Maynard Hiller for the preparation of the salts and Mr. E. Harrelson for carrying out many of the measurements.

REFERENCES

1. This work was supported by the Research Division of the U. S. Atomic Energy Commission.
2. H. R. Bronstein, A. S. Dworkin, and M. A. Bredig, J. Chem. Phys. 37, 677 (1962).
3. S. J. Yosim, L. F. Grantham, E. B. Luchsinger, and R. Wike, Rev. Sci. Instr. 34, 994 (1963).
4. L. F. Grantham and S. J. Yosim, J. Chem. Phys. 38, 1671 (1963).
5. L. F. Grantham, J. Chem. Phys. 43, 1415 (1965).
6. S. A. Rice, Discussions Faraday Soc. 32, 181 (1961).
7. D. O. Raleigh, J. Chem. Phys. 38, 1677 (1963).
8. A. S. Dworkin, H. R. Bronstein, and M. A. Bredig, Discussions Faraday Soc. 32, 188 (1961).
9. A. S. Dworkin, R. A. Sallach, H. R. Bronstein, M. A. Bredig, and J. D. Corbett, J. Phys. Chem. 67, 1145 (1963).
10. A. S. Dworkin, H. R. Bronstein and M. A. Bredig, J. Phys. Chem. 67, 2715 (1963).

11. To test their electron exchange hypothesis, Dworkin et al[10] substituted Sr^{+2} (which has similar size as Nd^{+2}) for Nd^{+2} and measured the conductivities of NdI_3-SrI_2 and found negative deviation from additivity rather than a maximum; these results therefore support electron exchange.

12. C. M. Fontana, E. Gorin, G. A. Kidder and C. S. Meredith, Ind. Eng. Chem. 44, 363 (1952).

13. L. F. Grantham, E. B. Harrelson, P. H. Shaw, and C. M. Larsen, Rev. Sci. Instr. 39, 699 (1968).

14. P. H. Dike, Rev. Sci. Instr. 2, 379 (1931).

15. L. Brewer and N. L. Lofgren, J. Am. Chem. Soc. 72, 3042 (1950).

16. M. A. Bredig, "Mixtures of Metals with Molten Salts" in "Molten Salt Chemistry" (Interscience, New York, 1964), M. Blander Ed., pp 367-425.

17. J. D. Corbett, "The Solution of Metals in Their Molten Salts" in "Fused Salts" (McGraw-Hill, New York, 1964), B. R. Sundheim, Ed., pp 341-407.

18. L. Yang, G. Pound, and G. Derge, Trans. AIME 26, 783 (1956).

19. L. F. Grantham, J. Phys. Chem., in press.

20. S. J. Yosim and S. W. Mayer, J. Phys. Chem. 64, 909 (1960).

21. L. F. Grantham, J. Chem. Phys. 44, 1509 (1966).

22. V. N. Fadeev and P. I. Fedorov, Russ. J. Inorg. Chem. 10, 788 (1965).

23. J. R. Chadwick, A. W. Atkinson and B. G. Huckstepp, J. Inorg. Nucl. Chem. 28, 1021 (1966).

24. A. S. Dworkin, H. R. Bronstein, and M. A. Bredig, J. Phys. Chem. 66, 1201 (1962).

25. A. S. Dworkin, H. R. Bronstein, and M. A. Bredig, J. Phys. Chem. 70, 2384 (1966).

26. H. H. Emons and D. Richter, Z. Anorg. Allgem. Chem. 339, 91 (1965).

27. H. H. Emons and D. Richter, Z. Anorg. Allgem. Chem. 353, 148 (1967).

28. J. A. Westenburg, Iowa State University of Science and Technology, Abstracts of Doctoral Theses, pp 1129-1130; University Microfilms, Inc., Ann Arbor, Mich., Order No. 64-9291.

29. K. S. Pitzer, J. Am. Chem. Soc. 84, 2025 (1962).

30. L. F. Grantham, to be published.

ELECTRICAL CONDUCTIVITY MEASUREMENTS IN MOLTEN FLUORIDE MIXTURES, AND SOME GENERAL CONSIDERATIONS ON FREQUENCY DISPERSION*

G. D. Robbins and J. Braunstein

Reactor Chemistry Division
Oak Ridge National Laboratory
Oak Ridge, Tennessee

INTRODUCTION

The electrical conductivities of molten salts are of fundamental importance both for the testing and development of theories of transport in ionic liquids and for practical applications in commercial processes. Since fluorides are among the simplest salts, their study is advantageous for the testing of theories. The fluorides are important in the electrolytic preparation of metals, such as aluminum, and are among the few salts suitable for use in molten salt nuclear reactors. In particular, mixtures of lithium fluoride and beryllium fluoride are used as the fuel salt solvent and coolant in the Molten Salt Reactor Experiment at Oak Ridge, and mixtures of lithium fluoride, beryllium fluoride, and thorium fluoride are proposed for use in a molten salt breeder reactor. Since the conductivities and viscosities of these mixtures vary over many orders of magnitude in the range between pure LiF and pure BeF_2, the characterization of these mixtures is of interest.

In this paper we first discuss problems relating to the measurement of conductance in highly conductive salt mixtures and then describe the measurement of electrical conductivities of some molten mixtures of LiF and BeF_2. We present results over the composition range 38 to 52 mole % BeF_2 and at temperatures between 400 and 530°C. A preliminary interpretation of the temperature dependence of the conductivities is made in terms of the zero mobility temperature concept of Angell.[1,2]

ELECTRICAL CONDUCTIVITY MEASUREMENTS

PRINCIPLES OF MEASUREMENT

Conductivity of Molten Salts

In highly conductive electrolytes, such as molten salts, the indiscriminate application of techniques developed for the measurement of electrical conductance in dilute aqueous solutions can cause errors. Often neglected is the origin and treatment of frequency dispersion of the measured resistance. Thus, in a recent review of electrical conductivity measurements in molten fluorides[3] it was shown that of the sixteen groups of investigators considered, ten did not address themselves to this question. The observations of the remaining authors on the presence, magnitude, and form of the frequency dependence of the measurements were contradictory. Accordingly, a brief discussion of electrical conductance measurements and some possible origins of frequency dispersion might be useful.

The electrical conductivity κ of a liquid and its resistance R are related by the expression

$$\kappa = \frac{1}{R}\,(\ell/a) \tag{1}$$

where (ℓ/a) is the cell constant, a characteristic of the conductivity cell. That the experimentally determinable quantity R_m, the measured resistance, can vary with the measuring frequency in dilute aqueous solutions was discussed in detail long ago by Grinnell Jones and co-workers,[4] and more recently by Ives and associates.[5] Variation of resistance with frequency has been observed also in organic solvents (e.g., methanol[6]) and in molten salts other than fluorides.[7-11]

Bridge Circuit and Balance Equations

One source of the variation of measured resistance with frequency can reside in assumptions associated with balance conditions for the measuring bridge. Often employed is some form of the basic Wheatstone bridge circuit, e.g., that shown in Figure 1. The upper arms are matched, standard resist-

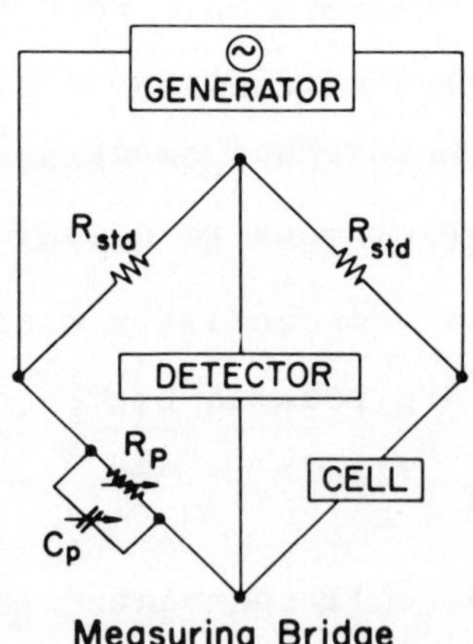

FIG. 1

Bridge circuit for conductance measurements.

ances, while one lower arm contains the cell; and the other, the balancing resistance and capacitance, connected in parallel. A generator impresses a sinusoidal potential across the bridge, and a phase sensitive null detector (preferably an oscilloscope) indicates bridge balance.

It is often stated or implied that when the bridge is balanced, the resistance R_p equals the resistance of the cell solution (for unmatched standard resistances, an equation incorporating their ratio is employed). As a result of electrode process studies, it is known that the most simple representation of the impedance measured in a cell is the series combination of solution resistance R_s and double-layer capacitance C_s[12] as shown in Figure 2a. Upon substituting this representation of the cell into Figure 1, the circuit analysis reveals that the two conditions which must be satisfied at balance are

$$\frac{R_s}{R_p} + \frac{C_p}{C_s} = \frac{R_{std}}{R_{std}} = 1 \qquad (2)$$

and

$$R_s C_s R_p C_p (2\pi f)^2 = 1 \qquad (3)$$

where f is the frequency of the sine wave. These may be combined to yield

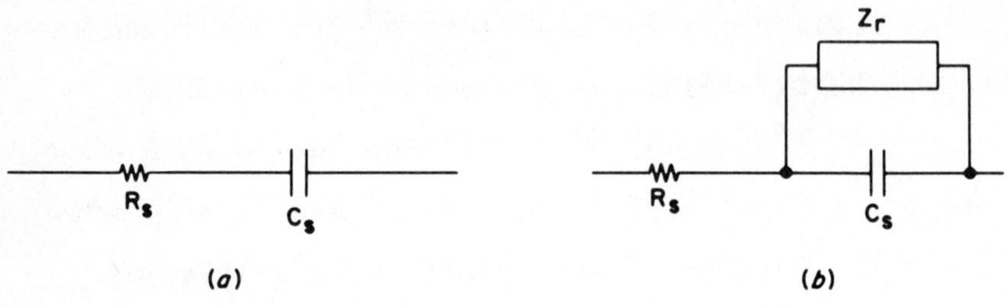

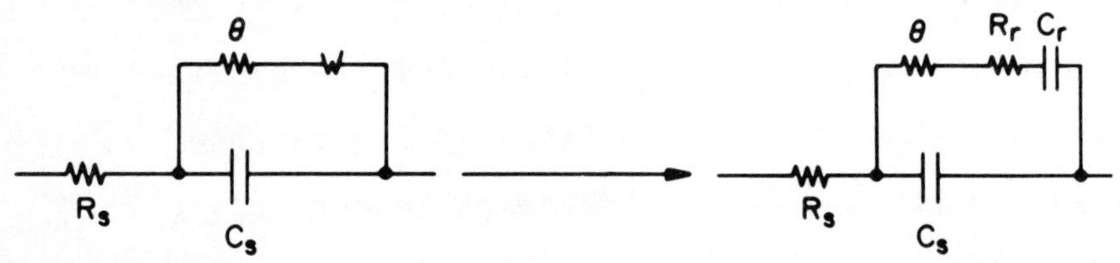

FIG. 2

Equivalent circuits of conducting cell.

$$R_p = R_s[1 + C_p{}^2R_p{}^2(2\pi f)^2] \qquad (4)$$

or

$$R_p = R_s + \frac{1}{R_sC_s{}^2(2\pi f)^2} \qquad (5)$$

Equation (5) demonstrates that R_p is not equal to R_s - that the reading on the resistance dials of such a conductivity bridge is not equal to the solution resistance - but is, in fact, an approximation in which the additive term $\frac{1}{R_sC_s{}^2(2\pi f)^2}$ is neglected.

This is usually a valid approximation in dilute aqueous solutions where the solution resistance is large. However, in molten salts R_s may be less by several orders of magnitude. To illustrate the effect of this approximation, if one assumes a resistance of 1000Ω in an aqueous solution, a capacitance of 20μf, and a measuring frequency of 1000Hz, the neglected term becomes 0.06Ω (compared to 1000Ω). In molten salts in cases where capillaries cannot be used to increase resistance (a condition often imposed in corrosive fluorides) values of 5Ω or less may be observed for R_s. Assuming similar values of capacitance[13] and measuring frequency, the neglected term in equation (5) becomes 12Ω (compared to 5Ω). For cases in which the second term in equation (5) is too large to have been neglected, the measured resistance (R_p in this case) will appear to be frequency dependent. Under such conditions one must either rely on equation (4) to calculated R_s or employ a different type bridge, e.g., with the variable components in series.

Although this approximation has been discussed previously,[5,14] it continues to be overlooked. It is usually not possible to assess the extent to which this factor may have contributed to published contradictory observations on frequency dependence[3] [via equation (4)] because the authors seldom report the values of capacitance required for bridge balance.

Interfacial Impedance

A second cause of the variation of measured resistance with frequency can arise from phenomena occurring at the electrodes. When the amplitude of an applied sinusoidal potential is less than the potential required for an electrochemical reaction, the representation of Figure 2a is approximately valid (C_s is potential dependent), and the metal/salt interfacial capacitance is charged and discharged with the potential variations. At applied potentials sufficiently large for reaction to occur, the transfer of charge across the solution/electrode interface may be represented (Figure 2b) as an impedance Z_r associated with this process and in parallel with the interfacial capacitance. Figure 2b can then be assumed to represent the conductivity cell during that part of each cycle that the potential exceeds the potential for reaction.

The form which frequency dispersion often assumes is $f^{-\frac{1}{2}}$. Although Faradaic admittance theory[15,16] predicts a similar dependence of impedances associated with an electrochemical reaction, the assumptions of this theory are sufficiently restrictive (e.g., that the amplitude of the sine wave approach zero) as to be of marginal applicability to most conductance apparatus. This theory represents the cell as consisting of a frequency-independent resistance Θ (Figure 2c) in series with the frequency-dependent Warburg impedance, -W-, which may be further represented at a single frequency as a resistance and capacitance, R_r and C_r, connected in series and associated with the electrode reaction. The mathematical analysis of this model predicts that R_r and the impedance associated with C_r vary as $f^{-\frac{1}{2}}$. Even though this is the form of some observed frequency dependence, the assumptions on which these results are based are generally not realized in electrical conductivity experiments, and Figures 2a and 2b remain the most useful equivalent circuit representations.

Assuming that frequency effects resulting from circuitry have been corrected, one can consider that the measured resistance is composed of the true solution resistance, R_s, and resistance associated with the electrode/solution interface, R_z.

$$R_m = R_s + R_z \qquad (6)$$

In order that the measured resistance represent the true resistance of the solution, R_z must either be corrected for or eliminated experimentally.

Jones and Christian[4] found that values of measured resistance varied linearily with $f^{-\frac{1}{2}}$ in aqueous solutions and employed this functional form to extrapolate resistance to infinite frequency, where R_z was assumed to be zero. Nichol and Fuoss[6] found a linear dependence of R_m on f^{-1} in methanol solutions. In molten nitrates,[10] chlorides and fluorides[11] measured resistance has been found to be nonlinear in $f^{-\frac{1}{2}}$, with resistance approaching a nearly constant value at high frequencies. DeNooijer[10] employed 20kHz as a measuring frequency while Winterhager and Werner[11] used 50kHz.

It is interesting to examine equation (6) and the equivalent circuit representation of Figure 2b in terms of extrapolating measured resistance to infinite- or very high-frequencies. As the frequency increases, the impedance of the capacitance path across the interface decreases relative to that of the reaction path, so that the equivalent circuit approaches that of Figure 2a. In addition, the practice of platinizing electrodes increases the interfacial capacitance, thereby reducing the effect of R_z.

Elimination of Spurious Frequency Dispersion

Several steps may be taken to help insure that the value of resistance employed in determining κ is solely a property of the conducting solution:

1. A test of the bridge circuit by substituting a calibrated resistance and capacitance, connected in series, having values similar to those expected from the conductivity cell indicates whether and within what frequency limits the measured resistance will represent solution resistance in a conductivity cell.

2. Reduction of the sine wave amplitude to the smallest value compatible with the measuring equipment (preferably below 100mV) reduces effects associated with any electrode reactions.

3. For the given experimental conditions, determination of the form of any frequency dispersion permits selection of either the best functional form for extrapolation to infinite frequency, or a frequency range in which the resistance is independent of frequency and is characteristic of the electrolyte solution.

EXPERIMENTAL

Conductance Bridge and Detector

The measuring bridge circuit was similar to that shown in Figure 1 with the exception of the balancing arm. This consisted of a resistance (General Radio, variable in steps of 0.1Ω from 0-10kΩ) and capacitance (specially constructed, variable in steps of 1μf from 0-110μf) connected in series. A General Radio decade capacitance box (0-1μf in steps of 100μμf) connected in parallel with the larger, variable capacitance increased the sensitivity of capacitive balance. The upper arms of the bridge contained matched, 100Ω(±0.5%) resistors. A Hewlett-Packard variable frequency oscillator supplied a sinusoidal potential across the bridge. The peak to peak voltage applied across the cell containing molten salt was approximately 30-50mV. An oscilloscope with two differential amplifiers (H.P., model 1400A) indicated impedance and phase balance.

This measuring assembly was tested employing a dummy cell consisting of a standardized resistance and capacitance connected in series. The accuracy of the bridge was better than ±0.5% for resistance of the order of 100Ω (capacitance of 21μf) in the frequency range 1-50kHz. The accuracy in determining capacitance was much less and depended on the magnitude of the resistance and on the frequency (±1% at 2.5kHz and ±10% at 8.5kHz). However, even though the value of capacitance required for bridge balance was not an accurate measure of the dummy capacitance at higher frequencies, at balance the accuracy of the resistance measurement was maintained.

Similar tests on a Jones bridge available in our laboratory (Leeds and Northrup),[17] which has a parallel-component balancing arm, revealed that while this bridge was found to be extremely accurate for measuring high resistances, the accuracy at 100Ω was only ±1% in the frequency range 1-50kHz. Furthermore, the product $R_s C_s R_p C_p (2\pi f)^2$, which should be constant and unity, varied by 15% over the range 1-10kHz.

Cell and Furnace Assembly

Figure 3 represents the silica conductance cell employed in these experiments. The relatively long conducting path (12cm) results in measured resistances of 100-500Ω. The platinized platinum electrodes, constructed from 20 gauge wire, are held immobile relative to the dip cell by silica hooks and glass tape. The L-shaped electrode design renders measured resistance independent of depth of

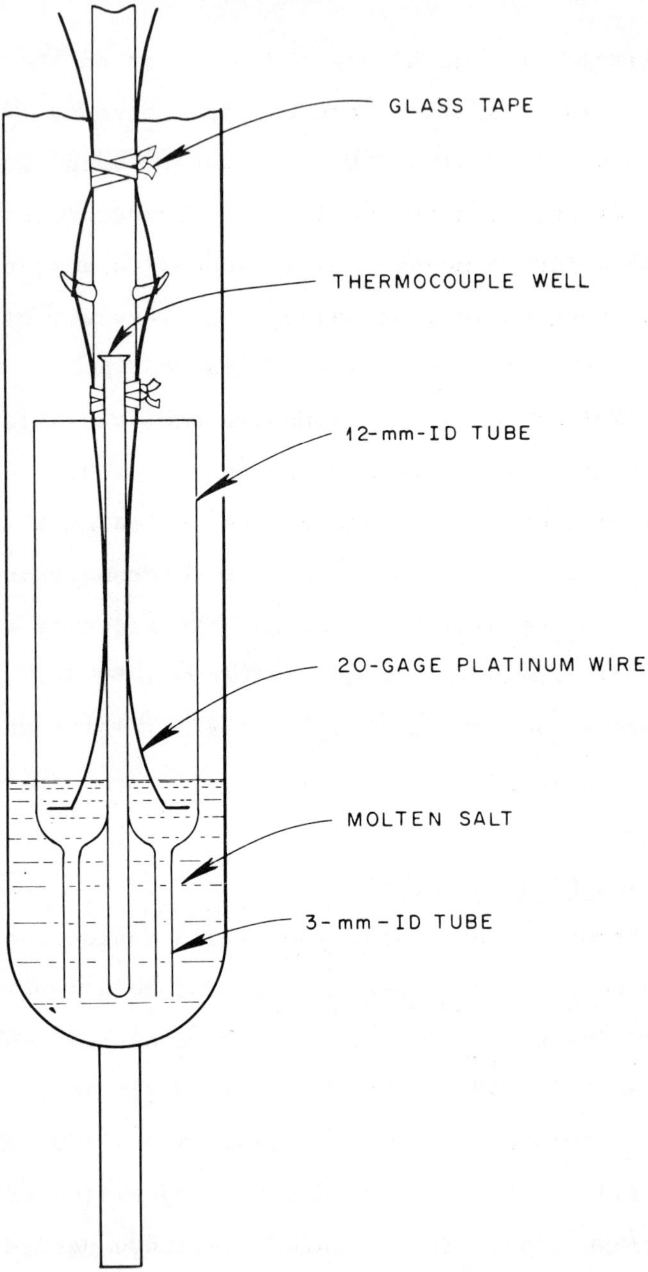

GLASS TAPE

THERMOCOUPLE WELL

12-mm-ID TUBE

20-GAGE PLATINUM WIRE

MOLTEN SALT

3-mm-ID TUBE

FIG. 3

Conductance cell assembly.

immersion (3-7mm salt height above electrodes). The entire cell assembly is immersed in a stirred, controlled temperature, potassium nitrate bath in the visual observation furnace[18] shown in Figure 4. Temperatures, measured with a calibrated chromel-alumel thermocouple and type K-4 potentiometer, are believed to be accurate to ±1°C.

Compatibility of Silica Cell with LiF-BeF_2 Mixtures

The molten lithium fluoride-beryllium fluoride mixtures are contained in the closed silica cell under a dried helium atmosphere (1 atm) and permitted to establish their equilibrium SiF_4 partial pressure via the reaction

$$2BeF_2 + \underline{SiO_2} \rightleftharpoons \underline{2BeO} + SiF_4\uparrow$$

The low equilibrium partial pressure of SiF_4 (<1mm at 500°C)[19] prevents major attack on the silica. The slow, solid-solid reaction $\underline{SiO_2} + 2\underline{BeO} \rightarrow \underline{Be_2SiO_4}$ and the low solubility of BeO and Be_2SiO_4 (<0.01 mole/kg at 500°C),[20,21] should not affect the conductance measurements.

Determination of Cell Constant

Based on availability, ease of handling, and liquid range, a reasonable choice among molten salts available for use in the determination of cell constants is potassium nitrate. However, in the light

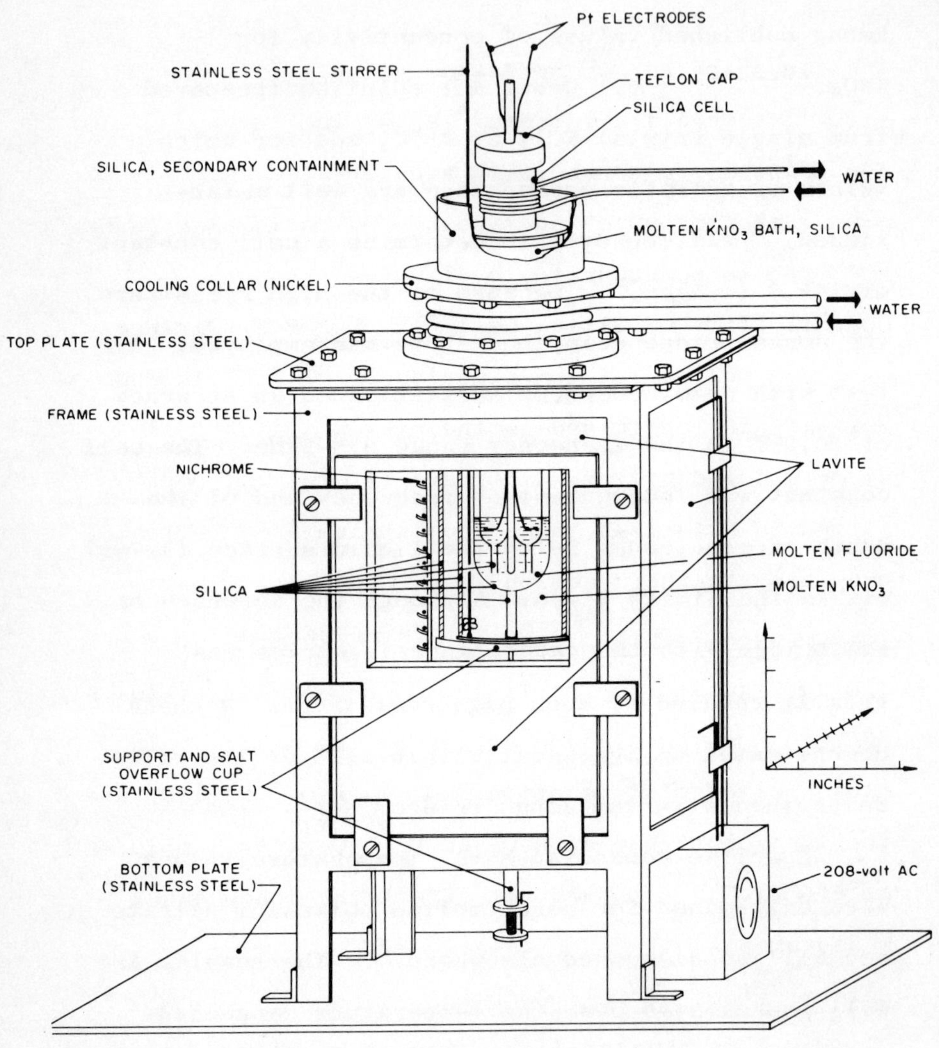

FIG. 4

Furnace with conductance cell in molten nitrate bath.

of the foregoing discussion and the discrepancies among published values of conductivity for KNO_3,[10,22-25] a 0.1 demal KCl solution (prepared from single crystal KCl) at 25°C, and for which values of specific conductance are well established,[26] was employed to determine a cell constant of 145.9 (±.3)cm^{-1}. Because of the high resistance, the Jones bridge described above was employed. A test with dummy components established an accuracy of ±0.05% in the frequency range 0.5-5kHz. The cell constant was independent of frequency and of the depth of electrodes below the liquid surface (3-7mm) within the limits given. Although the accuracy of the bridge with the capacitance in series was greatly reduced at such high resistances, a check demonstrated an agreement within 1.5% of the value determined with the Jones bridge.

Specific conductance vs. temperature values were determined for pure, molten potassium nitrate and will be discussed elsewhere.[27] The results are well represented over the temperature range 340-540°C by the equation (least squares, $\sigma = 0.0008$)

$$\kappa = -0.7098 + 4.6203 \times 10^{-3} t(^{\circ}C) - 2.0221 \times 10^{-6} t^2 \quad (7)$$

Experimental Procedure

Because of the high melting point of LiF and the temperature limitations of the furnace (550°C), $LiF-BeF_2$ (66-34 mole %) and BeF_2 were used as starting materials and handled exclusively in a dry box. The beryllium fluoride (Brush Beryllium Company) was vacuum distilled, and the $LiF-BeF_2$ (66-34 mole %) mixture was pretreated by hydrofluorination, followed by hydrogenation. (Analysis showed <30 ppm of Ni, Cr, or Fe and 4 ppm S.) The eutectic composition (52 mole % BeF_2)[28] mixture having been prepared and loaded in the cell in a dry box, other compositions were obtained by additions of chunks of LiF to the cell (sintered under an argon-hydrogen atmosphere) on successive days. Helium bubbling for several minutes was required to insure complete mixing. At a given temperature the measured resistance was stable with time and with depth of electrodes below the melt surface (3-7mm) and varied less than 0.5% over the frequency range 1-10kHz, the variation being approximately linear in $f^{-\frac{1}{2}}$. Values of measured resistance extrapolated to infinite frequency were less than 0.3% lower than the 10kHz values. Measurements at different temperatures were made in a random manner to test for possible time variations (which were not observed).

After four days of contact with the molten fluoride melt the dip cell was immersed in molten KNO_3, where no visible evidence of attack of silica could be seen below the surface of the fluoride melt. At the gas/melt interface some corrosion was evident. However, this area was not part of the conduction path.

In attempting to remeasure the cell constant in molten KNO_3, a significantly lower capacitance than that observed in the last fluoride measurement was noted. This indicated a reduction in the degree of platinization during soaking in KNO_3. Nevertheless, the cell constant agreed with that determined previously within 0.7%. The stability of the measured resistance with time and the absence of visible attack of the silica in conducting regions indicate successful employment of this material in these fluoride melts. We consider the overall accuracy of the fluoride data presented herein to be of the order of ±2%.

RESULTS AND DISCUSSION

Data and Empirical Equations

Values of specific conductance vs. temperature for lithium fluoride-beryllium fluoride mixtures of compositions 52, 47, 42, and 38 mole % BeF_2 are given in Table 1. In Figure 5 linear, computer-fitted,[29] least squares lines are drawn through the results for the two mixtures richer in lithium fluoride because of the limited temperature range of the measurements. Obvious curvature is observed for the two compositions of higher beryllium fluoride content, and several functional forms were least-squares fitted by computer. Two are given in Table 2, together with the linear equations for the two lithium-rich compositions. We present these different analytical expressions for representation of the data not because of any significance that might be attached to a particular form, but to permit alternative analytic evaluation of their derivatives. The standard errors of the coefficients and the standard error of fit for the equations are also given. The curves shown in Figure 5 for the two compositions richer in BeF_2 are of the form

$$\kappa = a + bt^2. \qquad (8)$$

TABLE 1

Specific Conductance vs. Temperature
$LiF-BeF_2$ Mixtures

Mole % BeF_2	t (°C)	κ (Ω^{-1} cm^{-1})
38	465.5	1.100
	486.1	1.226
	502.7	1.322
	533.2	1.504
42	455.3	0.844
	476.5	0.962
	501.4	1.097
	528.6	1.241
47	425.2	0.500
	446.2	0.587
	475.4	0.719
	499.7	0.830
	525.9	0.954
52	398.7	0.271
	418.4	0.329
	434.3	0.379
	449.5	0.427
	473.2	0.510
	498.6	0.602
	513.6	0.661
	528.5	0.715

TABLE 2

Computer-Fitted,[29] Least Squares Representation for κ vs. Temperature*

Mole % BeF_2	Temp. Range	Equation	a	Std Error of a	b	Std Error of b	c	Std Error of c	Std Error of Fit
38	465–535°C	$\kappa = a + bt$	-1.674	0.0212	5.961×10^{-3}	0.043×10^{-3}			0.0021
42	455–530°C	$\kappa = a + bt$	-1.615	0.0202	5.406×10^{-3}	0.041×10^{-3}			0.0023
47	425–525°C	$\kappa = a + bt^2$	-0.3566	0.0075	4.746×10^{-6}	0.033×10^{-6}			0.0025
	698–798°K	$\kappa = ae^{-b/(T-c)}$	9.159	0.900	1025	79.7	346.0	15.6	0.0010
52	400–530°C	$\kappa = a + bt^2$	-0.3211	0.0032	3.713×10^{-6}	0.015×10^{-6}			0.0017
	673–803°K	$\kappa = ae^{-b/(T-c)}$	8.790	0.974	1165	89.8	337.2	15.5	0.0014

*t, °C; T, °K.

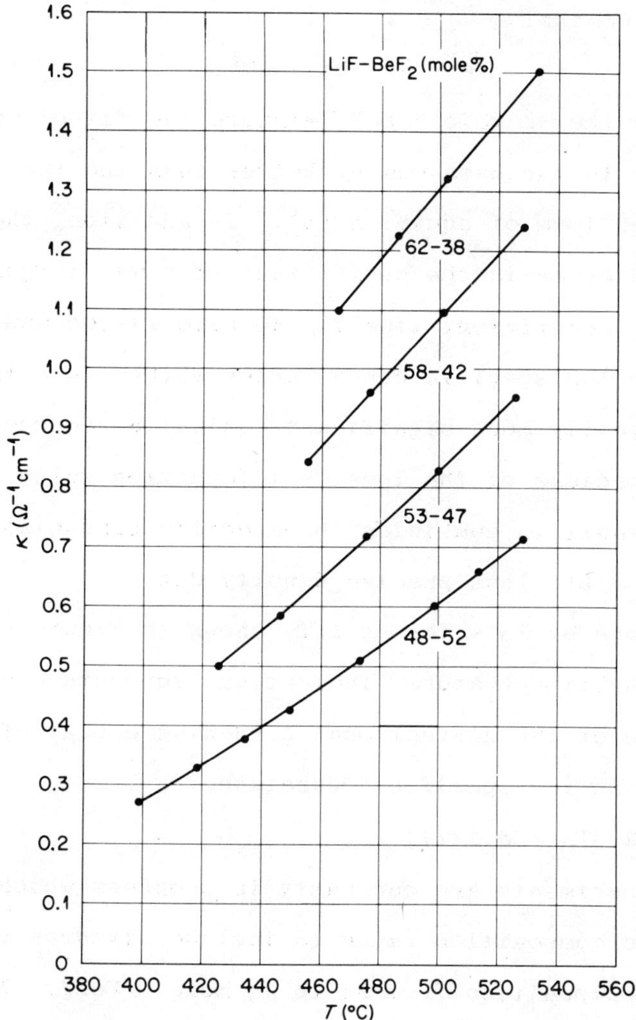

FIG. 5.
Specific conductance versus temperature for LiF-BeF_2 mixtures at compositions between 38 and 52 mole percent BeF_2.

Data for these two compositions were also fitted to the polynominal

$$\kappa = a + bt + ct^2. \qquad (9)$$

For the 52 mole % BeF_2 mixture the fit of equation (9) to the data was no better than the two-parameter form of equation (8). In addition, the standard error in the coefficient of t was larger than the coefficient itself. We have fitted equations to the specific conductances rather than the theoretically more significant equivalent conductances because of the loss of information which would result in combining the specific conductance with possibly less precise density data.

Plots of ln κ versus 1/T, shown in Figure 6, also exhibit curvature, indicating temperature dependence of the conventional Arrhenius energy of activation for specific conductance, E_κ. [$E_\kappa = -R \, d\ln \kappa/d(1/T)$]

Experiments are currently in progress which extend the composition range to include mixtures varying in composition from 34 to 70 mole % BeF_2. We await the completion of these experiments before presenting an interpretation of the composition dependence of the conductivities observed in this system.

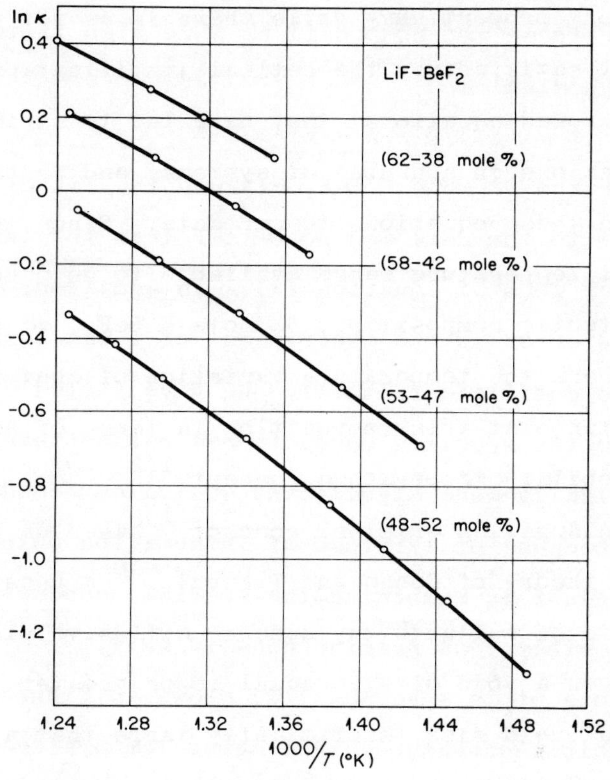

FIG. 6

Arrhenius plots of the specific conductance of LiF-BeF_2 mixtures.

Temperature Dependence and the Zero Mobility Temperature

The variation of the activation energy for conductance at low temperatures has been interpreted by Angell in terms of a model which incorporates a non-Arrhenius form for the temperature dependence of transport properties. While there is as yet no completely satisfactory theoretical justification for the proposed equations, they have led to striking correlations in a number of systems, and we have applied these equations to our data. Since the largest temperature range available to us occurs at the eutectic composition, 52 mole % BeF_2, we consider here the temperature variation of equivalent conductance at this composition in terms of Angell's zero mobility temperature concept.[1,2]

In Angell's original concept,[1] based on the free volume theory of Cohen and Turnbull,[30] a constituent particle of a liquid may undergo diffusive displacement when a void of size equal to or greater than a critical void size (sufficiently large that a neighboring particle may move into it) is created. Voids arise from the distribution of "free" volume, approximately equal to the total thermal expansion of the liquid above a temperature T_0 at which "free" volume begins to appear. By combining constants from an expression based on concepts of Cohen and Turnbull,[30] Angell[1] writes, for temperatures approaching T_0

$$D = AT^{\frac{1}{2}}e^{-k/(T-T_0)} \tag{10}$$

where D is the diffusion coefficient, and A and k are constants.

Angell[2] has subsequently pointed out that equation (10) can be obtained by an alternative theory proposed by Adam and Gibbs[31] which avoids introducing the concept of free volume. According to Adam and Gibbs, T_0 is that temperature at which the configurational entropy of the supercooled liquid becomes zero. In this theory translational motion occurs by cooperative rearrangement of a group of molecules, the temperature dependence of the process being related to the minimum size of cooperatively rearranging groups. T_0 is the zero mobility temperature, and k is related to the potential energy of the cooperatively rearranging groups. Angell has demonstrated the striking fit of this model to low temperature conductance data for a number of molten nitrates.[2] It also has been shown by Moynihan[32] and by Braunstein et al.,[33] that a single value of T_0 is consistent with conductance, viscosity, and solute diffusion in molten $Ca(NO_3)_2 \cdot 4H_2O$.

Application of the Model to Conductance Data

Combining the derivative of equation (10) with respect to 1/T with the similar derivative of the conventional Arrhenius relation for diffusion gives

$$E_D = \tfrac{1}{2}RT + kR\left(\frac{T}{T-T_0}\right)^2 \qquad (11)$$

where E_D is the activation energy for diffusion, and R is the gas constant. Thus a corrected energy for activation, E_{corr}, may be defined as

$$E_{corr} = E_D - \tfrac{1}{2}RT = kR\left(\frac{T}{T-T_0}\right)^2 \tag{12}$$

The relation

$$E_D = E_\Lambda + RT \tag{13}$$

where E_Λ is the activation energy for equivalent conductance, can be obtained from the Nernst-Einstein equation by assuming that the activation energies for ionic conductance of the different ions are either equal or sufficiently different in magnitude that one predominates.[2,34] The Nernst-Einstein equation is known not to apply to molten salts in the temperature region in which they are usually studied. Angell, however, employs this equation at temperatures in the region of the glass transition temperature, where the number of mobile particles and their interactions are so small that the conditions of validity of the Nernst-Einstein relation should apply.

The equivalent and specific conductance activation energies are related as

$$E_\Lambda = E_\kappa + \alpha RT^2 \tag{14}$$

α being the coefficient of expansion of the salt.

Combining equations (12)-(14)

$$E_{corr} = E_\kappa + \tfrac{1}{2}RT + \alpha RT^2 = kR\left(\frac{T}{T-T_0}\right)^2 \qquad (15)$$

Therefore, according to these concepts, a plot of E_{corr} vs. $\left(\frac{T}{T-T_0}\right)^2$ should be linear and pass through the origin.

Values of α were obtained from density data in fused fluoride melts.[35] Table 3 demonstrates the relative contribution of $\frac{1}{2}RT + \alpha RT^2$ for the LiF-BeF_2 (48-52 mole %) composition to the total E_{corr}. The agreement among values of E_κ obtained by several different methods is also shown. $E_\kappa^{(1)}$ was obtained graphically from Figure 6. $E_\kappa^{(2)}$ and $E_\kappa^{(3)}$ were calculated from the derivatives of the two analytic representations of the data shown in Table 2. The average value of the three methods was employed in computing E_{corr}. Plots of E_{corr} vs. $\left(\frac{T}{T-T_0}\right)^2$ for various values of T_0, as shown in Figure 7, resulted in a fit of the data corresponding to $T_0=325(\pm 25)^0K$. The value of k thus determined is $1400(\pm 150)^0K$. The temperature range of the measurements based on this value of T_0 is $2.1 \le T/T_0 \le 2.5$, which is slightly above the range found by Angell[1] over which equation (12) was found to fit the results for the molten $Ca(NO_3)_2$-KNO_3 system ($T \le 1.7T_0$). Angell estimated

TABLE 3

Calculated Values E_{corr} vs. Temperature

$LiF-BeF_2$ (48-52 mole %)

T (^0K)	$RT^2\alpha+\frac{1}{2}RT$ (kcal/mole)	$E_\kappa{}^{(1)}$	$E_\kappa{}^{(2)}$ (kcal/mole)*	$E_\kappa{}^{(3)}$	E_{corr} (kcal/mole)
685	0.88	9.3_0	8.9_7	9.2_3	10.0_5
704	0.91	8.5_7	8.5_2	8.5_5	9.4_6
725	0.94	8.0_0	8.0_8	8.0_2	8.9_7
746	0.97	7.8_0	7.7_0	7.6_2	8.6_8
769	1.01	7.3_0	7.3_3	7.3_1	8.3_2
794	1.05	6.5_5	6.9_9	7.0_6	7.9_2

*$E_\kappa{}^{(1)}$ by graphical methods

$E_\kappa{}^{(2)}$ from derivative of $\kappa = ae^{-b/T-c}$

$E_\kappa{}^{(3)}$ from derivative of $\kappa = a + bt^2$

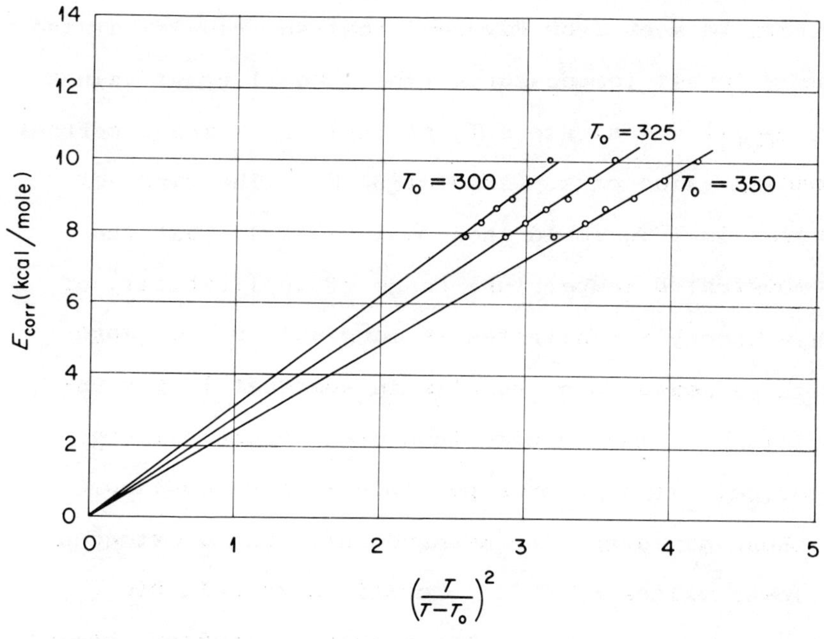

FIG. 7

Estimation of the zero mobility temperature.

an increase in k over the low temperature value (obtained below $1.7T_0$) of 12% at $2T_0$ and 39% at $3T_0$. If the value of k is found to vary in a similar manner in the fused fluoride system, a k of 1400 would be approximately 20% too large, indicating its correct value to be approximately 1170.

Since the divergence between experiment and equation (12) is greater at larger values of T/T_0, one might expect the data of the lower half of the temperature range to provide a better estimate of T_0.

It can be seen from Figure 7 that the results at the three lowest temperatures [the three largest values of $(\frac{T}{T-T_0})^2$] indicate a T_0 of 350(±25)°K and a corresponding value of k = 1200(±150)°K. (The range of values in T/T_0 would then be 1.9-2.1.) That the demonstrated temperature range of applicability of this theory for nitrates is marginal to the range of these measurement results in somewhat larger uncertainty in the results than might ordinarily be expected. If it proves possible to supercool some of these mixtures, the measurements can be extended to lower values of T/T_0 to obtain more reliable estimates of T_0 and k. The present estimates, however, appear to be consistent with results in other systems, e.g., T_0's of 200-330°K for the nitrates.[2] Values of k for the nitrates,[2] carbonates (estimated),[2] and a sodium silicate melt (viscosity)[2,36] are of the order of 700, 1400 and 5100°K, respectively.

CONCLUSION

An apparatus suitable for determining electrical conductivities in molten lithium fluoride-beryllium fluoride mixtures has been constructed and its use successfully demonstrated. Data presented herein characterize for the first time electrical conductivities of mixtures of lithium fluoride and beryllium fluoride. A value of the zero mobility temperature of the lowest melting mixture, estimated by the methods of Angell, is $T_0 = 350(\pm 25)^0K$.

ACKNOWLEDGMENTS

*Research sponsored by the U.S. Atomic Energy Commission under contract with the Union Carbide Corporation.

The authors wish to thank J. H. Shaffer and co-workers for preparation of the $LiF-BeF_2$ (66-34 mole %) starting material and B. F. Hitch for distillation of the beryllium fluoride.

REFERENCES

1. C. A. Angell, J. Phys. Chem., 68, 1917 (1964).
2. Ibid., 70, 2793 (1966).
3. G. D. Robbins, AEC Report ORNL-TM-2180, Oak Ridge, Tennessee, Mar 1968.
4. G. Jones and S. M. Christian, J. Am. Chem. Soc., 57, 272 (1935).
5. F. S. Feates, D. J. G. Ives, and J. H. Pryor, J. Electrochem. Soc., 103, 580 (1956).
6. J. C. Nichol and R. M. Fuoss, J. Am. Chem. Soc., 77, 198 (1955).
7. G. J. Hills and K. E. Johnson, J. Electrochem. Soc., 108, 1013 (1961).

8. D. L. Hill, G. J. Hills, L. Young, and J. O'M. Bockris, J. Electroanal. Chem., 1, 79 (1959).

9. E. R. Buckle and P. E. Tsaoussoglou, J. Chem. Soc., London, 667 (1964).

10. B. DeNooijer, Ph.D. Thesis, University of Amsterdam, The Netherlands, 1965.

11. H. Winterhager and L. Werner, Forschungsber. des Witschafts-u. Verkehrsministeriums Nordrhein-Westfalen, No. 438, 1957.

12. D. C. Grahame, J. Am. Chem. Soc., 63, 1207 (1941).

13. C. H. Liu, K. E. Johnson, and H. A. Laitinen, in "Molten Salt Chemistry," M. Blander, Ed., Interscience Publishers, New York, 1964, p. 715.

14. G. Jones and R. C. Josephs, J. Am. Chem. Soc., 50, 1049 (1928).

15. D. C. Grahame, J. Electrochem. Soc., 99, 370C (1952).

16. D. C. Grahame, Ann. Rev. Phys. Chem., 6, 337 (1955).

17. P. H. Dike, Rev. Sci. Instr., 2, 379 (1931).

18. K. A. Romberger, AEC Report ORNL-4254, Oak Ridge, Tennessee, Feb 1968.

19. C. E. Bamberger, R. B. Allen, and C. F. Baes, Jr., AEC Report ORNL-4229, Oak Ridge, Tennessee, Dec 1967.

20. B. F. Hitch and C. F. Baes, Jr., AEC Report ORNL-4076, Oak Ridge, Tennessee, Dec. 1966.

21. C. E. Bamberger, Oak Ridge National Laboratory, private communication; C. E. Bamberger, C. F. Baes, Jr., and J. P. Young, J. Inorg. and Nucl. Chem., 30, 1979 (1968).

22. H. Bloom, I. W. Knaggs, J. J. Molloy, and D. Welch, Trans. Faraday Soc., 49, 1458 (1953).

23. F. M. Jaeger and B. Kapma, Z. Anorg. Chem., 113, 27 (1920).

24. C. Kröger and P. Weisgerber, Z. Physik. Chem. (Frankfurt), 5, 192 (1955).

25. D. F. Smith and E. R. Van Artsdalen, AEC Report ORNL-2159, Oak Ridge, Tennessee, June 1956.

26. R. A. Robinson and R. H. Stokes, "Electrolyte Solutions," 2nd Ed., Butterworths, London, 1959, p. 462.

27. G. D. Robbins and J. Braunstein, Oak Ridge National Laboratory, to be published.

28. R. E. Thoma, H. Insley, H. A. Friedman, and G. M. Hebert, J. Nucl. Materials, 27, 166 (1968).

29. W. R. Busing and H. A. Levy, AEC Report ORNL-TM-271, Oak Ridge, Tennessee, 1962.

30. M. H. Cohen and D. Turnbull, J. Chem. Phys., 31, 1164 (1959).

31. G. Adam and J. H. Gibbs, J. Chem. Phys., 43, 139 (1965).

32. C. T. Moynihan, J. Phys. Chem., 70, 3399 (1966).

33. J. Braunstein, L. Orr, A. R. Alvarez-Funes, and H. Braunstein, J. Electroanal. Chem., 15, 337 (1967).

34. C. A. Angell, J. Phys. Chem., 69, 399 (1965).

35. B. C. Blanke, E. N. Bousquet, M. L. Curtis, and E. L. Murphy, AEC Report MLM-1086, Miamisburg, Ohio, 1956.

36. G. S. Fulcher, J. Am. Ceram. Soc., 8, 339 (1925).

THE DISTRIBUTION OF METALS BETWEEN MOLTEN FLUORIDES AND LIQUID METALS

D. M. Moulton, W. R. Grimes, and J. H. Shaffer

Reactor Chemistry Division
Oak Ridge National Laboratory
Oak Ridge, Tennessee

INTRODUCTION

Reprocessing of the MSBR

In the two-fluid region concept, the Molten Salt Nuclear Breeder Reactor will use as its fuel a solution of about 0.3 mole % uranium tetrafluoride in a molten lithium fluoride-beryllium fluoride solvent mixture. Fissionable uranium-233 will be produced by neutron bombardment of thorium-232 that will be dissolved in relatively large concentrations in a separate fluoride blanket salt mixture surrounding the reactor core. This fluid-fuel, fluid-blanket reactor concept readily lends itself to continuous reprocessing methods for recovering bred uranium from the blanket mixture and for removing objectionable fission products from the reactor fuel mixture. The present investigation has examined the reductive extraction of these materials from the respective molten fluoride mixtures into bismuth for application as a reprocessing method for molten salt breeder reactors.

Processes for recovering ^{233}U from the blanket mixture are directed toward the separation of ^{233}Pa, its precursor, on a relatively short cycle to reduce parasitic neutron absorption losses and fissioning events in the blanket mixture. Reprocessing methods for the fuel mixture are primarily concerned with the removal of rare earth fission products. By the reductive extraction process the fuel and blanket mixtures would be contacted, separately, with bismuth containing a dissolved reductant. The present work describes measurements of the equilibrium distributions between the salt and metal phases of the various species of interest.

Thermodynamic Considerations

Considering lithium as the reductant, the general equilibrium reaction for the process may be written as

$$MF_n(s) + n\,Li^o(m) \quad = \quad M^o(m) + nLiF(s), \qquad (1)$$

where "n" denotes the valence of species "M" and where the materials are dissolved in the salt (s) or bismuth (m) phase as indicated. This equation may be expressed as the sum of the six fundamental processes:

$$nLi(l) + n/2\,F_2(g) = nLiF(c) \qquad n\Delta G^o_{Li} \quad (2)$$

$$MF_n(c) = M^o(c) + n/2\,F_2(g) \qquad -\Delta G^o_m \quad (3)$$

$$nLiF(c) = nLiF(s) \qquad n\overline{\Delta G}_{LiF}(s) \quad (4)$$

$$MF_n(s) = MF_n(c) \qquad -\overline{\Delta G}_{MF_n}(s) \quad (5)$$

$$nLi^o(m) = nLi^o(l) \qquad -n\overline{\Delta G}_{Li}(m) \quad (6)$$

$$M^o(c) = M^o(m) \qquad \overline{\Delta G}_M(m) \quad (7)$$

Here the subscripts l, g, and c refer to pure liquid, gas, and crystal. Although we have thought it more convenient and appropriate to measure the equilibrium of (1) directly, the separate processes have been independently investigated and contribute to the rationale of the extraction process.

The free energy of formation of metal fluorides dissolved in LiF-BeF_2 (67-33 mole %) [(2) + (4) and (3) + (5)] has been tabulated for lithium, beryllium, and several other metals by Baes.[1] Foster, Wood, and Crouthamel[2] have measured $\overline{\Delta G}_{(m)}$ for lithium; $\overline{\Delta G}_{(m)}$ for uranium was found by Tien, Guion, and Pehlke.[3] Wiswall and Egan[4] have made some measurements of $\overline{\Delta G}_{(m)}$ of U, Th, Zr, and Ce and their analogous distribution between bismuth and a chloride melt. For all but lithium and uranium, however, the data are not extensive and some contain a large estimated error. Protactinium, which is one of the most important elements in this system, was not studied by any of these authors. The values of $\overline{\Delta G}_s$ (equations 4 and 5) are not very large (5-10 k cal) but those of $\overline{\Delta G}_m$ (equations 6 and 7) can be as much as -50 kcal since bismuth forms strong intermetallic compounds with all of the metals of interest except beryllium. These intermetallics are compounds of definite composition, not forming solid solutions, and many of them have the formal composition of salts with a Bi^{-3} anion, e.g., Li_3Bi, Ba_3Bi_2, LaBi, Th_3Bi_4, UBi_2. For beryllium there is no intermetallic and the solubility of the pure metal is negligible.

The Electromotive Series

If extraction is to be the basis of a separation process, it is helpful to express the results in a form that will show the separation of elements from each other. We have established an electromotive series for the reduction of metals in the salt-bismuth system. The ordinary standard half-cell reduction potential Eo for the reaction

$$M^{+n} + ne^- = M^o(c) \qquad (8)$$

is defined for the reduction of M^{+n} to pure metal M. Values of Eo for a number of elements may be obtained from Baes' compilation[1] in which the standard state for M^{+n} is the extrapolation of $X_{(s)}$ to unit mole fraction. We see that when the metal is being reduced into bismuth

$$E = Eo - \frac{RT}{nF} \ln \frac{\gamma(m) \; X(m)}{\gamma(s) \; X(s)} \tag{9}$$

(γ is the activity coefficient)

$\gamma(m)$ and $\gamma(s)$ may be measured separately but they always occur together so that we can define an alternate standard potential

$$Eo' = Eo - \frac{RT}{nF} \ln \frac{\gamma(m)}{\gamma(s)} . \tag{10A}$$

Hence

$$E = Eo' - \frac{RT}{nF} \ln \frac{X(m)}{X(s)} . \tag{10B}$$

This alternate standard potential, Eo', is determined directly from the distribution ratio between the two phases; the Eo's can then be used to establish an electromotive series. One important difference from the ordinary series should be noted: because the reduction is to a solution and not to a pure material, the metals will reduce over a range of potentials centered about Eo' with the width of the range less at higher valences. If the Eo's of two metals are close their extractions will naturally overlap. Within this limitation, however, the Eo's give a direct measure of the relative extractability of the various metals. This alternate standard potential is referred to the solution of the intermetallic in bismuth rather than to the pure metal and this avoids the sometimes confusing use of the term, $\gamma_{(m)} << 1$, at infinite dilution.

Valence

A complete description of the extractability of a metal requires a knowledge both of the potential Eo' and the valence of the ions in equilibrium with the reduced metal. Most of the elements being studied have a single stable oxidation state. Cerium and (probably) Pa were studied only in their lowest oxidation states (+3, +4). However, U, Sm, and Eu were added in higher valences (+4, +3, +3) which are the states in which they should be mainly found in the reactor. There is also the possibility that some metals will be reduced to intermediate states which are not very stable outside of the fused salt. When lower states are possible the distribution of the metal will vary over an unexpectedly wide range of potentials. It is possible to measure valences; we measure lithium distribution $(D_{Li}) = \frac{X_{Li}(m)}{X_{Li}(s)}$ and beryllium electrode potentials (E), and believe that lithium and beryllium have reliable valences of 1 and 2. If distributions are measured over a range of reductant concentrations, the valence, n, can be found from $\frac{d \log D}{d \log D_{Li}}$ or $\frac{d \log D}{dE}$. When two or more states are possible, this calculation yields a weighted average ionic valence n which will decrease with increasing D. If the lower oxidation state is normally stable, n will be near the lower valence during the experimentally observable part of the reduction. However, a solvent such as bismuth which forms strong intermetallics stabilizes the fully reduced metal and thus tends to promote disproportionation of the lower state. Hence it can be shown that when a lower state is only marginally stable in the salts alone it will not appear significantly in the presence of bismuth. Even then, of course, during a complete reduction, n will fall through the entire range of valences; but, when there is strong disproportionation, the lower state will not become predominant until D is so large that it is out of the range of practical measurement.

EXPERIMENTAL

Apparatus and Materials

The extractions were made in vessels constructed of 4-in., IPS stainless steel pipe with welded end closures of 1/4-in. stainless steel plate. These vessels were lined with either mild steel or graphite as the primary container for the molten salt and bismuth. The top plate had inlets for adding lithium metal directly to the molten metal phase, for gas sparging, for sampling, and for inserting a steel thermowell. When graphite was used as the primary container, pipes and tubes that were in contact with the liquid phases were either constructed of graphite or sheathed with graphite. Each experiment contained about 3 kg of bismuth and 2 to 4 kg of salt mixture.

Reagent grade bismuth was sparged with hydrogen at 700-900° in the extraction vessel to remove water and oxides from the metal and container. The molten salt mixture was prepared separately in nickel equipment by treating it with H_2-HF and then H_2 to remove oxide and structural metal impurities to a level of 100 ppm or less. Specific elements to be extracted were added to the salt mixture during its preparation - typically these were uranium (~0.5 wt %), rare earths (~100 ppm) with about 1-5 millicuries of its appropriate radiotracer, and ^{233}Pa (~1 millicurie or .05 micrograms) that was prepared by neutron irradiation of thorium. The prepared molten salt was piped to the extraction vessel as a liquid. The experimental procedure carefully guarded against atmospheric contamination of the extraction system at all times.

The extraction vessel was mounted vertically in a 2700 watt electrical resistance tube furnace. An auxiliary calrod heater was in contact with the bottom plate and was used to offset heat losses from the bottom of the tube furnace. Melt temperatures were adjusted and determined from comparative readings of a Pt-Pt (10% Rh) thermocouple and a Chromel-Alumel thermocouple which were in the thermowell of the extraction vessel. Melt temperatures were probably known to ±3 degrees.

The extraction experiment was started by adding lithium metal to the molten metal phase. After each successive addition of reductant (5 to 20 per experiment), the liquid phases were agitated by an inert gas sparge, allowed to coalesce under static gas conditions, and sampled. Samples of the salt phase were normally withdrawn into filter sticks of copper. Samples of the metal phase were withdrawn either into stainless steel filter sticks or in small graphite dippers. These graphite samplers were inserted through a tube that extended directly into the metal phase to avoid contamination by the salt. Although lithium was the usual reductant, thorium, beryllium and barium were used in some experiments.

Analytical

The distributions of elements of interest between the two liquid phases were determined by analyses of the samples withdrawn from the extraction system. Lithium, thorium, and barium in the bismuth phase were determined by a direct reading Paschen Emission Spectrometer. Barium in the salt was measured by neutron activation. Uranium in both phases was determined by wet chemical methods. Radiochemical methods were employed for analyzing ^{233}Pa and the various rare earths in the two liquid phases. Calculations based on these data yielded values for the half-cell potential for the individual species. In some of the later experiments a beryllium rod electrode was inserted into the salt phase to measure the potential of the liquid metal pool directly. These potentials are the basis of the data on cerium, europium, and lithium.

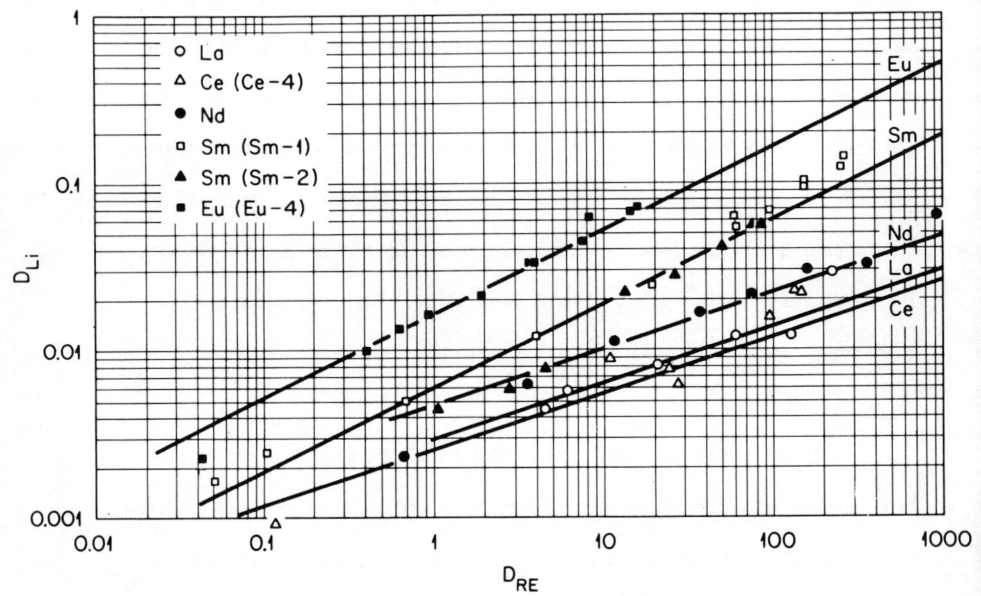

FIG. 1

Lithium distribution (D_{Li}) vs. rare earth distribution (D_{RE}) between bismuth and LiF-BeF_2 (66-34 mole %) at 600°. The slopes correspond to n = 3 (La, Ce, Nd) and n = 2 (Sm, Eu).

RESULTS

Figures 1-4 show most of the data used in deriving the reduction potentials listed in Tables 1 and 2. The potentials in Table 1 are based on the beryllium electrode

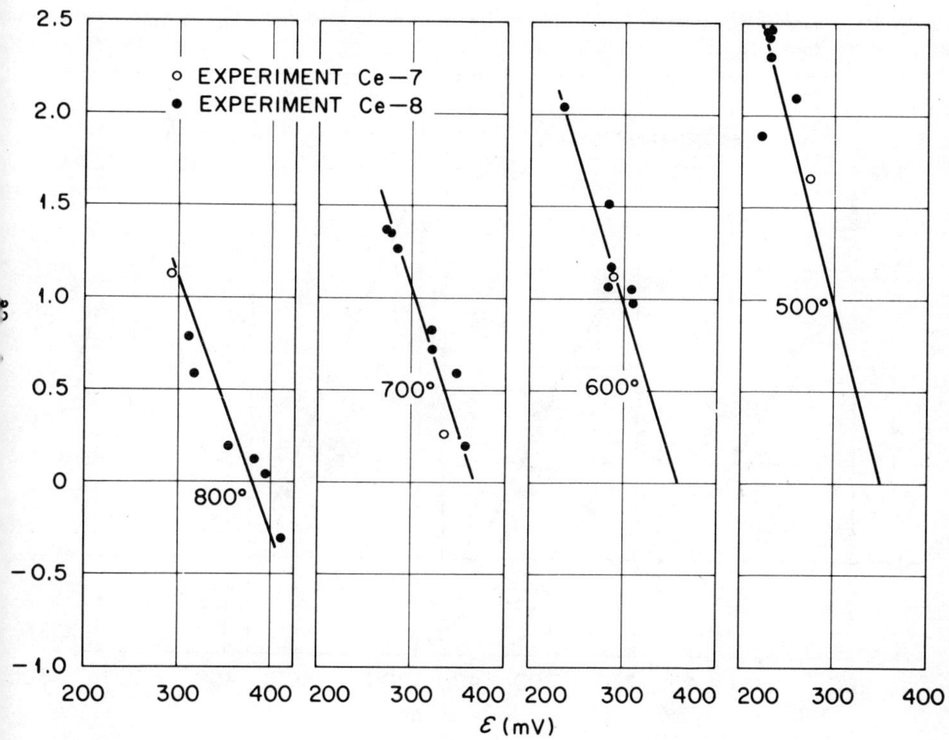

FIG. 2

Cerium distribution vs. beryllium electrode-metal pool potential between bismuth and LiF-BeF_2 (66-34 mole %). The slopes correspond to n = 3.

potential which is in turn based on a standard H_2-HF electrode at 0 volts.[1A] The values for Eo were taken from Baes.[1A, 1B] The uncertainties listed for Li, Eu, and Ce are the standard deviations of the measurements at each point. For the other elements the potentials are good to

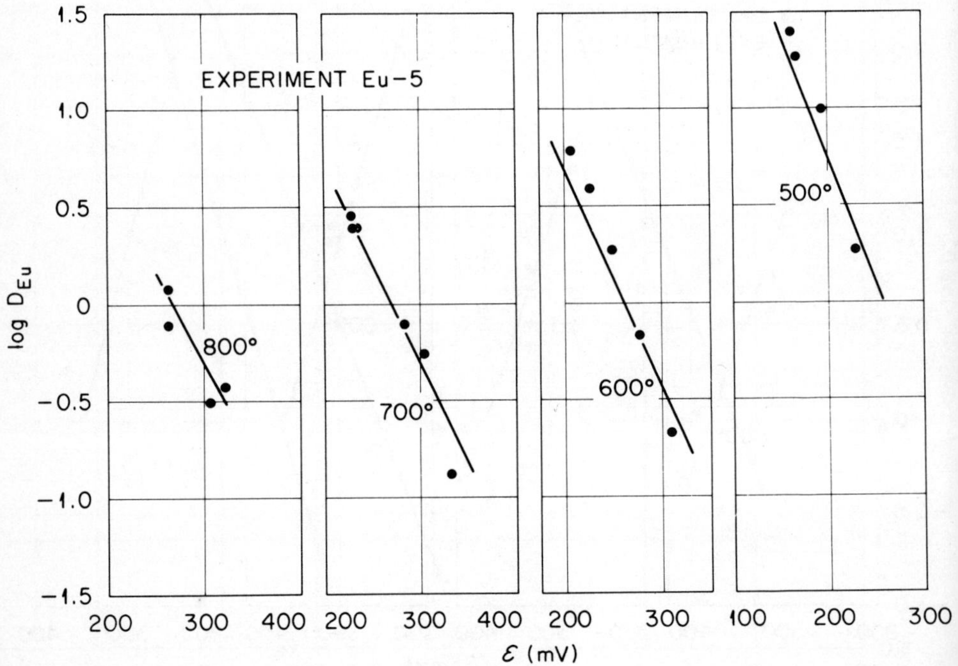

FIG. 3

Europium distribution vs. beryllium electrode-metal pool potential between bismuth and LiF-BeF_2 (66-34 mole %). The slopes correspond to n = 2.

perhaps ±0.02 volts. In Table 2 the lithium potential has been arbitrarily set at -1.860 and the others have been based on it; we have no absolute reference in this system.

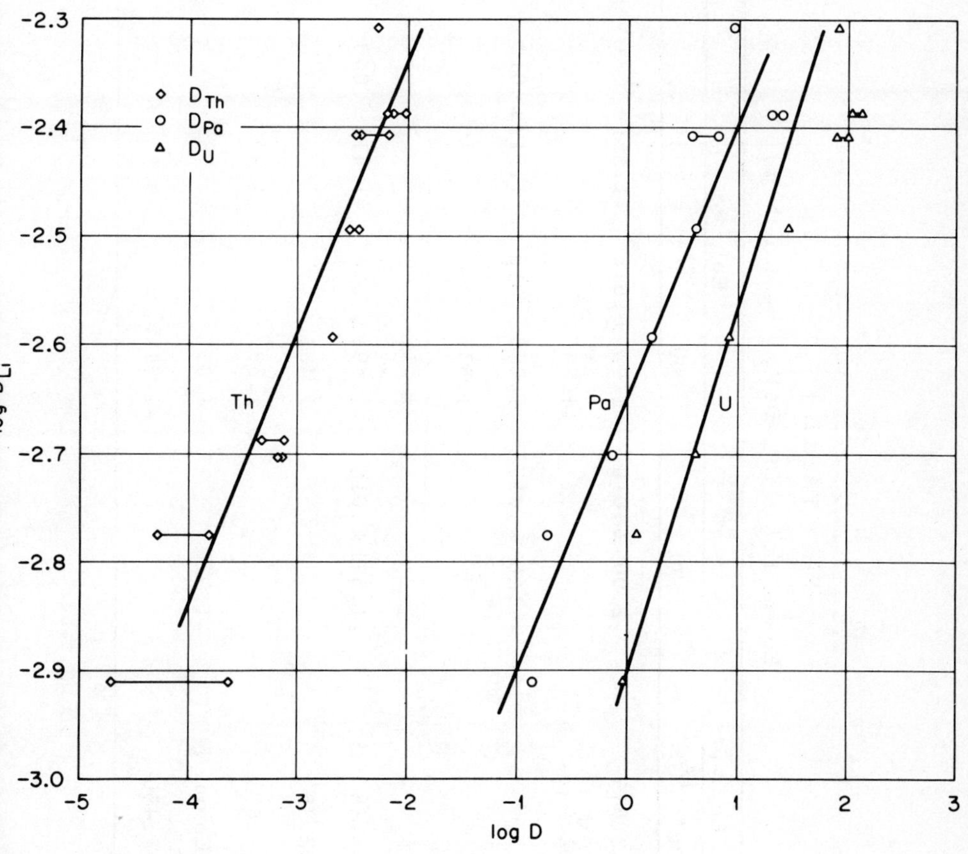

FIG. 4

Lithium distribution vs. actinide distribution between bismuth and LiF-BeF_2-ThF_4 (73-2-25 mole %) at 650°. The slopes correspond to n = 4 (Th, Pa) and n = 3 (U).

TABLE 1

Eo and Eo' in LiF-BeF_2 (66-34 mole %)

	Eo, v.	Eo', v.			
T °C	600[1, 1A]	500	600	700	800
Li^{+}	-2.645	-2.005 ± .010	-1.921 ± .016	-1.851 ± .021	-1.766 ± .012
Ba^{+2}	-2.4 (est.)		-1.80		
Eu^{+2}		-1.663 ± .013	-1.589 ± .016	-1.513 ± .012	-1.448 ± .010
Sm^{+2}			-1.537		
Nd^{+3}			-1.519		
Ce^{+3}	-2.28	-1.571 ± .015	-1.493 ± .008	-1.416 ± .009	-1.337 ± .016
La^{+3}	-2.35		-1.484		
Th^{+4} (1B)	-1.893		-1.58	-1.49	
Be^{+2}	(Eo only)	-1.920	-1.851	-1.783	-1.714
U^{+3}	-1.490		-1.34 ± .03		

TABLE 2

E_o' in $LiF-ThF_4-BeF_2$ (73-25-2 mole %) at 650°C

Li^+	-1.860
Th^{+4}	-1.520 ± .013
Pa^{+4}	-1.377 ± .011
U^{+3}	-1.320 ± .017

DISCUSSION

Rare Earths

The lithium and thorium values in $LiF-BeF_2$ (66-34 mole %) fall respectably close to those predicted from the literature values of Eo and γ(m). The cerium potential is also in agreement, but the cerium data of Ref. 4 and Ref. 1 allow a rather wide range to be predicted.

For rare earth and lithium fluorides the enthalpies and free energies of formation at 600°C (based on Eo for F_2 = 2.865 at 600° + 4.4 mv per 100°, Ref. 1) are

The estimated errors for Ce, Eu, and Li are about ±0.5 kcal. Sm, Nd, and La, which have poor tracers, were studied only briefly and without a beryllium electrode. The potentials were therefore calculated from the lithium distribution, and the greater uncertainties in both analyses mean that the errors in $\overline{\Delta G}$ are probably ±1 kcal or so.

TABLE 3

$\overline{\Delta G}$ and $\overline{\Delta H}$ for $M(m) + n/2\ F_2(g) = MFn(s)$ in LiF-BeF2 (66-34 mole %)

	$\overline{\Delta G}_{600}$	$\overline{\Delta H}$
NdF_3	-303	
CeF_3	-301.5	-346.1
LaF_3	-301	
EuF_2	-205.4	-232.4
SmF_2	-203	
LiF	-110.4	-125.7

It should be mentioned that measurements of the valence of rare earths frequently showed a value fractionally lower than the theoretical, relative to both the lithium distribution and the electrode. An accurate measurement of valence is not easy when there is much scatter in the data, so we are not convinced that it is a real effect. If it is real, it is probably due to a metal-phase activity effect and not to the formation of lower oxidation states in view of the above discussion.

Uranium

The reduction potential for uranium in $LiF-BeF_2$ is about 0.1 volts more negative than would be expected from the data in the literature. Uranium also seems to have the valence +4 rather than +3, whereas calculation indicates that n should approach 3 rapidly at $D_U > .01$. Furthermore, these two observations are inconsistent with each other since a high reduction potential means weak intermetallic formation which in turn implies a greater stability of low-valent states. While the difference between tri- and tetravalence is hard to measure, the difference in potential would require that D_U be in error by about 10^2. Similar anomalies were observed in the $LiF-BeF_2-ThF_4$ salt though here the magnitude of $Eo'_{Li}-Eo'_U$ could not be predicted exactly.

A complicating factor in these determinations is the possibility of solute-solute interaction in the metal phase. Although the activity coefficients in the particular systems of interest to us have not been studied in regard to solute interaction, it has been shown by Gluck and Pehlke[5] that a number of elements, including Cd, Sn, and Sb, can affect the activity of Zn in Bi at mole fractions of .01 or less for the solutes. Balzhiser and Ragone[6] have found that small amounts of copper raise the activity of uranium in bismuth. It is known (see, for example, Weinberg, et al.)[7] that Zr at a concentration of only 100 ppm or so will substantially decrease the solubility of uranium in bismuth. In some of our uranium studies zirconium was present and this may have affected the results. We have not yet seen any clear-cut indication of solute-solute interactions in these dilute systems, but we are continuing to study this matter since the practical application of the process will involve a solution of many elements at once.

CONCLUSIONS

The most remarkable finding in this work is that in LiF-BeF_2 mixtures the rare earth elements can be removed by extraction into bismuth without reducing the beryllium in the salt. If reduction to pure metal were tried, beryllium would reduce about 0.3 volts sooner, but its insolubility in bismuth together with the extreme stability of the rare earth bismuthides allows this order of reduction to be reversed. Bismuth is the only metal in which we have found this reversal; a few experiments in lead indicate that the intermetallics are not strong enough. A small amount of lithium is also removed, but this can be recovered in a further processing step (which is important since isotopically pure ^7Li must be used in a reactor). It is obvious that in any such processing scheme the first element to be removed is uranium, which is not otherwise necessary but presents no serious design problem. The position of protactinium between uranium and thorium means that it can be singled out and separated without major contamination by other materials, and the part of it that decays to uranium can be further separated and returned to the reactor. Finally, it may be inferred that rare earths can be separated to some extent from a melt containing large amounts of thorium. The separation in this case is limited by the fact that the solubility of the thorium-bismuth intermetallic is only about 0.25 mole % (Bryner and Brodsky[8]) so that the reduction potential cannot rise above this point. This particular system is of interest because of its importance for the single-fluid Molten Salt Breeder Reactor, an alternative to the two fluid reactor for which this study was made, in which both the thorium and the uranium with its fission products will be in the same salt. The separation of rare earths from thorium-bearing salts is presently being investigated as part of the program in support of such a reactor.

ACKNOWLEDGMENTS

This research was sponsored by the U. S. Atomic Energy Commission under contract with the Union Carbide Corporation.

REFERENCES

1. C. F. Baes, Jr., "The Chemistry and Thermodynamics of Molten-Salt-Reactor Fluoride Solutions," pp. 409-433 in Thermodynamics, vol I, International Atomic Energy Agency, Vienna, 1966.

1A. B. F. Hitch and C. F. Baes, Jr., An EMF Study of $LiF-BeF_2$ Solutions, ORNL-4257 (July 1968). (This work gives more accurate values for some of the quantities in Ref. 1. We are indebted to Dr. Baes and Mr. Hitch for discussions of their work.)

1B. C. F. Baes, Jr., personal communication.

2. M. S. Foster, S. E. Wood, and C. E. Crouthamel, Inorg. Chem. 3, 1428 (1964).

3. L. C. Tien, K. J. Guion, and R. D. Pehlke, pp. 501-514 in Thermodynamics, op. cit.

4. R. H. Wiswall, Jr. and J. J. Egan, "Thermodynamic Properties of Solutions of the Actinides and the Principal Fission Products in Bismuth," pp. 345-364 in Thermodynamics of Nuclear Materials, International Atomic Energy Agency, Vienna, 1962.

5. J. V. Gluck and R. D. Pehlke, Trans. Metallurgical Soc. A.I.M.E. 239, 36-47 (1967).

6. R. E. Balzhiser and D. V. Ragone, Ibid, 224, 485-490.

7. A. F. Weinberg, R. J. Van Thyne, and R. E. Steiner, Ibid, 221, 83-90 (1961).

8. J. S. Bryner and M. B. Brodsky, Proceedings, 2nd International Conference on Peaceful Uses of Atomic Energy 7, 209 (1958).

INVESTIGATION OF THE CAPACITY OF THE ELECTRICAL DOUBLE LAYER IN MOLTEN CHLORIDES USING A DROPPING METAL ELECTRODE

R. J. Heus, T. Tidwell,* and J. J. Egan

Brookhaven National Laboratory
Upton, New York

INTRODUCTION

Double layer capacitance measurements of molten salt–liquid metal interfaces are of interest for obtaining an understanding of the structure of the double layer in these systems as well as in aqueous systems where the compact double layer is not completely understood. Results are also needed if one is to make accurate measurements of electrode kinetics in molten salt systems.

---

*Present address: North Texas State University, Denton, Texas.

This paper describes a technique employing a dropping metal electrode (lead or bismuth) to measure the double layer capacitance of a molten salt (LiCl-KCl eutectic). The technique is similar to that first introduced by Grahame,[1] who employed a dropping mercury electrode to study the double layer in aqueous electrolytes. A constantly renewed metal surface was found to be a great advantage in his work and so it was hoped that it would also prove fruitful for high temperature systems.

Several workers have reported measurements on double layer capacitance and electrocapillary curves in molten salt systems using various techniques. This work has been reviewed by Laitinen and Osteryoung[2] and more recently by Graves, Hills, and Inman[3] as well as Delahay.[4] Many advances have been made by Russian workers and much of this work is summarized by Ukshe, Bukun, Leikis, and Frumkin.[5]

Still some difficulties remain. In particular, the capacitance is found to depend upon the frequency used for the measurement. This frequency dispersion for measurements reported by Ukshe et al. exists at frequencies below 20 kHz.

This is ascribed to the existence of electrochemical side reactions. To avoid the interference of these reactions only values measured at frequencies above 20 kHz are used.

In general, one tries to eliminate electrochemical side reactions when making double layer measurements. An electrode where such reactions are eliminated is called an ideal polarized electrode. Most accurate measurements of double layer capacitance in aqueous electrolytes have been made at ideal polarized electrodes. In molten salts this ideal has proved difficult to obtain.

Thus in the present work both the use of a dropping metal electrode and extensive pre-electrolysis of the melt have been employed to eliminate frequency dispersion and minimize side reactions.

EXPERIMENTAL

The apparatus used to obtain an adequate dropping metal electrode along with a suitable reference electrode and a means of obtaining the drop area is shown in Figure 1. The metal is loaded above the fritted glass filter (Pyrex) then melted and pushed through the filter with purified argon gas. Thus it enters the capillary free of oxide. The capillary itself is approximately 8 cm long with a 0.05 mm bore. Several of these

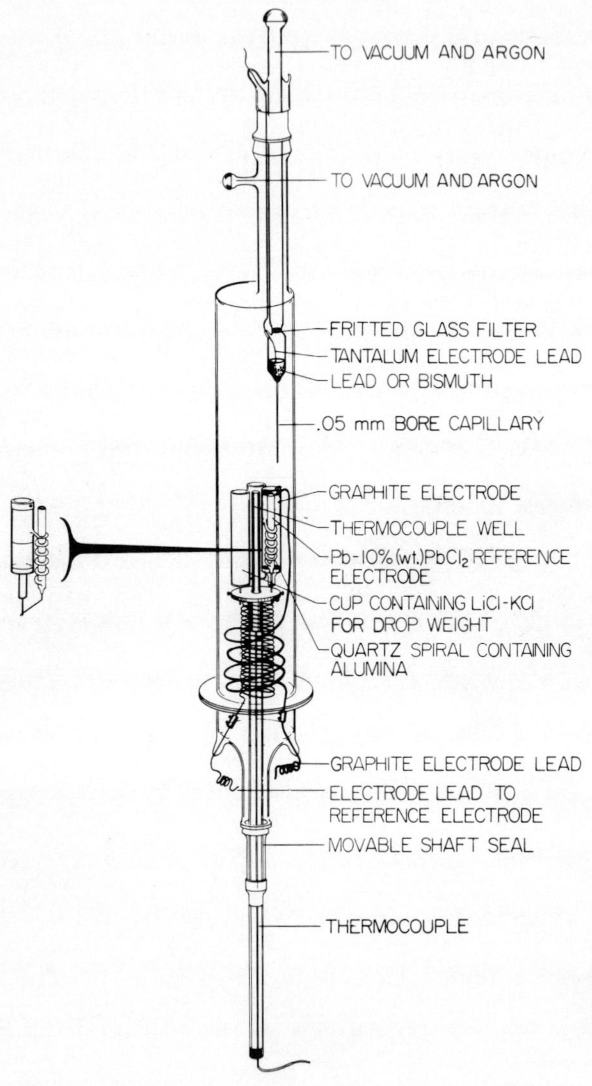

FIG. 1

Apparatus showing dropping electrode, reference electrode, and drop weight cups.

capillaries are drawn and the size selected under a microscope. After connecting the capillary to the filter this entire section is examined under the microscope for contamination. With such close examination, plugging of the capillary during a run is very rare.

The salt is contained in a quartz cylindrical cup and is shown enlarged at the left in Figure 1. A graphite cylinder which serves as the second electrode of large area fits closely inside the quartz cup. The reference electrode, Pb-$PbCl_2$ (10% by weight), is contained in the small tube to the side. These two compartments are joined by a spiral quartz tube filled with fine alumina powder (120 mesh). This type of junction has been proved very useful in previous transference experiments[6] and was employed here. It was very important that no $PbCl_2$ entered the large compartment, and this type of junction worked very well.

The cup containing the salt rested on a stand which could be raised and lowered as well as rotated. Thus, when the salt was melted, the cup was raised so that the capillary was immersed in the salt. A panel with appropriate pressure gages allowed one to regulate the pressure difference across the capillary. Two other Pyrex cups also on the stand and filled with salt were used to obtain the drop weight. The liquid metal flowed into these cups for a known time and a known drop time. A Chromel-Alumel thermocouple between the cups measured the temperature of the salt.

The entire apparatus (approximately 6.4 cm diameter and one-half meter high) is surrounded with a wire-wound resistance furnace which contains a suitable port hole to manipulate the three cups.

The capacitance of the dropping electrode—molten salt interface was measured with a General Radio 1650-A bridge in conjunction with a Tektronix oscilloscope. Since the drop is constantly changing its area and thus its capacitance, one transfers the degree of unbalance of the bridge onto the oscilloscope. With the help of a photograph of the envelope appearing on the oscilloscope screen one can calculate the exact area of the drop at the point of balance of the bridge. A Hewlett Packard 650A oscillator was used with the bridge.

Various potentials are imposed between the dropping electrode and the reference electrode by means of a potentiometer.

Purification of the salt was carried out in several steps. Weighed batches of the eutectic mixture were treated with HCl gas at 100°C for one hour then the temperature was raised to

500°C, where HCl gas was bubbled through the melt for several hours. Cl_2 gas was then bubbled through the melt for several hours followed by purified argon gas to remove excess HCl and Cl_2. The salt was then frozen and the resulting ingot transferred to an apparatus used for pre-electrolysis. This electrolysis was carried out between pyrolytic graphite electrodes (20 cm^2 each) under a high vacuum. It was found most helpful to electrolyze the salt at temperatures considerably above the temperature at which the capacitance measurements were made. Therefore, the salt was electrolyzed for two days at 700°C and one day at 800°C with 1.5 volts imposed across the electrodes. Currents of the order of 1 μA at 525°C still were observed at the end of the electrolysis. The salt was then transferred to another apparatus where it was filtered through a fritted quartz filter and cast into ingots which were the proper size to add to the cups used for capacitance measurements. During all of the above operations, care was taken to keep the salt from contact with the atmosphere.

RESULTS

The capacitance curves obtained with lead and bismuth are shown in Figure 2. These curves were measured at a frequency of 5 kHz, but the value of the capacitance did not change more than 3% by varying the frequency between 1 kHz and 20 kHz. The curves show results for lead at 390° and 480°C with very little change in capacitance with temperature. A second curve of Pb at 390°C is given to show the precision between different experiments. The results of Ukshe et al.[5] on lead in a LiCl-KCl (1:1) mixture are shown for comparison. The shape of the curve is similar to that obtained in this work but there is some discrepancy in the magnitude of the capacitance.

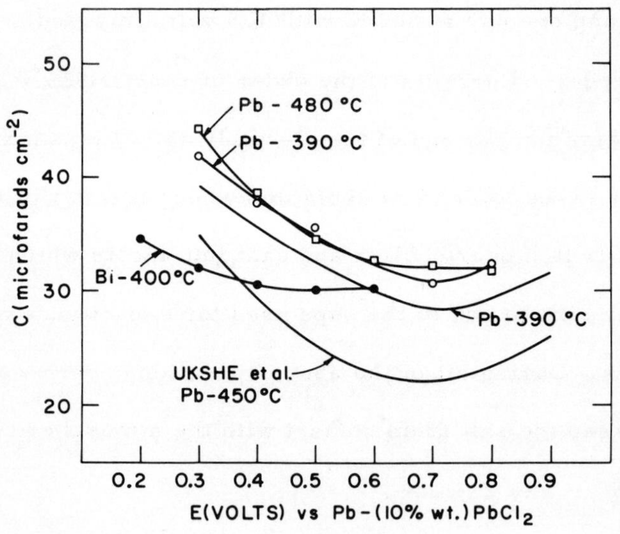

FIG. 2

Results of capacitance measurements using Pb and Bi in molten LiCl-KCl eutectic.

Also these authors report a temperature coefficient of capacitance of +5 μfd/100°C, a result which has not been confirmed in the present measurements.

It should also be mentioned that in measurements at high negative potentials (0.8 to 0.9 volts) a small current (up to 3 μA) was detected flowing between the dropping electrode and the reference electrode in our experiments. This was probably caused by the decomposition of the salt. In light of this, it is not certain that the upward trend of these curves at high negative potentials is real or due to an electrochemical reaction at the electrode.

In conclusion, it is believed that the use of dropping electrodes in high temperature systems has been useful for capacitance measurements as shown by the elimination of frequency dispersion. Further work is necessary on the purification of the salts used (a small current of approximately 1 μA always flowed through the salt during our measurements). Work is under way in our laboratory to extend these measurements to 800°C so that pure salts can be examined.

REFERENCES

1. (a) D. C. Grahame, J. Am. Chem. Soc., 63, 1207 (1941).
(b) D. C. Grahame, ibid., 68, 301 (1946).
(c) D. C. Grahame, ibid., 71, 2975 (1949).

2. H. A. Laitinen and R. A. Osteryoung in *Fused Salts*, B. R. Sundheim, ed., McGraw-Hill, New York, 1964, p. 255.

3. A. D. Graves, G. J. Hills, and D. Inman in *Advances in* Electrochemistry and Electrochemical Engineering, vol. 4, P. Delahay, ed., Interscience, New York, 1966, p. 117.

4. P. Delahay, *Double Layer and Electrode Kinetics*, Interscience, 1965.

5. E. A. Ukshe, N. G. Bukun, D. I. Leikis, and A. N. Frumkin, *Electrochim. Acta*, *9*, 431 (1964).

6. W. K. Behl and J. J. Egan, *J. Phys. Chem.*, *71*, 1764 (1967).

SOLUTIONS OF HALOGENS IN MOLTEN HALIDES

John D. Van Norman and Richard J. Tivers

Brookhaven National Laboratory
Upton, New York 11973

INTRODUCTION

It has been known for some time that gases dissolve to some extent in molten salts at elevated temperatures. As early as 1926, Wartenberg determined the solubilities of the halogen gases in a series of their respective alkali and alkaline earth halides.[1] With the advent of molten salt reactor programs and other applications of molten salt systems in modern technology, a number of recent investigations have been made concerning the dissolution of gases in molten salt media. Battino and Clever have summarized much of the work in a recent review article.[2] It has been found that the solubility of noble gases in molten salts is quite small and can be explained by a "hole" mechanism wherein the noble gas atom is accommodated in the molten salt by a "hole" of the same size.

The dissolution of halogens in molten halides, however, seems to involve an interaction between the dissolved gas and the molten halide. Spectral studies of Greenberg and Sundheim indicate the existence of trihalide ions in solutions of halogens in molten halides.[3] The solubilities of halogens in molten halides determined by Wartenberg,[1] Ryabukhin,[4] and Kowalski and Harrington[5] are greater by several orders of magnitude than those found for the noble gases, again suggesting some gas-melt interaction. Moreover, Kowalski and Harrington found evidence of a polychloride complex in molten chlorides using a radioisotope exchange technique.

The purpose of this investigation is to study the solution chemistry of halogens in molten halides. Electroanalytical techniques can be very useful in elucidating species in solution as well as determining various physical properties of such species. For example, Trusov and Borisova[6] have demonstrated that well defined polarograms of chlorine dissolved in molten lead chloride, silver chloride, and the lithium chloride-potassium chloride eutectic could be obtained at a pyrolytic graphite electrode. Potter and Del Duca[7] also used a pyrolytical graphite electrode in their chronocoulometric study of chlorine dissolved in the lithium chloride-potassium chloride eutectic. The technique of chronopotentiometry has been utilized by Ryabukhin[8] to measure diffusion coefficients of chlorine in molten potassium chloride, sodium chloride, and mixtures of potassium chloride-sodium chloride and potassium chloride-magnesium chloride. The platinum electrodes used in this study were attacked by the chlorine at the elevated temperatures, necessitating solution of diffusion equations for diffusion of chlorine toward the electrode and diffusion of platinum ions away from the electrode. An inert electrode would make interpretation of data much easier.

EXPERIMENTAL

Materials

The lead chloride and the silver chloride used in this investigation were of reagent grade and were further treated by passing HCl gas through the powder while the temperature was brought from room temperature to a point about 50°C above the melting point. After the salts became molten, generally requiring about four hours, HCl was passed for an additional two hours to remove the last traces of moisture. The dissolved HCl gas was removed by an argon purge and the molten chlorides were filtered through a quartz frit. The lithium chloride–potassium chloride eutectic was prepared in the manner described by Laitinen, Ferguson, and Osteryoung.[9] Since chlorine was later passed through all the chloride melts, the salts would be further purified as pointed out by Maricle and Hume.[10] The lead bromide and the lithium bromide were treated in a manner similar to the chlorides, HBr treatment and filtration.

Various forms of carbon were tried as an electrode material. The anodes used were a high density graphite, density 1.9, of spectroscopic grade made by the Ultra-Carbon Co. Three forms of carbon were tried as possible cathodes. The high density graphite proved useless as chlorine gas enters the pores of the graphite, affecting measurements. Pyrolytic graphite blades were tried but fabrication problems were encountered. A satisfactory cathode material was found to be a vitreous carbon made by the Beckwith Carbon Corporation.

The vitreous carbon is impermeable to gases, resists attack by chlorine at the temperatures studied, and has a coefficient of thermal expansion matched by some Pyrex glass. Carbon to Pyrex seals were made that were leak tight and were compatible with the molten chlorides studied.

Apparatus and Measurements

Solubility measurements of chlorine in molten chlorides were made in the cell shown in Figure 1. The melt level was above the female joint. Chlorine was bubbled through the lower section of the cell, as shown in the figure, for a minimum of 12 hours, found to be sufficient time to achieve equilibrium. The bubbler was then raised out of the way and the solution allowed to settle to eliminate any bubbles trapped in the lower section. The Pyrex male joint was lowered into place and the sample well was then checked visually for any trapped chlorine. The argon bubbling tube was pushed through the membrane down into the sample well. Argon was bubbled through the chlorine saturated molten chloride and the off-gases were passed through a series of gas bubblers containing an aqueous potassium iodide solution. The chlorine reacted with the KI solution forming iodine. Argon was bubbled until no more chlorine was detected in the off-gas stream, usually requiring

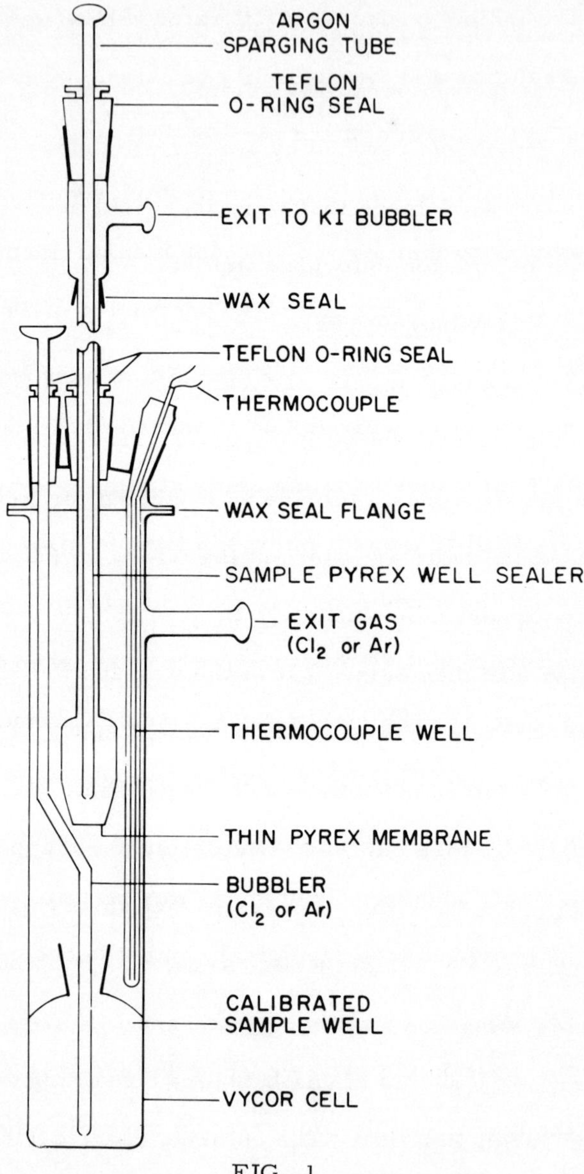

FIG. 1

Apparatus for the measurement of chlorine solubility in molten chlorides.

1 hour. The iodine produced was titrated with a standard thiosulfate solution and the solubility of chlorine was calculated knowing the volume of the sample well.

Solubilities of bromine in molten bromides were measured in a cell similar to that for the chlorine measurements except that the top of the bromine cell was constructed of Teflon and all seals were O-ring seals. The cell top, all gas lines, and the bromine generator were inside a transite box maintained at 61° to $63^{\circ}C$ to insure bromine being at atmospheric pressure. For solubility measurements in lithium bromide the male joint with the membrane was made of quartz.

Chronopotentiometric measurements were taken in the cell shown in Figure 2. The cathode was a small vitreous carbon crucible sealed into Pyrex. A graphite rod extended down the Pyrex tube to make contact with the vitreous carbon crucible in which a small amount of lead metal was placed. Upon melting the lead provided good electrical contact between the rod and the crucible.

Once the salts in the cells reached temperature, as measured by the thermocouple immersed in the melt, the melt was saturated with chlorine gas by bubbling through the melt for a day and over the melt for the remaining time of the run.

The electrodes were then lowered into the melt and cathodic chronopotentiograms taken by passing a constant current between the electrodes while measuring the potential between

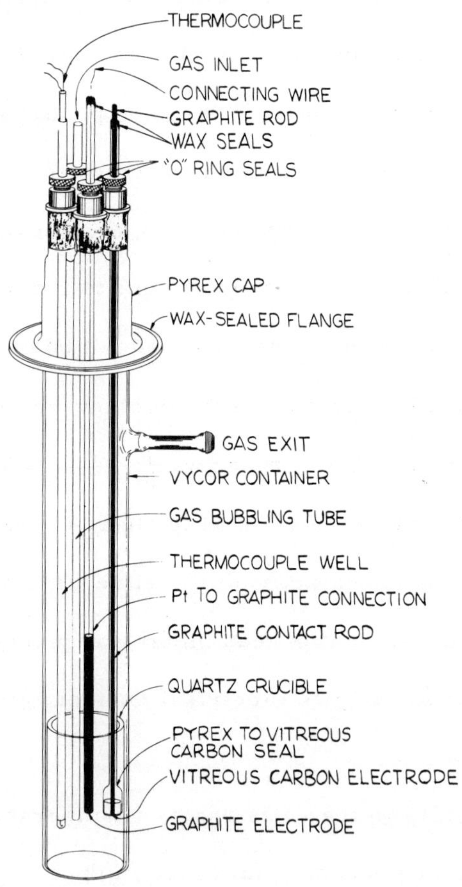

FIG. 2

Chronopotentiometric cell for molten chlorides.

them as a function of time. The anode served as the reference electrode, a chlorine electrode that is nonpolarizable. The constant current regulator was built by the Instrumentation Division of Brookhaven National Laboratory. The chronopotentiograms were recorded on a L&N High Speed Speedomax G recorder that had a pen speed of 0.25 sec full scale and a chart speed of 2.0 or 4.0 ips. Transition times were read directly from the chart as previously described,[11] with an estimated accuracy of $\pm$0.025 sec. Transition times were generally kept between 0.5 and 1.5 sec.

RESULTS AND DISCUSSION

Solubility Studies

The solubilities of chlorine at one atmosphere pressure in lithium chloride–potassium chloride eutectic, lead chloride, and silver chloride as well as the solubilities of bromine at one atmosphere pressure in lead bromide and lithium bromide are listed in Table 1. The solubility values for chlorine are averages of at least two determinations in each of two separate batches of salt. The solubility values for bromine are averages of one or two determinations in two separate batches of salt. The estimated experimental error is $\pm$5%.

It can be seen from Table 1 that the solubility of chlorine becomes greater as the temperature is increased. This increase at higher temperatures has been noted in chlorides and chloride mixtures by Ryabukhin,[4] by Lukmanova and Vil'nyanski,[12] and by Kowalski and Harrington[5] for certain mixtures of lead chloride–potassium chloride. A plot of log X_{Cl_2} versus 1/T should be linear with a slope equal to $-\Delta\overline{H}^o/2.303\,R$. Figure 3 shows such a plot for the LiCl-KCl eutectic and $PbCl_2$.

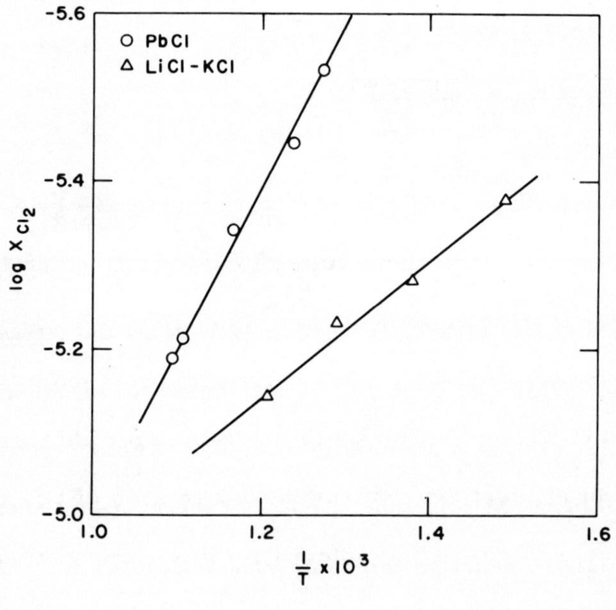

FIG. 3

Temperature dependence of chlorine solubility in molten $PbCl_2$ and the LiCl-KCl eutectic.

TABLE 1

Solubility of Halogens in Molten Halides

Salt	Gas	Temperature, °C	Solubility, moles/cc x 10^7	Solubility, mole fraction x 10^6
LiCl-KCl eutectic	Cl_2	400	1.26	4.19
		450	1.54	5.21
		500	1.70	5.83
		550	2.05	7.14
$PbCl_2$	Cl_2	513	0.52	2.95
		535	0.64	3.62
		585	0.78	4.45
		630	1.04	6.12
		639	1.10	6.47
AgCl	Cl_2	518	2.31	6.95
$PbBr_2$	Br_2	430	13.9	92.1
		519	12.3	83.2
LiBr	Br_2	617	17.5	61.0
		709	10.4	37.2

The solubility of bromine becomes less as the temperature increases. This can probably be attributed to the lesser stability of Br_3^- ion at higher temperatures. Table 2 gives the calculated heats of solution, $\Delta\overline{H}^o$, for the systems studied. The estimated errors in the heats of solution for the chlorine systems are $\pm 10\%$. In view of the fact that results were obtained at only two temperatures for the bromine systems, the results must be considered tentative.

TABLE 2

Calculated Heats of Solution

Salt	Gas	Temperature range	$\Delta \bar{H}^{o}$, kcal/mole
LiCl-KCl eutectic	Cl_2	400° to $550^{o}C$	+3.7
$PbCl_2$	Cl_2	513° to $639^{o}C$	+9.4
$PbBr_2$	Br_2	430° to $519^{o}C$	−1.3
LiBr	Br_2	617° to $709^{o}C$	−9.5

The solubility results obtained, in general, are quite comparable to results obtained in most other investigations cited. However, Olander and Camahort[13] investigated the solubility of chlorine in the lithium chloride— potassium chloride eutectic at 400° and $500^{o}C$. They found no measurable quantity of chlorine dissolved in the molten eutectic with the lower limit of detection quoted at 4×10^{-9} moles/cc, two orders of magnitude less than the values found in the present investigation. The reason for this difference is unknown at the present time.

HALOGEN SOLUTIONS IN MOLTEN HALIDES

Electrochemical Studies

Cathodic chronopotentiograms were obtained for chlorine dissolved in all three chloride systems studied and as a function of temperature for the lithium chloride–potassium chloride eutectic. As previously mentioned, vitreous carbon proved to be the best cathode material. Caution had to be taken to make sure that there was no bubble formation on the surface of the electrode. In chlorine or bromine saturated melts, bubbles form at any point or rough surface. Electrodes were visually examined for any bubbles that might interfere with proper diffusion conditions to the electrode surface.

Figure 4 shows a typical cathodic chronopotentiogram of chlorine obtained at a vitreous carbon electrode in molten chlorides. The transition time, τ, shown on the curve, is taken as that point at which the potential rise becomes linear. At the potential at τ, the surface concentration of chlorine has been reduced by a factor of over 1000 and the basic equation of chronopotentiometry

$$i_o \tau^{\frac{1}{2}} = \frac{nF\pi^{\frac{1}{2}}D^{\frac{1}{2}}C}{2} \qquad (1)$$

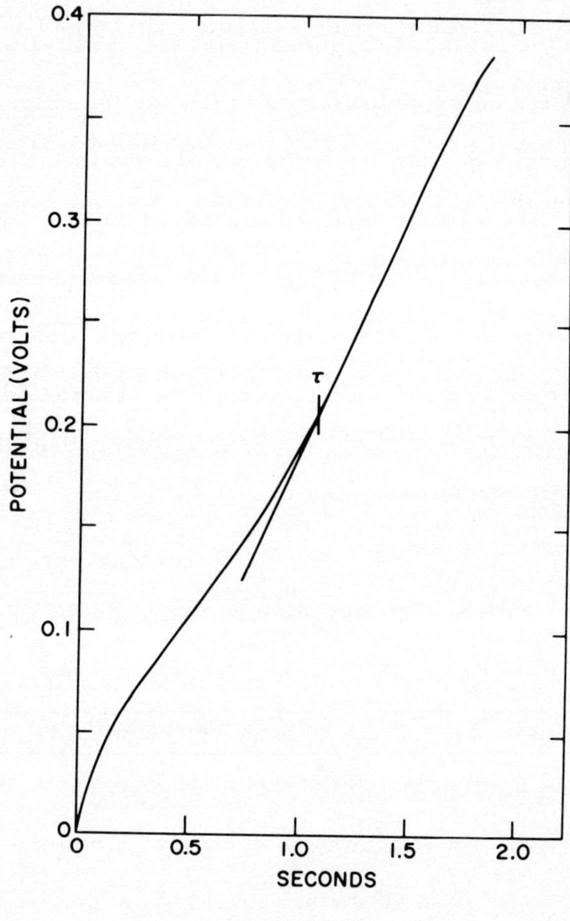

FIG. 4

Typical cathodic chronopotentiogram of chlorine dissolved in molten chloride; LiCl-KCl eutectic at 450°C, current density of 0.27 mA per cm^2.

is obeyed. The product $i_0\tau^{\frac{1}{2}}$ should be constant at any given concentration and temperature; this was verified experimentally within the limits of the measurements, $\pm 5\%$. Diffusion coefficients for chlorine dissolved in the molten chloride were calculated from equation 1 and the results obtained are given in Table 3. The results obtained are based on an average of 10 or more measurements in each of two separate batches of salts. Because of the possibility of undetected bubbles being on the electrode surface and the variations between separate batches of salt, the estimated error in the diffusion coefficient values is $\pm 20\%$.

TABLE 3

Diffusion Coefficients of Chlorine in Molten Chlorides

Salt	Temperature, °C	$D \times 10^4$ cm^2/sec
LiCl-KCl	400	0.60
eutectic	450	1.6
	500	1.7
	550	1.5
$PbCl_2$	580	3.5
AgCl	518	3.7

As can be seen from Table 3 the diffusion coefficients measured are greater by an order of magnitude than those normally found for ions in molten salts ($\approx 10^{-5}$ cm^2/sec). These results agree with those found by Ryabukhin[8] who measured diffusion coefficients of chlorine in KCl, NaCl, and NaCl-KCl, NaCl-$MgCl_2$ mixtures finding values of 10^{-3} to 10^{-4} cm^2/sec. The value of 0.60×10^{-4} cm^2/sec found in this investigation for Cl_2 dissolved in LiCl-KCl eutectic at 400°C agrees well with that calculated from the data of Potter and Del Duca.[7] A further check on this value was made by performing a polarographic "pilot ion" experiment in which a polarogram of Cl_2 was obtained in an LiCl-KCl melt to which a known amount of silver chloride had been added. By comparing the wave heights obtained with the relative amounts of each electroactive species, Cl_2 and Ag^+, it was found that the diffusion coefficient of chlorine was greater than that of Ag^+ ion in the eutectic, $D_{Ag^+} = 2.4 \times 10^{-5}$ cm^2/sec,[14] by a factor of 3 or 4.

The reason for the high diffusion coefficients of chlorine in molten chloride is unknown. One appealing explanation takes into account the fact that chlorine dissolves in the melt forming Cl_3^- ions, the trichloride ion. A chain conduction of the Grotthus type is then possible wherein a Cl_3^- ion transfers a Cl_2 molecule to an adjoining Cl^- ion which then becomes the

Cl_3^- ion and the process repeating itself. Diagrammatically shown it would be

$$Cl-Cl-Cl^- + Cl^- \longrightarrow Cl^- + Cl-Cl-Cl^-$$

with the net effect being transport of chlorine molecules. Such a mechanism requires exchange of chloride ions with the chlorine gas. This exchange has been studied by Kowalski and Harrington[5] where they found a rapid exchange between Cl^{36} ions in molten $PbCl_2$-KCl mixtures and chlorine gas bubbled through the melt.

With the exception of the diffusion coefficient of Cl_2 in LiCl-KCl eutectic at 400°C, the D_{Cl_2} in the eutectic remains fairly constant which implies a small value for the energy of activation of diffusion. The same result was reported by Potter and Del Duca.[7] For a process that is diffusion controlled, a positive energy of activation for diffusion would be expected. However, for a process such as a Grotthus type of conduction, a small value for the energy of activation for diffusion is reasonable.

It is evident that the chemistry of solution of halogens in their molten halides is complex and that further studies are needed to more fully understand such systems.

ACKNOWLEDGMENTS

The authors would like to express their appreciation to Drs. James J. Egan and Richard H. Wiswall, Jr., for many helpful discussions and suggestions. This work was performed under the auspices of the U. S. Atomic Energy Commission.

REFERENCES

1. H. v. Wartenberg, Z. Elektrochem., 32, 330 (1926).

2. R. Battino and H. L. Clever, Chem. Rev., 66, 395 (1966).

3. J. Greenberg and B. R. Sundheim, J. Chem. Phys., 29, 1029 (1958).

4. Y. M. Ryabukhin, Russ. J. Inorg. Chem. (English Transl.), 7, 565 (1962).

5. M. Kowalski and G. W. Harrington, Inorg. Nucl. Chem. Letters, 3, 121 (1967).

6. G. N. Trusov and S. R. Borisova, Elektrokhim., 1, 709 (1965).

7. A. E. Potter and B. S. Del Duca, Electrochemical Reduction of Chlorine on Carbon in Molten Lithium Chloride-Potassium Chloride, NASA TN D-4175, National Aeronautics and Space Administration, 1967.

8. Y. M. Ryabukhin, Russ. J. Phys. Chem. (English Transl.), 39, 1563 (1965).

9. H. A. Laitinen, W. S. Ferguson, and R. A. Osteryoung, J. Electrochem. Soc., 104, 516 (1957).

10. D. L. Maricle and D. N. Hume, *ibid.*, *107*, 354 (1960).

11. J. D. Van Norman, *ibid.*, *112*, 1126 (1965).

12. T. L. Lukmanova and Ya. E. Vil'nyanski, *Izv. Vyssh. Ucheb. Zaved., Khim., Khim. Tekhnol.*, *9*, 567 (1966).

13. D. R. Olander and J. L. Camahort, *A. I. Ch. E. J.*, *12*, 693 (1966).

14. C. E. Thalmayer, S. Bruckenstein and D. M. Gruen, *J. Inorg. Nucl. Chem.*, *26*, 347 (1964).

ELECTRODE REACTIONS IN MOLTEN FLUORIDES

G. Mamantov

Department of Chemistry
University of Tennessee
Knoxville, Tennessee 37916

INTRODUCTION

Molten fluorides are of considerable importance in nuclear technology, production of aluminum and fluorine, electrodeposition of refractory metals in coherent form, formation of corrosion-resistant diffusion coatings, and electrochemical fluorination of organic and inorganic compounds; they also show promise as the electrolytes for thermal batteries. From the standpoint of basic research, molten fluorides are a class of nonaqueous solvents in which interesting redox and acid-base chemistry may be pursued; unusually low and high oxidation states have been claimed to exist in molten fluorides.

Most of the studies of electrode reactions in molten fluorides are quite recent; for example, comprehensive reviews of electrochemistry in molten salts,[1,2] published in 1964, do not discuss studies in molten fluorides. The purpose of this chapter is to summarize the present status of studies of electrode processes in molten fluorides. Since the emphasis to date has been on applied research, little may be said at present about the structure of the double layer and electrode kinetics in pure fluoride melts. The work in cryolite (Na_3AlF_6) melts, related to the production of aluminum, will not be covered, since recent reviews are available.[3-6]

EXPERIMENTAL TECHNIQUES

Melts employed. Three types of melts have been predominantly employed for electrochemical studies: 1) the relatively low-melting acid fluorides, such as KHF_2,[7-11] and KF·2HF, used primarily for the production of fluorine[12] and preparative electrochemical fluorination.[13,14] These melts have obviously much in common with anhydrous hydrogen fluoride, an electrochemical solvent of increasing importance.[14-16] 2) The alkali fluoride mixtures, such as the NaF-KF eutectic[17] (m.p. 710°C), and the LiF-NaF-KF eutectic[18-29] (46.5-11.5-42.0 mole %, m.p. 454°C) (commonly known as FLINAK) are important in fundamental studies. FLINAK is used for electrodeposition of refractory metals in coherent form.[22-27,30,31] A similar composition of alkali fluorides, sometimes mixed with the alkaline earth fluorides, is used for the formation of diffusion coatings[32-36] (General Electric "metalliding" process). FLINAK has also been used in an experimental high energy density thermal battery, using K_3CuF_6/inert electrode as the cathode and magnesium as the anode.[37] 3) Molten fluoride compositions, containing lithium fluoride, beryllium fluoride, and frequently, zirconium(IV) fluoride and/or uranium(IV) fluoride and thorium(IV) fluoride, are used in the molten salt reactor experiment (MSRE), in operation at the Oak Ridge National Laboratory. Among these, LiF-

BeF_2 (66-34 mole %, m.p. 454°C) and LiF-BeF_2-ZrF_4 (65.6-29.4-5.0 mole %, m.p. 434°C) have been used for electrochemical studies.[19,20,38-46]

The fluoride melts are normally purified by: a) sparging the melt with an H_2/HF mixture[47] (HF reacts with the oxides;[48] H_2 reduces metallic impurities, such as Ni(II), Fe(II), Cr(II)); b) fusion with NH_4HF_2,[23] c) drying under high vacuum (10^{-5} torr) prior to melting;[28,29] d) addition of an active metal, such as beryllium or zirconium,[49] and e) pre-electrolysis.[10,36,49] None of the above purification procedures <u>alone</u> apparently result in melts that are comparable in purity to the commonly used LiCl-KCl eutectic, purified by a vacuum-HCl treatment.[50,51] Pizzini and coworkers[29] estimate that the content of hydrolysis products (as OH^-) in FLINAK after the vacuum treatment is in the range 10^{-2}-10^{-1} formal; they believe that this impurity content is of the same order of magnitude or smaller when compared to FLINAK pretreated with HF/H_2 or NH_4HF_2. Purer melts, as far as hydrolytic contamination is concerned, are those containing ZrF_4. In these melts, precipitation of the fairly insoluble ZrO_2 takes place upon contamination with water. ZrO_2 normally settles to the bottom of the container, because of its high density; it may, however, float on the surface of the melt because of lack of aggregation and surface tension effects. Additional work on the purification of fluoride melts, to be used in electrochemical studies, is highly desirable.

Containers. Glass or quartz cannot be used to contain fluoride melts for long-term experiments, since traces of H_2O will result in the formation of HF and attack on the container. In the case of LiF-BeF_2 (66-34 mole %), it has been shown,[52] however, that the reaction $2BeF_{2(d)} + SiO_{2(s)} = SiF_{4(g)} + 2BeO_{(s)}$ may be reversed by operating in the presence of SiF_4; thus, in the absence of water, quartz is a satisfactory container material for Li-Be fluorides, at least for short-term experiments. A very common material for containing molten fluorides is graphite or carbon. It is desirable to use a higher density grade material, such as ATJ graphite,[53] in order to minimize the penetration by the melt. In addition, it is desirable to heat the graphite crucible _in vacuo_ to a temperature, at least as high as the intended experimental temperature, for an extended period of time ($>$ 1 day). Cleaning the graphite container with alcohol in an ultrasonic bath prior to bakeout or immersing it in liquid nitrogen is desirable to remove finely divided graphite particles. This latter problem should be greatly minimized by the use of pyrolytic graphite[54] (or a coating of it on ordinary graphite) or glassy carbon[55] containers; unfortunately, these materials are fairly expensive.

Metals or alloys such as molybdenum[23] and monel,[35,36] have also been used satisfactorily as containers. According to Cook,[87] monel containers in an inert atmosphere are preferable to graphite. Possible interaction with solutes and the effect of stray electrical potentials, resulting in anodic dissolution, should be considered when using metallic containers.

The crucible (graphite or metal) containing the salt is invariably enclosed in another container; in this manner, the operations are conducted either in an inert atmosphere (for example, high purity helium passed through a molecular sieve trap cooled to 77°K) or under vacuum (~ 10^{-2} torr or better). It should be realized that operation under vacuum in a system that is not leak-free, will result in a rapid contamination of the melt.

<u>Indicator (Working) Electrodes.</u> Almost exclusively, solid electrodes, such as platinum,[26,39] molybdenum[25] and

pyrolytic graphite[20] have been employed. For best reproducibility, electrodes should not be subjected to the extremes of anodic and cathodic potentials. With platinum, at anodic potentials with respect to the Ni/Ni(II) couple, electroactive films, such as chemisorbed hydrogen and oxygen, may be produced.[56] At sufficiently high cathodic potentials, alkali metals, such as lithium, are evolved, which alloy with platinum. The anodic dissolution of the alloy results in a rough surface (increased area).

Pyrolytic graphite is a particularly good indicator electrode material for the study of the reduction of metallic ions to metals (Ni(II), Fe(II), Cr(II)) in LiF-BeF_2-ZrF_4 (65.6-29.4-5.0 mole %).[20,56,57] In FLINAK, pyrolytic graphite has a much smaller useful cathodic potential range; a large reduction peak at ~ -1.3 V (vs. Ni/NiF_2 ($X_{Ni(II)} = 1$)) is observed,[67] believed to be due to the predeposition of potassium and the formation of a graphite-potassium intercalation compound.[59] The solid electrodes may be sheathed in an insulator[20,28] or immersed directly into the melt.[23,26,39] Boron nitride is the most common insulating material for use in molten fluorides. It has been noted[86] that boron nitride is oxidized by the more noble metal ions, such as Mo(III); at temperatures near 500°C we[49] have not observed this to be the case. Boron nitride is slowly penetrated by the melts;[44] the extent of penetration depends on the type of melt, grade of boron nitride, and the treatment to which BN was subjected in machining and other handling. The use of pyrolytic BN[54] has not been explored as yet.

Bismuth pools (in boron nitride,[56] or quartz[60]) have been used as working electrodes in Li-Be fluorides. The useful potential range is considerably smaller than with platinum, presumably because of dissolution of bismuth on the anodic side and the predeposition of lithium, due to the formation of a Li-Bi alloy, on the cathodic side. The evaluation of bismuth electrodes is continuing.

Reference Electrodes. The earlier work in this area is summarized in two reviews.[61,62] Recently the usefulness of HF/H_2,[45] Be(II)/Be,[46] and Ni(II)/Ni[17,23,44] couples as reference electrodes in molten fluorides has been demonstrated. The last two couples may be separated from the bulk of the melt by means of a wetted boron nitride membrane.[44] Very fine metallic frits fabricated from a relatively noble metal have been used to separate electrode compartments in other melts;[63] attempts to use copper frits for this purpose in molten fluorides were not successful,[49] apparently because electrode reactions involving adsorbed species were occurring at the copper surfaces.

It is frequently convenient to use a quasi-reference electrode[39] - usually a platinum rod with a large surface area. Such an electrode is equivalent to the mercury pool in conventional polarography; however, since the quasi-reference electrode is used in a three electrode system, (either in linear sweep voltammetry or chronopotentiometry), the current passing through the reference electrode is extremely small ($\lesssim 10^{-8}$ A). Therefore, the potential of the quasi-reference electrode under most operating conditions remains remarkably constant (± 10 mV for a period of months), provided strong reducing agents, such as U(III) (and probably strong oxidizing agents), are not added to the melt. In that case, the potential shifts in the direction expected from the Nernst equation. The quasi-reference electrode in Li-Be fluorides is believed to function as HF/H_2 electrode (both HF and H_2 are apparently adsorbed on the electrode).

<u>Electrochemical Methods.</u> EMF measurements,[44-46] current-potential curves (determined by the galvanostatic

method),[9-11,28,29] chronopotentiometry,[23-27,20,39,42] chronoamperometry,[39] linear sweep voltammetry[38,39,42,43] and derivative voltammetry[56] have been employed for studies of electrode reactions in molten fluorides. In our experience, the linear sweep voltammograms are usually better defined than the chronopotentiograms obtained under the same conditions (usually with wire electrodes), and yield more readily qualitative information. On the other hand, the theory of more complex cases, such as disproportionation of the product,[39,42] has been or can be readily worked out for chronopotentiometry, as compared to linear sweep voltammetry.

EMF MEASUREMENTS

Electrode potentials in molten fluorides have been measured by several groups.[17,44-46,49] These quantities have also been calculated from thermodynamic data[64,65] and distribution equilibria using liquid bismuth.[66] A summary of the results to date is given in Table 1. Electrode potentials have also been estimated by chronopotentiometry[30] and voltammetry;[56] a number of such results are given in the text.

TABLE 1

Electrode Potentials in Molten Fluorides

	HMR[a]	B[b]	G[c]	MS[e]	JMM[f]
Li(I)	-2.674	-3.012		-2.29	
Ba(II)	-2.657			-2.17	
La(III)	-2.518	-2.68		-1.85	
Ce(III)	-2.445	-2.61		-1.86	
Sm(III)	-2.323	-2.48			
Th(IV)	-1.675	-2.232		-1.95	
Be(II)	-1.517	-2.211		-2.22	-2.120
Zr(IV)	-1.352	-1.772			-1.742
U(IV)	-1.327	-1.752		-1.7	
U(IV)/U(III)		-1.512			
Al(III)	-0.977		-1.5		
Mn(II)	-0.680		-1.04		
Cr(II)	-0.510	-0.789			-0.754
Cr(III)	-0.377		-0.70[d]		
Fe(II)	-0.204	-0.413			-0.408
Co(II)	-0.032		-0.07		
Ni(II)	0	0	0	0	0
Fe(III)	0.058		-0.12[d]		
Cu(I)			0.48		
Ag(I)	1.216		0.64		
F_2	2.890	2.219			

[a]Theoretical EMF series (V vs. Ni/Ni(II)) for solid fluorides at 500°C calculated by Hamer, Malmberg and Rubin.[64]

[b]Electrode potentials in LiF-BeF_2(67-33 mole %) at 500°C calculated by Baes.[65, 85] The standard state of the ions is the hypothetical mole fraction solution in LiF-BeF_2 (67-33 mole %), with the exception of Li(I), Be(II), and F^-, for which the standard state is LiF-BeF_2 (67-33 mole %).

[c]EMF measurements by Grjotheim[17] in NaF-KF (40-60 mole %) at 850°C. Standard state is hypothetical unit mole fraction solution.

[d]Mellors and Senderoff[84] believe that these potentials correspond to Cr(III)/Cr(II) and Fe(III)/Fe(II) couples, respectively.

[e]Potentials calculated in LiF-BeF_2 (67-33 mole %) at 600°C from distribution equilibria studies using liquid bismuth by Moulton, et al.[66] Standard states in both salt and metal (in bismuth) phases are the hypothetical unit mole fraction solutions (except for beryllium metal).

[f]Unpublished EMF studies in LiF-BeF_2-ZrF_4 (65.6-29.4-5.0 mole %) at 500°C by Jenkins, Mamantov and Manning.[49] Standard state in f) same as in b).

ELECTROCHEMICAL STUDIES RELATED TO ATMOSPHERIC CONTAMINANTS

Water, oxygen, and even carbon dioxide are electroactive under some (or most) experimental conditions in molten salts. In molten fluorides, species derived from these atmospheric contaminants, either by chemical or electrochemical reactions, such as OH^-, HF, O^{-2}, O_2^{-2}, H_2, may themselves be reduced and/or oxidized at proper potentials. Complications, resulting from chemisorption on electrodes and passivation, frequently enter the picture.

Since, to date, most fluoride melts have been seldom free of traces of moisture (for FLINAK, $[OH^-]$ of 10^{-2} formal is not unusual[29]), the effects of atmospheric contamination need to be considered. However, aside from the work of Pizzini and coworkers,[9-11, 28, 29] little has been done in this area. Pizzini, Sternheim and Barbi[9] and Pizzini and Magistris[10] studied hydrogen evolution in molten KHF_2 at 250°C by the galvanostatic method. They believe that two electrode reactions are involved in the presence of water.

$$HF + e = 1/2\ H_2 + F^- \quad (1)$$

$$H_3O^+ + e = 1/2\ H_2 + H_2O \quad (2)$$

Reaction (1) occurs at ~0 to -0.1 V (with respect to HF/H_2 reference electrode); reaction (2) at ~-0.4 V. HF arises from the slow dissociation of HF_2^-

$$HF_2^- = HF + F^- \quad (3)$$

and H_3O^+ by the reaction

$$H_2O + HF = H_3O^+ + F^- \quad (4)$$

Pizzini and Morlotti[28] investigated hydrogen and oxygen evolution in molten FLINAK, at 600°C at platinum electrodes. They assumed the occurrence of reactions

$$H_2O + F^- = OH^- + HF \quad (5)$$

$$H_2O + 2F^- = O^{-2} + 2HF \quad (6)$$

to yield a relatively high HF_2^- concentration in the melt (equilibria (5) and (6) have been studied thoroughly in $LiF-BeF_2$ (66-34 mole %) by Mathews and Baes[48]). HF_2^- results in HF (reaction (3)) which is reduced to H_2 (reaction (1)). A second voltage plateau (~ -0.5 V with respect to HF/H_2

electrode) was attributed to hydrogen discharge directly from water. On the anodic side, two voltage plateaus were observed at current densities higher than the anodic limiting current for the hydrogen oxidation reaction (H_2 saturated melts were used). The first reaction at ~+0.5 V (with respect to HF/H_2 reference electrode) was believed to be

$$2OH^- + 2F^- = O_2^{-2} + 2HF + 2e \quad (7)$$

The second anodic reaction at ~+1.0 V was attributed to

$$O^{-2} = 1/2\, O_2 + 2e \quad (8)$$

Stability of peroxides in FLINAK was demonstrated by dissolving Na_2O_2 in the melt at 600°C and titrating the peroxide in the aqueous solution of the solidified mixture. The reduction of O_2 and O_2^{-2} was not studied; O_2^{-2} was not considered.

We have observed[56,67] the reduction of HF, and OH^- (added as LiOH), and the oxidation of H_2 and O^{-2} (added as K_2O or BeO). The reproducibility of voltammograms is poor; adsorption is believed to be present. Addition of large amounts of LiOH to FLINAK results in the passivation of the platinum, but not the pyrolytic graphite, electrode.[67] Additional systematic studies of this general area are definitely desirable.

Pizzini and coworkers observed the passivation of aluminum in molten KHF_2,[11] lead in molten PbF_2-NaF (33-67 mole %),[68] nickel in FLINAK, NiF_2-KF (10-90 mole %) and related melts,[29] and iron in FLINAK[81] in the presence of traces of water or oxides.

ELECTROCHEMISTRY OF TRANSITION METALS

In this section, work on all group B elements, including the actinides, will be reviewed; thus, the term "transition metals" is used quite broadly.

a) Iron and Nickel. The reduction of Fe(II) to the metal has been studied in FLINAK[18,20] and LiF-BeF_2 (66-34 mole %).[20] Well-defined voltammograms (at scan rates > 1 V/min) and chronopotentiograms were obtained at the sheathed pyrolytic graphite electrode. Anodic stripping voltammograms were shown to be more reproducible than cathodic voltammograms; both techniques, however, are useful for in situ determinations. The slow scan rate experiments at the platinum wire electrode[18] indicated the applicability of the Heyrovsky-Ilkovic equation;[1] this result, as well as the stripping experiments,[20] indicate that the process proceeds reversibly (at least at slower scan rates). This conclusion was also reached by Senderoff.[30] The diffusion coefficients at 500°C, calculated from the modified Randles-Sevcik equation, were 1×10^{-6} and 5×10^{-6} cm^2/sec in

FLINAK and $LiF-BeF_2$, respectively. The oxidation of Fe(II) to Fe(III) has been observed in FLINAK.[30] Pizzini and coworkers[81] have very recently reported on their measurements of anodic and cathodic overvoltage in FLINAK at iron electrodes in the interval 500-750°C. At oxide-free electrode surfaces, anodic dissolution and cathodic deposition of the metal occur without apparent overvoltages.

EMF measurements on the Ni(II)/Ni couple in FLINAK and $LiF-BeF_2-ZrF_4$ (65.6-29.4-5.0 mole %) have shown[44] the applicability of the Nernst equation and the usefulness of this couple as a reference electrode. The kinetics of the $Ni^{+2} + 2e = Ni$ process at ~500°C have been studied by the voltage step method.[41,49] The preliminary values of the standard rate constant, transfer coefficient and the molar exchange current density in FLINAK and $LiF-BeF_2-ZrF_4$ (65.6-29.4-5.0 mole %) at 500°C are $0.6-8.2 \times 10^{-4}$ cm/sec, 0.64-0.44, and 0.9-3.2 A/cm^2, respectively. The extreme values for both melts are given above; the wide range is caused primarily by the poor linear dependence of the logarithm of the exchange current on the logarithm of the concentration of Ni(II). The values are believed to be more reliable than those reported previously.[41] The above value of the molar exchange current density for the Ni(II)/Ni couple is considerably smaller than the value obtained for the same couple in molten LiCl-KCl at 450°C (110 ± 20).[82] Nearly all of the values of the standard rate constant obtained in molten salt systems,[69] are considerably larger than the value given above. Additional work is in progress to determine the importance of adsorption of Ni(II) at the electrode. Much additional information is needed to determine the relative contributions of charge transfer, reaction, and crystallization overvoltages[70] to the irreversibility of some electrode processes[22-27,30,31] (see below) in molten fluorides. The importance of reaction overvoltage was demonstrated for the deposition of molybdenum from a chloride melt;[71] a similar mechanism is believed likely[27] for the deposition of molybdenum and tungsten from fluoride melts.

The voltammetry of nickel in FLINAK has been studied.[21] The utility of anodic stripping technique to determine as little as 1 ppm of nickel was demonstrated. As in the case of Fe(II), the diffusion coefficient is ~1 x 10^{-6} cm^2/sec at 500°C. The activation energy for diffusion, calculated from a plot of log D vs. 1/T, was found to be 18 kcal/mole.

The anodic dissolution of nickel in several fluoride melts was studied by Pizzini, Morlotti and Roemer.[29] In oxide-free melts, no activation overvoltage for the anodic and cathodic (in molten NiF_2-KF) reactions was observed. Saturation of the melt with NiO and addition of water led to the open circuit passivation of the electrode. Oscillations of the potential (in chronopotentiometry) or current (in chronoamperometry) were also observed. The observed behavior is explained in terms of mass transport and charge transfer reactions across the phase boundaries Ni/NiO and NiO/solution, as well as across the bulk of NiO. The anodic activation of a passive nickel electrode is attributed to the formation of a higher nickel oxide, namely Ni_3O_4.

b) Elements of Groups IVB, VB, AND VIB. Most of the studies in this area have been done by Senderoff and Mellors,[22-27,30,31] in developing a process for the electrodeposition of these metals (titanium excluded) in coherent form. Since reviews of this work are available,[30,31,72] only a brief summary is given here.

The electrode reactions for the reduction of tantalum,[23] zirconium,[25] niobium,[26] molybdenum and tungsten[27] in FLINAK were studied by chronopotentiometry. The results are summarized in Table II. In all cases, the step involving

TABLE 2

Chronopotentiometry of Refractory Metal Ions in FLINAK at 500-860°C

System and Reference	Proposed Mechanism	$E_{\tau/4}$, V (vs. Ni/NiF_2; $X_{Ni(II)}=10^{-2}$) 750°C	Concentration (formal)
Tantalum[23] (added as K_2TaF_7)[a]	Ta(V) → Ta(II) Ta(II) → Ta	$E_1 = -1.03 \pm 0.05$ $E_2 = -1.26 \pm 0.06$	$2 \times 10^{-2} - 10^{-1}$
Niobium[26] (added as K_2NbF_7)[b]	Nb(V) → Nb(IV) Nb(IV) → Nb(I) Nb(I) → Nb	$E_1 = -0.11$ $E_2 = -0.756 \pm 0.015$ $E_3 = -1.02 \pm 0.03$	$2.5 - 8 \times 10^{-2}$
Zirconium[25] (added as K_2ZrF_6)[c]	Zr(IV) → Zr	$E \approx -1.5$	7.2×10^{-2}
Molybdenum[27] (added as solution of Mo(III-IV) in FLINAK)[d]	Mo(III) → Mo	$E \approx -0.3$	$\sim 10^{-1}$
Tungsten[27] (added as solution of W(IV-V) in FLINAK)[e]	Single step if mean valence <5	$E = -0.34 \pm 0.04$ (at 600°C)	$\sim 3 \times 10^{-2} - 10^{-1}$

[a] E_1 and $i\tau_1^{1/2}/C$ are independent of concentration and current density. $i\tau_2^{1/2}/C$ decreases with increasing concentration. n's deduced from theory of stepwise reductions and the shape of the first potential-time curve.[73] First product (TaF_2) slightly soluble (reverse chronopotentiometry). 2nd step is not reproducible and apparently irreversible. $D_{Ta(V)} = 1.5 \times 10^{-5}$ cm^2/sec (750°C). $E_a = 8.5$ kcal/mole (650-800°C).

[b] For the combined step (1 and 2) $i\tau^{1/2}/C$ is independent of concentration and current density. n_2 deduced as in the case of Ta. Nb(I) + 3Nb(V) = 4 Nb(IV) occurs rapidly (must convert Nb to (IV) state in the plating bath). Nb(I) somewhat soluble. The last step apparently irreversible. $D_{Nb(V)} = 2.0 \times 10^{-5}$ cm^2/sec (750°C). $E_a = 8.75$ kcal/mole (650-800°C).

[c] Process apparently irreversible. No evidence for lower valent zirconium fluorides. At higher temperatures (~800°C) Zr reacts with KF to produce volatile K metal.

[d] Solution lost 20% of Mo during 1-2 hrs. (mean valence rose from 3.3 to ~4). Nernstian plots not linear at 600-700°C; at 800°C, $n \approx 2$. Large difference between cathodic and anodic $E_{1/4}$'s at 600-700°C (less at 800°C). Irreversibility ascribed to slow dissociation of a polynuclear complex anion to a mononuclear anion which disproportionates.

[e] W crucibles used (initial oxidation state unknown). Rapid change in solution composition. System more irreversible than Mo (similar mechanism as for Mo).

the reduction to the metal is not a simple diffusion-controlled reaction. In the case of tantalum and niobium complications result from the presence of a slightly soluble intermediate; in the case of zirconium the deposit is apparently attacked by the solvent. In the case of molybdenum and tungsten, the irreversibility is ascribed to the slow dissociation of a polynuclear complex anion. From the practical standpoint, it appears that these complications, resulting in reaction overvoltage, are essential for the production of coherent deposits of metals. Neither iron nor nickel have been successfully deposited as coherent coatings, but only as dendrites or powder;[30] for both of these systems the electrode reactions are simple, diffusion-controlled steps.

Attempts to verify the reported oxidation of Ta(V) to the hexavalent state[24] in FLINAK have been inconclusive.[67]

Reduction of Cr(II) to the metal in LiF-BeF_2-ZrF_4 (65.6-29.4-5.0 mole %) is apparently quasi-reversible.[57] Chromide diffusion coatings have been obtained by depositing chromium from a chromium(II) solution in FLINAK (or a similar molten fluoride bath).[35] Cr(III) must be converted to Cr(II), either by reaction with metallic chromium or pre-electrolysis, in order to obtain the chromide coating.

The operating temperatures range from 700 to 1200°C, depending upon the diffusion rate of chromium in the substrate metal. The latter factor also determines the desired current density (0.5-2.5 mA/cm^2). Concentrations range from 0.2 to 5 mole %. The process operates satisfactorily by shorting the chromium anode and the substrate metal (steel, for example) cathode; however, a small voltage (0.1-0.3 V) may be applied to increase the deposition rate of chromium.

c) Uranium and Thorium. The reduction of U(IV) in LiF-BeF_2-ZrF_4 (65.6-29.4-5.0 mole % and similar compositions) at 500°C has been shown to be a reversible one-electron process[39] occurring at -1.5 V (vs. Ni/NiF_2, $X_{Ni(II)}$ = 1). The system was studied by linear sweep voltammetry, chronopotentiometry, and chronoamperometry. The voltammograms were best defined at unsheathed platinum electrodes; several other electrode materials were also employed. Disproportionation of the product, U(III),[74] was not observed at 500°C in the short time span of the experiments; at higher temperatures this effect should be noticeable, particularly with a slow electrochemical technique, such as controlled potential coulometry. The diffusion coefficient of U(IV) at 500°C was found to be 2 x 10^{-6} cm^2/sec; the activation energy for diffusion (480-600°C) was determined as 10.5 kcal/mole, of the same order of magnitude as E_a for other species in molten fluorides[18,21,57] (also see Table 2), but somewhat higher than E_a for diffusion of U(IV) in LiCl-KCl eutectic (7.7 kcal/mole in the interval 420-620°C)[83] or most of the values of E_a of *self-diffusion* coefficients in pure molten salts.[75]

Voltammetric studies of U(IV) at fast scan rates (up to 100 V/sec) indicate the presence of weak adsorption of U(IV) at platinum electrodes.[43] This effect is not noticeable at scan rates < 2V/sec.

The electrode reaction U(IV) → U(III) should be useful for the *in situ* determination of U(IV) in Li-Be fluorides as indicated by the linear dependence of peak current on the concentration of U(IV) (0.1-0.9 mole % or 0.07-0.55 formal) and reasonably good reproducibility (relative standard deviation of 2.0% for 41 determinations during 36 days), even at scan rates of 1 V/min (at this slow scan rate effects of convection and cylindrical diffusion are noticeable[39]). On the other hand, the oxidation of U(IV) to U(V)[42] is not analytically useful since the electrode reaction is followed by the disproportionation of the electrochemically produced U(V) and the attack of the working electrode (pyrolytic graphite or platinum-10% rhodium) by U(V) or U(VI).

The reduction of Th(IV) in FLINAK by linear sweep voltammetry is being investigated.[76] The process is apparently a single 4 electron step that occurs at ~-1.8 V (vs. Ni/NiF_2, $X_{Ni(II)} = 1$). The study has been complicated by the presence of traces of atmospheric impurities. This results in the precipitation of ThO_2 and voltammetric waves due to impurities near the potential region where Th(IV) is reduced.

ELECTROCHEMISTRY OF GROUP A ELEMENTS

Outside of studies in cryolite melts, little fundamental work has been done on these elements in molten fluorides. The process of applying corrosion-resistant diffusion coatings, previously discussed for chromium, has also been developed for beryllium,[32] boron[33] and silicon[34] (see Table 3). The review article describing the process[36] states that aluminum, scandium, titanium, vanadium, manganese, nickel, germanium, yttrium, zirconium, and the rare earths have also been plated to form diffusion coatings. The usefulness of molten fluorides (as compared to other molten salts) is determined[36] by their fluxing action (oxide films

TABLE 3

Typical Characteristics of the Diffusion Coating Process in Molten Fluorides for Beryllium, Boron and Silicon

System and Reference	Operating Temperature	Concentration of Solute	Current Density (A/cm^2)
Beryllium[32] (added anodically and as BeF_2) in LiF-NaF-KF, (45-10-45 mole %)[a]	700 - 800°C	33.3 mole % (min.) BeF_2	10^{-3} - 3×10^{-2}
Boron[33] (added anodically and as KBF_4) in LiF-NaF-KF[b]	700 - 800°C	1 - 5 mole % KBF_4	10^{-3} - 3×10^{-2}
Silicon[34] (added anodically and as K_2SiF_6 or SiF_4) in LiF-Na-KF[c]	600 - 800°C	1 - 5 mole % K_2SiF_6	Depends on the substrate metal and the temperature.

[a] Be anode used. Inert atmosphere (N_2, forming gas (90% N_2, 10% H_2), Ar) more desirable than vacuum operation.[87] Purity of salts critical. Impressed voltage: 0.1-0.3 V (may be operated by shorting cathode and anode). Resulting coatings ~0.1-10 mils thick.

[b] B anode (pieces of B in a graphite basket with holes) used. Also see comments under Be.

[c] Si pieces in a graphite basket used as anode. Also see comments under Be.

are dissolved from cathode surfaces), high boiling points, and high thermodynamic stability.

The production of fluorine by the electrolysis of molten KF• 2HF at ~100°C, using carbon anodes, has been reviewed by several authors[12, 77, 78] and will not be discussed here. At much higher temperatures (> 600°C), in alkali metal fluorides (or other nonvolatile fluorides), the anodic reaction at carbon electrodes may result in the formation of fluorocarbons, such as CF_4, C_2F_6, and C_3F_8.[14] The "anode effect"[79] in fluoride melts is believed to involve the formation of CF_4 or $(CF)_x$ giving rise to a film which covers the anode and hinders the wetting by the melt. We have found[58] that graphite electrodes are passivated using linear sweep voltammetry at slow or moderately rapid scan rates (< 100 V/sec), at potentials more anodic than ~ +1.5 V (vs. Ni/NiF_2 ($X_{Ni(II)}$ = 1)). At rapid scan rates (≳ 100 V/sec), 2 current peaks are observed prior to the anodic decomposition potential. Although the origin of the 2 current peaks has not been established, formation of $(CF)_x$ and fluorocarbons is not unlikely.

ACKNOWLEDGMENTS

I would like to thank N. C. Cook, D. L. Manning, S. Senderoff, J. D. VanNorman, and J. P. Young for valuable comments on the manuscript, C. F. Baes and D. M. Moulton for electrode potential data, and my students, F. R. Clayton, H. W. Jenkins and F. Whiting, who contributed to some of the work reviewed in this chapter. The support of the Atomic Energy Commission under Contract AT-(40-1)-3518 is gratefully acknowledged.

REFERENCES

1. C. H. Liu, K. E. Johnson and H. A. Laitinen in *Molten Salt Chemistry*, M. Blander, ed., Interscience, New York, 1964.

2. H. A. Laitinen and R. A. Osteryoung in *Fused Salts*, B. R. Sundheim, ed., McGraw-Hill, New York, 1964.

3. K. Grjotheim and C. Krohn, *Chem. Zvesti*, *21*, 762 (1967).

4. N. E. Richards in *Proceedings of the First Australian Conference on Electrochemistry*, J. A. Friend and F. Gutmann, eds., Pergamon Press, Oxford, 1965, p. 867.

5. N. E. Richards and B. J. Welch, *ibid.*, p. 901.

6. R. Piontelli, *ibid.*, p. 932.

7. A. J. Arvia and J. B. deCusminsky, *Trans. Far. Soc.*, *58*, 1019 (1962).

8. A. J. Arvia and J. B. deCusminsky, *J. Chem. Phys.*, *36*, 1089 (1962).

9. S. Pizzini, G. Sternheim and G. B. Barbi, Electrochim. Acta, 8, 227 (1963).

10. S. Pizzini and A. Magistris, Electrochim. Acta, 9, 1189 (1964).

11. S. Pizzini, A. Magistris, and G. Sternheim, Corr. Science, 4, 345 (1964).

12. A. J. Rudge, The Manufacture and Use of Fluorine and Its Compounds, Oxford University Press, London, 1962.

13. J. Burdon and J. C. Tatlow in Advances in Fluorine Chemistry, Vol. I, M. Stacey, J. C. Tatlow and A. G. Sharpe, ed., Butterworth, London, 1960, p. 129.

14. S. Nagase, Fluorine Chemistry Reviews, 1, 77 (1967).

15. B. Burrows and R. Jasinski, J. Electrochem. Soc., 115, 348 (1968).

16. L. G. Spears and N. Hackerman, J. Electrochem. Soc., 115, 452 (1968).

17. K. Grjotheim, Z. Phys. Chem., N.F., 11, 150 (1957).

18. D. L. Manning, J. Electroanal. Chem., 6, 227 (1963).

19. D. L. Manning and G. Mamantov, J. Electroanal. Chem., 6, 328 (1963).

20. D. L. Manning and G. Mamantov, J. Electroanal. Chem., 7, 102 (1964).

21. D. L. Manning, J. Electroanal. Chem., 7, 302 (1964).

22. G. W. Mellors and S. Senderoff, J. Electrochem. Soc., 112, 266 (1965).

23. S. Senderoff, G. W. Mellors and W. J. Reinhart, J. Electrochem. Soc., 112, 840 (1965).

24. G. W. Mellors and S. Senderoff, J. Electrochem. Soc., 112, 642 (1965).

25. G. W. Mellors and S. Senderoff, J. Electrochem. Soc., 113, 60 (1966).

26. S. Senderoff and G. W. Mellors, J. Electrochem. Soc., 113, 66 (1966).

27. S. Senderoff and G. W. Mellors, J. Electrochem. Soc., 114, 586 (1967).

28. S. Pizzini and R. Morlotti, Electrochim. Acta, 10, 1033 (1965).

29. S. Pizzini, R. Morlotti, and E. Roemer, J. Electrochem. Soc., 113, 1305 (1966).

30. S. Senderoff, Metall. Reviews, 11, 97 (1966).

31. S. Senderoff and G. W. Mellors, Science, 153, 1475 (1966).

32. N. C. Cook, U. S. Patent 3,024,175 (March 6, 1962).

33. N. C. Cook, U. S. Patent 3,024,176 (March 6, 1962).

34. N. C. Cook, U. S. Patent Re. 25,630 (Aug. 4, 1964).

35. N. C. Cook, U. S. Patent 3,232,853 (Feb. 1, 1966).

36. N. C. Cook, J. D. Evans, and B. A. Fosnocht in Proceedings, International Conference on Protection Against Corrosion by Metal Finishing, Basel, 1966.

37. C. B. Root and J. B. Hunt, Abstract No. 16, J. Electrochem. Soc., 114, 194C (1967).

38. D. L. Manning, J. M. Dale, and G. Mamantov in Polarography, 1964, G. J. Hills, ed., vol. 2, Interscience, New York, 1966, p. 1143.

39. G. Mamantov and D. L. Manning, Anal. Chem., 38, 1494 (1966).

40. J. P. Young, G. Mamantov, and F. L. Whiting, _J. Phys. Chem._, _71_, 782 (1967).

41. G. Mamantov, H. W. Jenkins, and D. L. Manning, Preprints, Division of Fuel Chemistry, American Chemical Society, Miami Beach, Florida, _11(1)_, 147 (1967).

42. D. L. Manning and G. Mamantov, _J. Electroanal. Chem. and Interf. Electrochem._, _17_, 137 (1968).

43. G. Mamantov and D. L. Manning, _ibid._, _18_, 309 (1968).

44. H. W. Jenkins, G. Mamantov, and D. L. Manning, _ibid._, (in press).

45. G. Dirian, K. A. Romberger and C. F. Baes, Jr., U.S.A.E.C. Report ORNL-3789, 76, (1965).

46. B. F. Hitch and C. F. Baes, Jr., U.S.A.E.C. Report ORNL-4257 (1968).

47. J. H. Shaffer, U.S.A.E.C. Report ORNL-3708, 288 (1964).

48. A. L. Mathews and C. F. Baes, Jr., _Inorg. Chem._, _7_, 373 (1968).

49. H. W. Jenkins, G. Mamantov and D. L. Manning, unpublished work.

50. H. A. Laitinen, W. S. Ferguson and R. A. Osteryoung, _J. Electrochem. Soc._, _104_, 516 (1957).

51. H. A. Laitinen, _Pure Appl. Chem._, _15_, 227 (1967).

52. C. E. Bamberger, J. P. Young and C. F. Baes, Jr., _J. Inorg. Nucl. Chem._, (in press).

53. L. D. Loch in _Reactor Handbook_, 2nd ed., C. R. Tipton, Jr., ed., vol. I, Interscience, New York, 1960, p. 888.

54. Prolytic graphite and boron nitride are available from High Temperature Materials, Inc., Lowell, Mass.

55. Glassy carbon is available from the International Carbon Co., 500 Fifth Avenue, New York, N.Y., and Beckwith Carbon Corporation, Van Nuys, Calif.

56. G. Mamantov and D. L. Manning, unpublished work.

57. D. L. Manning and J. M. Dale in Molten Salts: Characterization and Analysis; G. Mamantov, ed., M. Dekker, 1968.

58. G. Mamantov, F. R. Clayton and F. Whiting, unpublished work.

59. W. Rudorff in Advances in Inorganic Chemistry and Radiochemistry, vol. I, H. J. Emeleus and A. G. Sharpe, ed., Academic Press, New York, 1959, p. 223.

60. C. Romberger and J. Braunstein, unpublished work.

61. R. Laity in Reference Electrodes, D. J. G. Ives and G. J. Janz, eds., Academic Press, New York, 1961.

62. A. F. Alabyshev, M. F. Lantratov and A. G. Morachevskii, Reference Electrodes for Fused Salts, Sigma Press, Washington, D.C., 1965.

63. H. R. Bronstein, J. Electrochem. Soc., 112, 1032 (1965).

64. W. J. Hamer, M. S. Malmberg and B. Rubin, J. Electrochem. Soc., 112, 750 (1965).

65. C. F. Baes, Jr. in SM-66/60, Thermodynamics, vol. I, IAEA, Vienna, 1966.

66. D. M. Moulton, W. P. Teichert, W. K. R. Finnell, W. R. Grimes, J. H. Shaffer, U.S.A.E.C. Report ORNL-4229, 39 (1968).

67. F. Whiting, G. Mamantov, and J. P. Young, unpublished work.

68. S. Pizzini and L. Agace, Corr. Sci., 5, 193 (1965).

69. A. D. Graves, G. J. Hills, and D. Inman in Advances in Electrochemistry and Electrochemical Engineering, vol. 4, P. Delahay, ed., Interscience, New York, 1966.

70. K. J. Vetter, Electrochemical Kinetics, Academic Press, New York, 1967.

71. S. Senderoff and G. W. Mellors, J. Electrochem. Soc., 114, 556 (1967).

72. A. D. Graves and D. Inman, Electroplating and Metal Finishing, 314 (1966).

73. P. Delahay, New Instrumental Methods in Electrochemistry, Interscience, New York, 1954.

74. G. Long, U.S.A.E.C. Report, ORNL-3789, 72 (1965).

75. A. Klemm in Molten Salt Chemistry, M. Blander, ed., Interscience, New York, 1964.

76. F. R. Clayton and G. Mamantov, unpublished work.

77. G. H. Cady in Fluorine Chemistry, vol. I, J. H. Simons, ed., Academic Press, New York, 1950.

78. C. L. Mantell, Electrochemical Engineering, 4th ed., McGraw Hill, New York, 1960.

79. G. Kortum, Treatise on Electrochemistry, 2nd ed., Elsevier, Amsterdam, 1965, p. 538.

80. Yu. V. Baimakov and M. M. Vetyukov, Elektroliz Rasplavlennykh Solei, Metallurgiya, Moscow, 1966, p. 195.

81. S. Pizzini, E. Roemer, and R. Morlotti, Electrochim. Metall., 3, 151 (1968).

82. H. A. Laitinen, R. P. Tischer and D. K. Roe, _J. Electrochem. Soc._, _107_, 546 (1960).

83. C. E. Thalmayer, S. Bruckenstein, and D. M. Gruen, _J. Inorg. Nucl. Chem._, _26_, 347 (1964).

84. G. W. Mellors and S. Senderoff, in _Applications of Fundamental Thermodynamics to Metallurgical Processes_, G. R. Fitterer, ed., Gordon and Breach, New York, 1967.

85. C. F. Baes, Jr., private communication.

86. S. Senderoff, private communication.

87. N. C. Cook, private communication.

VOLTAMMETRIC STUDIES OF CHROMIUM(II) IN MOLTEN $LiF-BeF_2-ZrF_4$ AT 500°C

D. L. Manning and John M. Dale
Oak Ridge National Laboratory
Oak Ridge, Tennessee

INTRODUCTION

At the Oak Ridge National Laboratory interest in molten fluorides has increased immensely due to their application as reactor fuels. As a result, considerable research is being conducted to better characterize this important class of molten salt systems. We are investigating electroanalytical methods as applied to molten fluorides with the hope of adapting such techniques to in-line methods of analyses. This paper is concerned with a voltammetric study of the reduction of chromium(II) in molten $LiF-BeF_2-ZrF_4$ (65.6-29.4-5.0 mole %). The composition is the same as the solvent salt used in the Molten Salt Reactor. Chromium is usually encountered as a corrosion product in molten fluorides.

EXPERIMENTAL

Details of the experimental set-up are presented elsewhere.[1] The melt (~40 ml volume) is contained in a graphite cell (~ 2" dia x 4" long) which is then enclosed in a quartz jacket to maintain a vacuum or controlled atmosphere. The stationary dip-type electrodes are inserted into the melt through the "Cajon" O-ring fittings which are located on the cap of the enclosure. Usually the melt is maintained in an inert atmosphere. The cell can be evacuated, however, when necessary. Disassembly of the cell, salt additions, etc., are carried out in a dry box.

Pyrolytic graphite indicator electrodes were prepared as previously described.[2,3] The unsheathed electrodes were approximately 1 millimeter in diameter and the electrodes which were sheathed in BN were one-eighth inch in diameter.

Glassy carbon indicator electrodes were prepared from glassy carbon rods (3 mm dia x 25 mm/long), available from

International Carbon Corp., 500 Fifth Avenue, New York, 36, New York. A small pin-hole was drilled through the top of the rod using an "Elox" electrical discharge process. The glassy carbon rod was sheathed with BN as previously described[2] and then fastened to a graphite holder (~ 1/4" dia x 1" long) with a small metal pin. The graphite holder was tapped with a 3/48 hole and the electrode was then joined to a 1/8" nickel rod threaded with 3/48 threads. These electrodes and the PGE electrodes are extremely fragile and utmost care is utilized in handling.

The controlled potential-controlled current cyclic voltammeter is described elsewhere.[4] With this instrument, scan rates from 0.05 to 500 V/sec are available and cell currents to 100 mA can be measured.

A Tektronix Type 549 storage oscilloscope with a type 1A7 pre-amplifier was used for read-out of the curves. Photographs were made with a Tektronix Type C-12 Camera attachment and Type 42 polaroid film.

For read-out of voltammograms at slow scan rates a Moseley Type 2D-2A X-Y recorder was used.

The nickel-nickel fluoride reference electrode is described elsewhere.[5] It consists of a thin-walled boron nitride compartment containing the same fluoride solvent and a fixed concentration of dissolved NiF_2 ($\sim 10^{-3}$ mole fraction in this case). A nickel electrode is inserted into this compartment. Electrical contact is made when the BN (1/32" wall thickness) becomes wetted by the melt. Hot pressed boron nitride is ordinarily an insulator in molten fluorides although it is slowly penetrated by the melt. Generally, about a week is required for the resistance through the BN to drop to ~1000 ohms or less.

The solvent used was LiF-BeF_2-ZrF_4 (65.6-29.4-5.0 mole %). Chromium(II) was added as CrF_2. The experimental temperature was 500°C unless otherwise stated.

RESULTS AND DISCUSSION

Typical voltammograms for the reduction of chromium(II) to the metal at a sheathed pyrolytic graphite electrode are shown in Figure 1. The reduction of chromium(II) occurs at approximately -0.9 volts versus a Ni/NiF_2 reference electrode. Well defined peak-shaped curves are obtained at scan rates to ~10 V/sec whereas at faster rates of voltage scan, the curves lose definition rapidly. The same behavior was observed at the glassy carbon indicator electrode. It appears that either electrode material can be used for

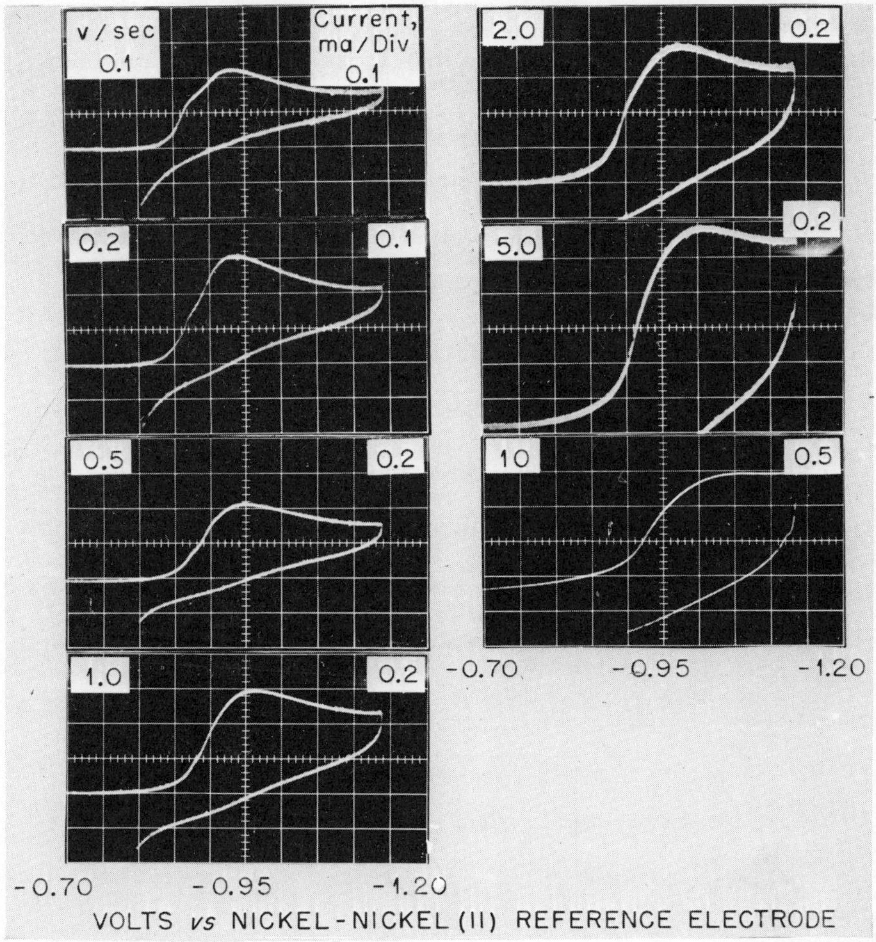

FIG. 1

Voltammograms for reduction of Cr(II) in $LiF-BeF_2-ZrF_4$ at 500°C. Formality of Cr(II), 0.018; PG Electrode area ~0.08 cm^2.

measurements in molten fluorides. In either case, the curves were better defined than those recorded at platinum indicator electrodes.

The Randles-Sevcik equation for the reversible deposition of an insoluble substance at a planar electrode at 500°C is $i_p = 2.28 \times 10^5\ n^{3/2}\ A\ D^{1/2}\ C\ v^{1/2}$ where i_p is peak current, (A), n = electron change, C = concentration (moles/cm^3), A = electrode area (cm^2), D = diffusion coefficient (cm^2/sec) and v is the scan rate (volts/sec). A plot of i_p versus $v^{1/2}$ is shown in Figure 2. In general, the plot was linear to about 1 v/sec and at faster scan rates a pronounced downward curvature was noted. From the slope

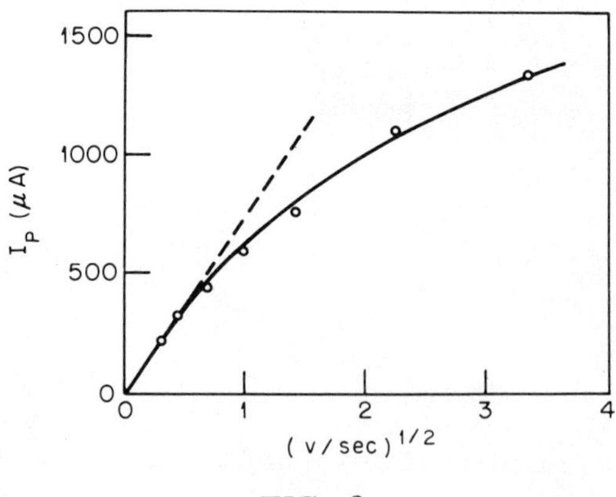

FIG. 2

Plot of peak current vs. $v^{1/2}$ for Cr(II) reduction.
Cr(II) = 0.018 F; PGE ~ 0.08 cm^2 A.

of the line over the linear portion of the graph a D value of $\sim 1 \times 10^{-6}$ cm^2/sec was calculated. This is about the same order of magnitude as previously determined D values of electroactive species in fluoride melts.[1,2,6,7]

The downward curvature of the i_p vs. $v^{1/2}$ plots at the faster scan rates suggests that the rate of the electrode reaction is the principal factor controlling the peak currents as pointed out by Delahay.[8] It appears, therefore, the system may best be described as "quasi-reversible," namely, the kinetics of electron transfer become competitive with the rate of potential change. This situation was treated by Matsuda and Ayabe[9] for the case where both oxidized and reduced forms are soluble. Although the case for the insoluble reduced species was not treated by Matsuda and Ayabe, it appears reasonable that the experimental observations for the two cases would be similar.

The dependence of peak current on the concentration of Cr(II) was investigated to ascertain the analytical utility of these measurements from the standpoint of analysis. This was done at a scan rate of 0.1 v/sec and a linear plot was obtained over the concentration range of Cr(II) investigated;

namely, about 0.0083 to 0.26 F. To ascertain the reproducibility of the voltammetric measurements at the micro-ampere level (ppm range of diffusing species) and over a relatively long period of time, the consistency of the peak current was observed at frequent intervals for approximately 12 weeks. The results are tabulated in Table 1.

The same electrodes were used throughout. The measurements were somewhat more reproducible at the sheathed electrode; however, these measurements eventually became large and erratic. It appears that the useful life of sheathed electrodes can be estimated as a few weeks at a working temperature of 500°C.

An analytical determination of chromium directly in a molten salt reactor fuel would involve obtaining the measurement in the presence of uranium(IV). A voltammogram of chromium(II) and uranium(IV) at a uranium concentration level expected in molten salt breeder reactor fuels is shown in Figure 3. The Cr(II) - Cr(0) reduction wave occurs at the foot of the U(IV) - U(III) reduction wave. As a result, the peak current for chromium is enhanced. An example of enhancement is the measurement of ~80 ppm of Cr(II) in the absence and presence of 0.3 mole percent of UF_4. In the

TABLE 1

Reproducibility of the Chromium Reduction Wave in $LiF-BeF_2-ZrF_4$ at 500°C
Concentration Cr(II) 0.0083 F

Pyrolytic Graphite Electrodes, Area ~0.08 cm^2

Scan Rate	PG in BN		PG unsheathed	
V/sec	i_p, μA	Std. Devia., μA*	i_p, μA	Std. Devia., μA*
0.02	60	± 9	58	± 11
0.05	90	± 11	81	± 16
0.10	125	± 15	110	± 21
0.20	187	± 20	145	± 31

* 30 Determinations at each scan rate.

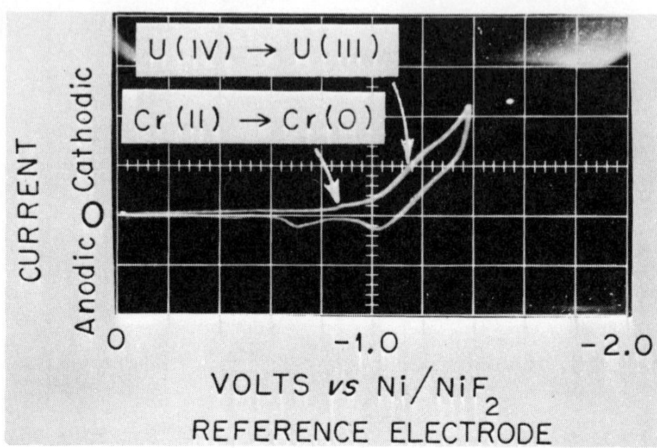

FIG. 3

Cyclic voltammogram for reduction of U(IV) and Cr(II) in $LiF\text{-}BeF_2\text{-}ZrF_4$ at 500°C.
U(IV) ~0.192 F; Cr(II) ~0.008 F; Scan 0.1 v/sec; current scale, 0.5 mA/Div.; Glassy Carbon Electrode.

absence of U(IV), the chromium i_p is about 42 μA whereas in the presence of U(IV) an increase to 164 μA was observed. This is discouraging from the standpoint of an analytical determination of chromium in the presence of uranium. However, in cases where the chromium wave can be resolved, the voltammetric approach may still be useful for the detection of chromium directly in the melt, and may provide a semi-quantitative measurement on an empirical basis. In the absence of interferences, it appears that controlled potential voltammetry would be a good choice as a means of monitoring chromium(II) directly in molten fluorides.

ACKNOWLEDGMENT

This research is sponsored by the U. S. Atomic Energy Commission under contract with the Union Carbide Corporation.

REFERENCES

1. D. L. Manning, J. Electroanal. Chem., 6, 227 (1963).

2. D. L. Manning and Gleb Mamantov, J. Electroanal. Chem., 7, 102 (1964).

3. D. L. Manning and Gleb Mamantov, J. Electroanal. Chem., 13, 137 (1968).

4. T. R. Mueller and H. C. Jones, Ann. Prog. Rept., Anal. Chem. Div., ORNL-4039, January 1967, p. 1.

5. H. W. Jenkins, Gleb Mamantov, and D. L. Manning, J. Electroanal. Chem., (in press).

6. D. L. Manning, J. Electroanal. Chem., 9, 302 (1964).

7. Gleb Mamantov and D. L. Manning, Anal. Chem., 32, 1494 (1966).

8. P. Delahay, J. Phys. Coll. Chem., 54, 630 (1960).

9. H. Matsuda and Y. Ayabe, Z. Elektrochem., 59, 494 (1955).

THERMOCHEMISTRY, COMPLEXATION, ELECTRON AND OXYGEN TRANSFER IN FUSED NITRATES

J. Jordan, W. B. McCarthy, and P. G. Zambonin

Department of Chemistry
The Pennsylvania State University
University Park, Pennsylvania

INTRODUCTION

This paper is a review of a novel type of chemistry which has transpired as a result of recent studies in our laboratories. Published results are reviewed critically and complemented by a preliminary report of work in progress. We feel that our findings compel a drastic reappraisal of the applicability of conventional Lux-Flood acid-base concepts to fused nitrates. Instead of the oxide ion, $O^{=}$, the superoxide ion, O_2^{-}, emerged from our studies as a chemical entity of paramount importance. This finding is related to remarkable features of redox chemistry which are peculiar to molten nitrates. Indeed, fused nitrate solvents stand out by unique electron transfer (redox) properties. Likewise, complexation chemistry exhibits some interesting facets. In contradistinction, precipitation processes appear reminiscent of conventional inorganic chemistry, except for the effects of enhanced ion-pairing on the solubility of precipitates.

HISTORICAL CONTEXT

Electron Transfer and Acid-Base Reactions

Generally oxide transfer was considered by most authors as the analogue to acid-base reactions in accordance with the Lux-Flood (L-F) formalism[1] which has been very successful in silicate melts and is exemplified in Equation 1:

$$\underset{}{O^{=}} + \underset{\text{L-F acid}}{SiO_2} = \underset{\text{conjugate L-F base}}{SiO_3^{=}} \qquad (1)$$

Virtually all authors took it for granted that the nitrate melt *per se* was similarly self-dissociating yielding oxide, in accordance with one of the following equilibria[2,3]

$$NO_3^- = NO_2^+ + O^= \qquad (2)$$

$$2NO_3^- = 2NO_2 + 1/2O_2 + O^= \qquad (3)$$

Nitrate melts have been known to contain a very strong oxidant, capable e.g. of converting chromium(III) to chromium(VI),[4] and bromide to bromine.[2] The controversy as to the prevalence of Reaction 2 *versus* Reaction 3 revolved mainly around the question whether NO_2^+ or NO_2 served as the electron acceptor in such processes.

A similar unresolved controversy prevailed in the literature concerning the nature of the so-called "oxygen electrode" in fused nitrate solvents. Results have been obtained by some workers which they interpret as consistent with the two electron transfer[5]

$$O^= = 1/2O_2 + 2e^- \qquad (4)$$

Others, however, have reported a puzzling Nernst slope which suggests a one-electron transfer.[6] A related electrode reaction[7,8] has also been postulated, viz.:

$$4OH^- = O_2 + 2H_2O + 4e^- \qquad (5)$$

Prevalence of Reaction 5 <u>versus</u> Reaction 4 is evidently contingent on the equilibrium

$$O^{=} + H_2O = 2OH^{-}; \; K_E = (OH^{-})^2/(O^{=})(H_2O) \qquad (6)$$

The confusion in the field has been compounded by extreme but conflicting assumptions implying that the value of K_E was approximately zero[9] or approximately infinity.[7,8]

Ion-Association and Complexation

There has been much interest in the formation of complexes in molten salts in recent years and a variety of theories have been presented to predict their properties. Of these, two have received much attention. The first is the essentially thermodynamic quasi-chemical theory proposed by Flood, Førland and Grjotheim.[10] This theory uses a thermodynamic treatment based on nearest neighbor interaction to discuss complexes. The statistical mechanical quasi-lattice theory proposed by Blander[11] is the second model which has received attention. It relies on statistical-mechanical considerations for calculating thermodynamic parameters.

Halide complexes of various metals, including silver,[12] nickel,[13] and cadmium,[14] have been investigated by various electrochemical methods. Experimental results have been compared with corresponding theoretical predictions. In most cases, the agreement between theory and experiment was quite satisfactory as far as formation constants are concerned. However, it has not been possible to discriminate between the two theories in this manner because they predict the same formation constants (within the limits of experimental error).

EXPERIMENTAL

Electrochemical Studies

The details of the experimental apparatus and procedure have been described elsewhere.[15,16] Two important points should be emphasized here: (1) since silica containing materials interfere with the results, platinum containers were used and all glass surfaces were covered with platinum foil. (2) Water was carefully removed from the melt by degassing with nitrogen dried over magnesium perchlorate.

Measurement of K_E

Pure sodium hydroxide was added to a dried nitrate melt in a closed, evacuated system at a constant temperature of 250°C. After equilibrium had been attained, the partial pressure of the water evolved was measured. From previous studies[17] giving the concentration of water as a function of its equilibrium pressure (i.e. Henry's Law Constant), the amount of water formed by Reaction 6 was calculated.

Thermometric Titration

The procedure used for the performance of the titrations in an adiabatic cell has been described previously.[18] The apparatus was slightly modified as illustrated in Figure 1: an improved cover for the dewar consisting of a tightly fitting stainless steel-asbestos-aluminum foil "sandwich" was used. Because hydroxide will attack glass under the experimental conditions, all glass surfaces were covered with platinum foil. All titrations were performed at 300°C, using equimolar sodium-potassium nitrate as the solvent.

RESULTS AND DISCUSSION

Redox Couples Involving $O^{=}$, $O_2^{=}$, O_2^{-}, O_2 and OH^{-}

Recent studies in our laboratories have yielded conclusive evidence that the oxide ion is capable of existence in nitrate melts only at negligible concentrations under normal experimental conditions.[16,15,19] The reason is inherent in the following equilibria

Oxygen transfer: $NO_3^{-} + O^{=} \rightleftharpoons NO_2^{-} + O_2^{=}$ (7)

Oxygen and electron transfer: $2NO_3^{-} + O_2^{=} \rightleftharpoons 2NO_2^{-} + 2O_2^{-}$ (8)

Disproportionation: $2O_2^{-} \rightleftharpoons O_2^{=} + O_2$ (9)

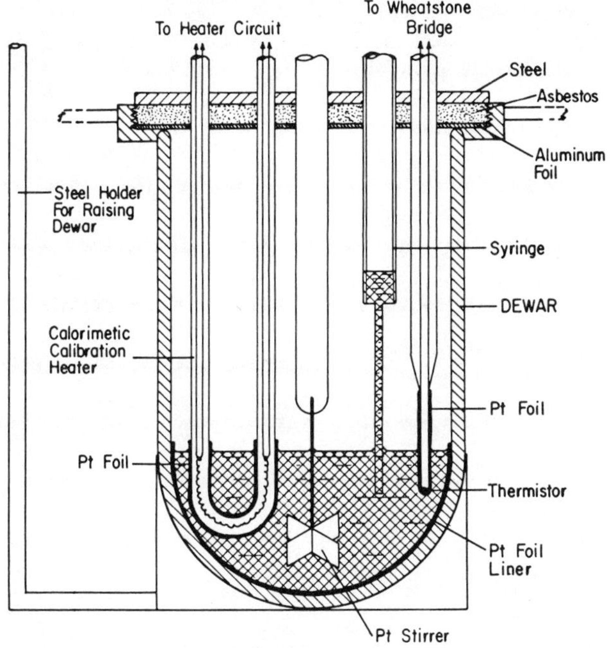

FIG. 1

Thermometric Titration Cell.

Disproportionation: $$3O_2^{=} = 2O_2^{-} + 2O^{=} \qquad (10)$$

The equilibrium constant for the disproportionation of peroxide into superoxide plus oxide (Reaction 10) was calculated from electrochemical data reported in the literature,[20] obtained in a molten alkali hydroxide solvent in which the ion O^{--} is stable. Expressed in molality units, the following assignment transpired:

$$K_{10} = \frac{[O_2^-]^2[O^{--}]^2}{[O_2^{--}]^3} = 10^{-11} \text{ at } 229^\circ C \qquad (11)$$

From voltammograms, recorded with the aid of Levich's rotated platinum disk[21] indicator electrode in molten sodium-potassium nitrate at the same temperature, we have estimated the standard potentials of the half reactions $O_2 + e^- = O_2^-$ and $O_2^- + e^- = O_2^{--}$. Combining these, the disproportionation constant of superoxide (into peroxide and oxygen) was evaluated as:

$$K_9 = \frac{[O_2^{--}][O_2]}{[O_2^-]^2} = \exp\left[\frac{F}{RT}\left(E^o_{O_2^-/O_2^{--}} - E^o_{O_2/O_2^-}\right)\right] = 3.5 \times 10^{-6} \qquad (12)$$

By reacting KO_2 with potassium nitrite in our fused nitrate solvent and determining the corresponding equilibrium concentrations, we have deduced an assignment for the thermodynamic constant which governs the Oxidation-Reduction Process 8, viz.:

$$K_8 = \frac{[NO_2^-]^2[O_2^-]^2}{[O_2^{--}][NO_3^-]^2} = 6.7 \times 10^{-11} \qquad (13)$$

Finally, for the Oxygen Transfer Reaction 7, it is readily apparent that:

$$K_7 = \frac{[NO_2^-][O_2^{--}]}{[NO_3^-][O^{--}]} = \left(\frac{K_8}{K_{10}}\right)^{1/2} \approx 3 \qquad (14)$$

The pertinent data make it quite clear that the important oxygen anions in the nitrate melts are superoxide and peroxide rather than oxide. Figure 2 shows the equilibrium concentrations of the various species that result when a given amount of sodium oxide is added to the nitrate solvent in the absence of oxygen. In the presence of oxygen at

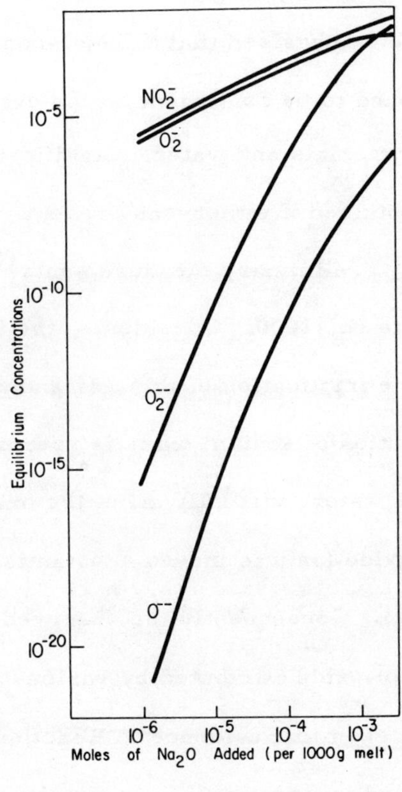

FIG. 2

Equilibrium concentrations of several ions in a nitrate melt.

appreciable partial pressures, as in an oxygen electrode, any oxide or peroxide is converted virtually completely to superoxide as a result of Equilibrium 9. This accounts for the experimental findings[6,19] that the oxygen electrode in molten nitrates has the Nernstian slope expected of the one-electron transfer process

$$O_2^- = O_2 + e^- \tag{15}$$

It should be emphasized that the self-consistency of our results was found to be contingent on the exclusion of silica containing materials and water. Significantly different results were obtained if either was present. Regarding the effect of water, preliminary measurements[17] have yielded the estimate $K_E > 1000$. Considering the interrelated equilibria (6-10) the implication of this assignment is that if a millimolal solution of sodium oxide is prepared in the presence of excess water, virtually all of the oxide will be converted to hydroxide (as was indeed substantiated directly in our experiments). Consequently the observed two-electron transfer per mole of oxide attributed by various authors on the basis of electrochemical evidence to Reaction 4 can plausibly be ascribed to Reaction 5. As for the effect of silica, it is reasonable to assume that its presence will yield spurious results due to the abstraction of oxide ion as in Reaction 1.

Hydroxo-Complexes of Calcium

It appears likely that complexation in molten salts might be generally enhanced due to relatively low dielectric constants. These are of the order of 4 in nitrate melts. Consequently, an attempt was made to explore the possible prevalence of hydroxo-complexes of alkaline earth cations, such as the series $CaOH^+$, $Ca(OH)_2$, $Ca(OH)_3^-$, and $Ca(OH)_4^=$.

In *ad hoc* preliminary experiments, it was ascertained that a solid phase which analyzed as $Ca(OH)_2$ was precipitated from fused nitrate solvents. The solubility of the pure compound was on the order of 10^{-3} molal at 300°C. However, attempts to evaluate an invariant solubility product

$$K_{sp} = (Ca^{2+})(OH^-)^2 \qquad (16)$$

were abortive. At a given temperature the value of K_{sp} varied by as much as a factor of a hundred, depending on the relative excess (with respect to the stoichiometry of $Ca(OH)_2$) of either Ca^{2+} or OH^- in the melt. These findings are not inconsistent with complexation. In order to elucidate the matter further, thermometric titrations were selected as a preferred methodological approach, since indicator electrodes for either calcium or hydroxide ion in molten salts are not available. Furthermore, thermometric titrations yield direct information regarding the heat of reaction.

Figure 3 presents several typical thermometric titration curves of sodium hydroxide with calcium nitrate. These curves confirm the existence of complexes of calcium in the fused nitrate solvent system. There are two obvious features of the curves that suggest that reactions other than precipitation were occuring. First, the apparent endpoint

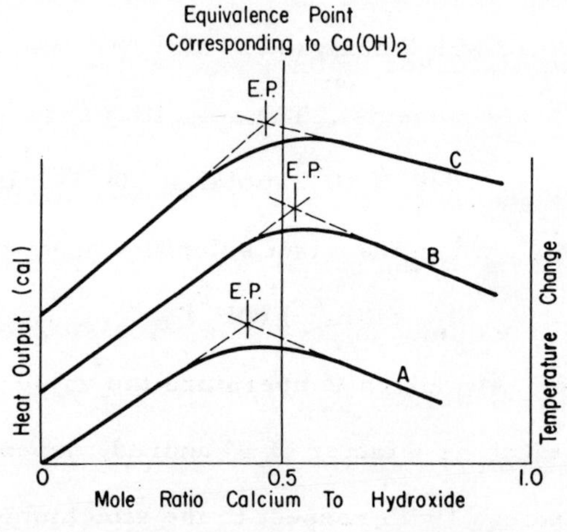

FIG. 3

Thermometric titration curves.
E.P.: Extrapolated experimental (apparent) end-point.
Initial concentration of OH^-: A, 1.26×10^{-2} molal; B, 0.873×10^{-2} molal; C, 1.41×10^{-2} molal.
Extraneous effects (stirring, dilution, etc. ....) negligible.

did not correspond to the theoretical endpoint for a calcium hydroxide precipitate. Second, past the apparent endpoint the slope was much greater than would be expected for a trivial excess reagent line. When the curves were analyzed mathematically in an attempt to calculate a solubility product defined by Equation 16 (assuming no complexation occurred) and a corresponding heat of precipitation, ΔH_p, it became evident that the curves could not possibly be accounted for by the precipitation of $Ca(OH)_2$ alone.

If precipitation were the only reaction occurring, the curve could be described by the equation

$$-q = \Delta H_p \, m_p \tag{17}$$

where q is the heat corresponding to a given abscissa point in Figure 3 and m_p is the number of moles precipitated at that point. The value of m_p could be calculated if K_{sp} were known. However, neither K_{sp} nor ΔH_p are known. Even so, it is possible to calculate them from the curve. At a point significantly past the theoretical endpoint, the value of m_p is approximately independent of the value of K_{sp}. Thus, a preliminary value of ΔH_p was calculated using an estimated K_{sp}. This ΔH_p assignment was used to calculate a value of m_p at, or prior to, the theoretical endpoint. From this value of m_p an improved estimate of K_{sp} was obtained. An improved value of ΔH_p was then computed using again the point past the theoretical endpoint. This iterative process was continued until K_{sp} converged towards a constant. A computer program was written to perform the necessary calculations. It was found that the values of K_{sp} showed a definite trend as various points were chosen along the curve. This indicates that there were more equilibria involved than just precipitation. Also the corresponding values of ΔH_p were much larger than would be expected for a precipitation reaction. The most reasonable explanation for these results is the presence of calcium hydroxide complexes. One further conclusion can be drawn. The fact that the experimental curves had a large negative slope after the endpoint, suggests that the complex $CaOH^+$ plays an important role.

When the possibility of complexes is considered Equation 13 becomes

$$-q = \Delta H_p \, m_p + \sum_n \Delta H_n \, \beta_n \, (OH^-)^n \tag{18}$$

where ΔH_n is the heat of formation and β_n is the overall formation constant of the nth complex. Unfortunately, this equation alone is indeterminate. It is possible to obtain a given curve using several alternative assignments for the various unknowns. Therefore, some additional information is necessary in order to interpret our experimental titration curves unambiguously. Measurements of an "apparent solubility product," aK_{sp}, defined in the following equation:

$$aK_{sp} = (Ca^{2+})_{total}\,(OH^-)^2_{total} \qquad (19)$$

where

$$(Ca^{2+})_{total} = \sum_n (Ca(OH)_n) \qquad (20)$$

and

$$(OH^-)_{total} = (OH^-) + \sum_n n\,\beta_n (Ca(OH)_n) \qquad (21)$$

is expected to yield the desired information. The total concentration of calcium in solution, $(Ca^{2+})_{total}$, in equilibrium with the solid can be expressed as a function of the free hydroxide concentration (OH^-) as follows:

$$(Ca^{2+})_{total} = K_{sp} \sum_n \beta_n (OH^-)^{n-2} \qquad (22)$$

If the free hydroxide concentration were known the values of the solubility product and the formation constants of the complexes could be calculated. Since we can only measure the total hydroxide concentration and not the free hydroxide concentration, we must work under conditions where they are approximately equal. From the preliminary measurements of the solubility product, it is possible to calculate the conditions necessary to meet this requirement. Appropriate analytical procedures are being developed.

This discussion has not mentioned the possibility of polynuclear complexes. The measurements of the apparent solubility product do not allow the determination of the presence of such complexes. However, when the thermometric titration data are considered, the presence of these complexes will be detected and, if they exist, the stability constants can be estimated.

ACKNOWLEDGMENTS

Research described in this paper was supported in part by the United States Atomic Energy Commission under contract AT(30-1)-2133 with the Pennsylvania State University. William B. McCarthy held a National Science Foundation Graduate Fellowship and a fellowship from the Esso Education Foundation during this work. Pier Giorgio Zambonin was on leave from the University of Bari, Bari, Italy.

REFERENCES

1. H. Lux, Z. Electrochem., 45, 303 (1939).

2. F. R. Duke and M. L. Iverson, Anal. Chem., 31, 1233 (1959).

3. L. E. Topol, R. A. Osteryoung and J. H. Christie, J. Phys. Chem., 70, 2857 (1966).

4. M. Y. Spink, Ph. D. Thesis, The Pennsylvania State University, 1965.

5. R. N. Kust and F. R. Duke, J. Am. Chem. Soc., 85, 3338 (1963).

6. A. M. Shams El Din and A. A. A. Gerges, Proc. Australian Conf. Electrochem., 1st, Sydney, Hobart, Australia, 1963, 562, (Publ. 1965).

7. M. Francini and S. Martini, EURATOM Document 2496. e, (Ispra), 1965.

8. K. Romberger, Ph. D. Thesis, The Pennsylvania State University, 1967.

9. A. M. Shams El Din, Electrochim. Acta, 7, 285 (1962).

10. H. Flood, T. Førland and K. Grjotheim, Z. anorg. allgem. Chem., 276, 289 (1954).

11. M. Blander, J. Chem. Phys., 34, 432 (1961).

12. M. Blander, F. F. Blankenship and R. F. Newton, J. Phys. Chem., 63, 1259 (1959); D. L. Manning, R. C. Bansal, J. Braunstein and M. Blander, J. Am. Chem. Soc., 84, 2028 (1962).

13. J. H. Christie and R. A. Osteryoung, J. Am. Chem. Soc., 82, 1841 (1960).

14. D. Inman, Electrochim. Acta, 10, 11 (1965); J. Braunstein and A. S. Minano, Inorg. Chem., 3, 219 (1964).

15. P. G. Zambonin and J. Jordan, J. Am. Chem. Soc., 89, 6365 (1967).

16. P. G. Zambonin and J. Jordan, Anal. Letters, 1, 1 (1967).

17. William B. McCarthy, unpublished work in progress.

18. J. Jordan, J. Meier, E. J. Billingham, and J. Pendergrast, Anal. Chem., 31, 1439 (1959).

19. P. G. Zambonin and J. Jordan, J. Am. Chem. Soc., in press (1969).

20. J. Goret and B. Tremillon, Bull. Soc. Chim. France, 67 (1966).

21. V. G. Levich, Physicochemical Hydrodynamics, Prentice-Hall, Inc., Englewood Cliffs, N. J., 1962.

AUTHOR INDEX

Numbers in parentheses are reference numbers and indicate that an author's work is referred to although his name is not cited in the text. Numbers that are underlined show the page on which the complete reference is listed.

AUTHOR INDEX

I

J

SUBJECT INDEX

SUBJECT INDEX

Z